Miners, Quarrymen and Saltworkers

History Workshop Series

General Editor

Raphael Samuel, *Ruskin College, Oxford*

Already published

Village Life and Labour

In the press

Childhood (2 vols)

Routledge & Kegan Paul
London, Henley and Boston

edited by

Raphael Samuel

Tutor in Social History and Sociology
Ruskin College, Oxford

Miners, Quarrymen and Saltworkers

First published in 1977
by Routledge & Kegan Paul Ltd
39 Store Street,
London WC1E 7DD
Broadway House,
Newtown Road,
Henley-on-Thames,
Oxon RG9 1EN and
9 Park Street,
Boston, Mass. 02108, USA
Set in Photon Bembo
and printed in Great Britain by
The Camelot Press Ltd, Southampton

ISBN 0 7100 8353 X (c)
ISBN 0 7100 8354 8 (p)

Contents

Illustrations

Plates

between pages 176 and 177

Figure

Notes on contributors

Merfyn Jones was brought up in the slate-quarrying district of North Wales and is the grandson of a quarryman on both sides of his family. An earlier version of his chapter was given at the History Workshop on workers' control in the nineteenth century, held at Ruskin College in 1971; it is part of a larger and continuing work. He did postgraduate work at Warwick University and has been for the past three years co-director of the SSRC coalfield project at Swansea, which has been rescuing miners' libraries, gathering lodge minute books, and recording memories of the South Wales coalfield.

Brian Didsbury was brought up in Northwich, Cheshire. He left school at fifteen and was for twelve years a chemical worker at the ICI factory in Winnington. He was a student at Ruskin College from 1972 to 1974. He is now a trade union organizer for the Transport and General Workers' Union in Cheshire.

Dave Douglass was brought up in Felling, Co. Durham. He went down the pit at sixteen to work at Wardley colliery; when the pit closed he transferred to the Doncaster coalfield, where he now works as a ripper at Hatfield Main colliery. He was a student at Ruskin College from 1970 to 1972, working mainly on the history of which Part 4 in the present volume is a part. He is editor of *The Mineworker*, a rank-and-file paper circulating chiefly in the Doncaster coalfield and South Wales. He is a singer and a collector of lore, language and songs.

Raphael Samuel has been tutor in social history and sociology at Ruskin College since 1964.

General editor's introduction

Capitalism did not grow up all of a piece, and its nineteenth-century development, though swift, was also highly uneven. In ironmaking the giant furnaces of the Black Country existed cheek-by-jowl with thousands of back-yard smithies, where metals were hand-hammered at the hearth. In Sheffield, some thirty or forty rolling mills supplied the working materials for sixty handicraft trades, most of them, in the 1860s, being conducted on the basis of sub-contract and out-work, with journeymen masters working double-handed or alone. Textiles, by mid-Victorian times, were largely mechanized, but dressmaking (where employment increased three-fold between 1841 and 1861) depended on the poor needlewoman's fingers. In agriculture the living-in farm servant gave way to daywagemen, monthmen, and job hands, to itinerant harvesters and travelling gangs; but on the railways a whole army of job-for-life men was being recruited, with pensions and promotion to tie them to their employers, company houses, and regular, all-the-year-round wages. Another major line of cleavage – and certainly the most enduring – was between men and women. Even where they worked together at the same occupation, and on the same premises, they belonged to different worlds: bookbinding, for instance (to be discussed in a subsequent *History Workshop* volume), was a male craft, but a woman's sweated trade, with large numbers of season hands who could only expect employment for four to five months of the year.

These disparities were nowhere more extreme than among mineral workers, the subject of this book. Some were employed in mammoth capitalist enterprises, such as the Dowlais iron works of Sir Josiah Guest, or the slate quarries of Lord Penrhyn. Others, such as the Whitby jet diggers, the Brandon flint knappers, or the lead miners of Derbyshire and North Wales, worked in 'poor men's ventures' in which production was in the hands of self-governing workers' companionships. A coal mine could vary in size from a day-hole pit, worked by a pair of men, to a multi-recessed catacomb; a brickworks could be a factory or a shed. Conditions of employment were exceedingly various, depending upon local tradition as well as upon the state of technology, or the size of the individual plant. Brickmaking, for instance, was a woman's trade in the Black Country, where there were employment opportunities for men in the ironworks and coal-pits; but in the industrial towns of Lancashire, where women found employment in the cotton mills, while openings for men were scarce, it was a mainly male preserve. Patterns of settlement were various too. At one extreme there were more or less closed, hereditary occupational communities, enjoying ancient privileges and rights, like the Free Miners of the Forest of Dean, the lead miners of Derbyshire, or the Purbeck marblers (quarrymen on the Isle of Purbeck claimed to be descended from the Phoenicians). At the other extreme there were mushroom settlements which sprang up in the new mining districts of mid-Victorian times, such as Cleveland, Hodbarrow and the Rhonddas, where strangers were recruited from all parts.

Miners and quarrymen do not fit easily into the conventional categories of Victorian work. They were neither exactly labourers, nor yet artisans. Their work was highly skilled, yet it carried no formal apprenticeship, and was supported by no formal restriction on entry to the trade. Their wages were comparatively high, yet they were almost totally non-unionized. In Marxist terms their work was 'handicraft' in character even when it was being carried on in the service of large-scale industry. They had to perform a series of more or less autonomous operations with little co-ordination between them. They were all-round men within a particular class of work; their earnings depended on both physical strength and manual dexterity and skill; they had to combine the labourer's muscle power with the craftsman's ability to work to fine limits. Unlike the factory, there was no machinery to regulate working pace and dictate production routines; nor were there premises to police, or valuable raw materials to guard against depradators, as there were in the manufacturing workshop. Management was comparatively undeveloped. In a Cornish tin or copper mine there was, practically speaking, none at all – the work was put up for auction each month, and let out to the 'tributers' – the men who drilled the rock singly or in pairs – to exploit to the best of their ability. The 'bargain' system which Merfyn Jones describes in the slate quarries of North Wales was similar: once the price for the job had been fixed, the workers were left to get on with it. Coal miners had few such formal autonomies but, in the days of the pillar and stall system, a man might spend the best part of the week at the face without being visited by a gaffer. In Durham, the development of the 'longwall' system of coal-getting increased the need for supervision, by grouping larger numbers of men together, but even so the work was to a large extent regulated by the system of 'marras' or workers' companionships, which Dave Douglass describes in this book. Another system of indirect employment was that of the family berth, which Brian Didsbury refers to in his chapter on Cheshire saltworkers. It was also to be found on many of the southern and midland brickfields, as also in some quarries. In fact sub-contract, in mid-Victorian times, was perhaps the characteristic form of employment in mineral work; management and supervision were at a minimum: ownership in greater or less degree was dissociated from control.

Workers' combinations in mining and quarrying were in many ways distinct from those in manufacturing industry, being at most times invisible. Amongst Cornish miners there was practically no trade unionism at all: the 'bargain' system, and the ease of moving from mine to mine, seems to have served as an alternative to it. In brickmaking the migratory and seasonal character of the labour force made permanent association impossible; but there was a formidable and to some extent successful resistance to machinery and a good deal of local job monopoly; while in mid-Victorian Lancashire there were a formidable series of local combinations which used intimidation and terror to supplement bargaining strength and keep machinery at bay. Trade unionism among coal miners was, down to the 1890s, an on-and-off affair. From time to time powerful county unions appeared and recruited members at a terrific pace; but their triumphs were short-lived, and membership melted away when conditions were unfavourable or the coalowners made a concerted attack. In Scotland trade unionism in the coalfield had been endemic from the time of the Napoleonic wars, but the first permanent county union did not appear until 1871; and as late as 1894 little more than 30,000 of the 70,000 miners who came out on strike were trade unionists. The North Wales Quarrymen's Union which Merfyn Jones writes about was exceptional among mineral workers for its longevity and steadiness.

Among coal miners the absence of formal trade unionism was to some extent compensated for by alternative forms of organization – secret solidarities which the owners were powerless to break even when their rule was nominally unopposed. Among the Scottish miners one of the most frequent methods of resistance was the 'darg', or restriction on output, which was imposed whenever there was a threat of falling earnings, or as a reprisal for unsettled grievances. Amongst the miners of South Wales trade union membership was fickle, but the slightest change in working arrangements was liable to become a flashpoint of struggle, and the progress of production was punctuated by day-to-day bargaining at the face. 'They can do nothing without having a general talk over it first', *The Times* Commissioner complained after a visit to the Rhonddas in 1873, 'and 1,000 men occasionally wait upon a manager to ask a question or make a request, when the business might be quite as well done by one. Even after they descend the pit in the morning "confabulation" precedes the operation of the mandril and the pick.' Wider solidarities, too, would spring to life with seeming spontaneity, as Dave Douglass describes in his account of the Felling strike of 1887 and the Wardley Funeral Strike of the following year. An earlier glimpse of them is given in a report from North Wales which appeared in the *Mining Journal* for 1859, and which shows how action could be mysteriously concerted from nowhere.

> STRIKE AT THE COLLIERIES AND IRON-WORKS IN THE WREXHAM MINING DISTRICT. For the last three weeks a ruinous strike, to both employers and employed, has been raging at the following works: Brynmally (Mr. Clayton), Brymbo (Messrs. Darby), Vron (Messrs. Maurice and Lowe), Frood (Messrs. Sparrow and Poole), and Westminster (Marquis of Westminster). The men at first demanded sixpence a day extra, but the employers stated that they were not willing to give the sum required until the employers in South Staffordshire and Shropshire should set them an example. The men afterwards reduced their demand to fourpence per day. One of the principal employers of the district, Mr. W. Darby, says the reason why an advance is not given is owing to the very low price at which the produce is now selling. The way the notice was given demands a passing remark, as being disgraceful in the extreme. An anonymous, ill-worded notice was one night posted upon the various pits' heads by a man from a distance, with his face covered with crape, stating that a rise of sixpence a day would be expected after Wednesday, Aug. 10, the pay-day being the following Saturday. Nothing was personally said on the subject to the employers, but on the Thursday morning the works were all at a stand-still, in consequence of the men staying away. No application was made to the employers, no committee was deputed to wait upon them to ascertain their intentions, or to argue the matter over, but the men simply gave an anonymous notice, waited till the period had expired, and then, without explanation or further warning, stayed at home. Could any mode of proceeding be more unwise? Why not at first, when dissatisfied with their pay, have met,

> and some half-dozen appointed as a deputation to the employers? A large open-air demonstration was held on the Wrexham race-course by the 'turn-outs' on Monday afternoon, for the purpose of trying to come to some arrangements with the employers. The meeting was called by a placard signed 'A collier'. For a time no-one came forward to commence the proceedings, and the 'turn-outs' began to question whether or not they had been made the subject of a hoax; but soon after this the 'Collier' came forward in the person of Mr. J. S. Joseph, mining engineer, who addressed the men at some length, arguing most forcibly the propriety of a deputation waiting upon the masters, for the purpose of coming to some arrangements, at the same time undertaking to be a member of such deputation. He also stated that the men in that district were getting before the strike sixpence per day more than those in the Rhos and Ruabon district. Several of the men here spoke at some length, and it was finally carried that a deputation, consisting of 25 men, an equal number from each of the works, should be appointed to wait on a meeting of the employers at Brymbo, with full powers to make terms as regards wages. One of the speakers, named Lloyd, a collier, in addressing the meeting, made a calculation with regard to wages, in which he stated it was now impossible on the low scale of wages to make both ends meet, and proved, in his manner, that at the week's end the men who now received 17s per week, would be 1s. 1d in debt.

Whatever the original characteristics of its work force, mining and quarrying made an essential contribution to the industrial revolution, and to the shape of the capitalist economy. The acrid smell of burning clay was not the least of the smokes that hung over Sheffield and the Black Country (in the 1870s some 40,000 tons of clay a year were consumed in the crucibles of the Sheffield steelworks); 100,000 char-stones from Duke's quarries, Derbyshire, lined the track of the London and Birmingham railway; copper and tin from Cornwall charged the furnaces of the Swansea valley. Quarrying provided the raw material for a range of mineral-based industries – chemicals, brickmaking, glasswork, pottery; building materials for construction work of every kind; limestone for the improving agriculturist, fettling materials for the ironworks, sand for the moulder's bed. Coal mining supplied motive power to industry, and domestic comfort to the home. Minerals also took their place in overseas trade, more especially in the early and middle years of the nineteenth century, when Britain was the Chile of Europe as well as the Workshop of the World (in the 1850s well over half the world's copper was being produced in the Redruth copperbelt and the Tamar Valley, while the Devon Great Consols mine at Gunnislake accounted for half the world's production of arsenic). There was a huge overseas trade in Cheshire salt and quite a large one in fire-bricks ($9\frac{1}{2}$ million of them were exported from Newcastle upon Tyne in 1864). Welsh slate travelled from Portmadoc all over the world, 'the chief trade being perhaps with France, Belgium, Holland and the ports of the Baltic generally'; granite paving setts were exported from Galloway to Tsarist Russia; granite statuary from Aberdeen to Buenos Aires; and 'an almost world-wide fame' was claimed for the hydraulic cement of Barrow-on-Soar. These essays are offered as a contribution to economic history as well as to that of British labour. But like their predecessors and those to come the chief focus is on the workplace. They are concerned with job control as well as with trade unionism, with social relations both at the point of production, and in their family and community setting. Dave Douglass's 'Pit talk in county Durham', with which the volume concludes, extends the discussion to the cultural repercussions of work and to the way in which the miner's life is recorded in speech and song. Merfyn Jones, in his study of the 'Slate quarrymen of North Wales', looks at the religious and national dimensions of class struggle; while the opening essay on 'Mineral workers' tries to relate the experience of work to the development of the capitalist economy as a whole.

A note on the History Workshop

The History Workshop started at Ruskin College, Oxford – a trade union college for adult men and women – in 1967, and has existed for some years as a loose coalition of worker-historians and full-time teachers and researchers. These books draw on its work. Their aim – as set out in the introduction to *Village Life and Labour* – is to foster a 'people's history' which will bring the boundaries of history closer to those of people's lives. This volume is the second of a series on 'Work' and will be followed by volumes on The Workshop Trades, The Uniformed Working Class, and Women's Work. The Workshop now publishes the twice yearly *History Workshop Journal* (subscriptions and information from PO Box 69, Oxford), and workshop meetings, part-festival, part work-in-progress, have now spread from Ruskin to other parts of the country, drawing on the strengths of local experience and research (for help or advice in organizing a local workshop, write to 19 Elder Street, Spitalfields, London E.1).

Part 1

Mineral workers

Raphael Samuel

I

Coal miners are the only class of mineral worker to have lodged themselves in the historian's consciousness. This is partly because of numbers, partly because of records, partly because of trade union strength. It owes a good deal to the years from 1888 to 1914, when miners acquired a common political and trade union consciousness, and when the regional peculiarities of the individual coalfields – so marked in earlier years that it is difficult to speak of an industry at all – lost some of their original force. Amongst other classes of mineral worker no such national profile appears. Some groups were too localized, others too dispersed, and many (in mid-Victorian times) had auxiliary or alternative occupations which obscured their group identity. China clay diggers were peculiar to Cornwall and Lee Moor; arsenic miners to Devon, saltworkers to Cheshire, Droitwich and the Tees. Jet diggers may have been important in the economy of Whitby – indeed the making of jet ornaments in the 1860s was the most important trade in the town[1] – but with the exception of some nearby Cleveland villages[2] they do not appear in any strength elsewhere. The same may be said of the flint knappers of Brandon,[3] the plumbago miners of Borrowdale or the hydraulic cement makers of Barrow-on-Soar. Gypsum, which like jet, was responsible for the mass production of cheap ornaments (today's 'Victoriana'), was only worked extensively in Derby and Notts.[4] The Purbeck Marblers do not figure in the annals of the building trade, though their kerbstones and sills found their way to towns all over southern England; nor do the sett-makers of Mountsorrel or Paenmaenmawr. Groups of workers like these may attract the notice of the antiquarian – and latterly the industrial archaeologist – but it is difficult for the economic historian to assign them to a graph.

So far as metalliferous miners are concerned, one influence making for their neglect by posterity is the sharp chronological break which took place in the 1860s and 1870s, when foreign competition destroyed the profitability of copper, tin and lead. Copper was the first to be affected, and according to Robert Hunt, Keeper of Mining Records at the Museum of Geology, some 7,380 of 11,321 copper miners left Devon and Cornwall in the eighteen months ending December 1867.[5] (The number of copper mines in Cornwall decreased from ninety in 1863 to forty in 1873: in 1893 only six were still working.)[6] A comparable disaster overtook Cornish tin in the following decade: the number employed in the Cornish mining industries declined from 26,814 in 1873 to 5,193 in 1898.[7] In the coal industry, by contrast, the number of miners went on increasing right down to 1921; the pit villages were becoming more sharply defined culturally and politically, more settled in their population, and more organized, precisely at the time when their metalliferous counterparts were in decay. The Rhondda Valley stamped itself on twentieth-century British consciousness; Devon Great Consuls – during the 1850s the largest copper mine in the world – faded into grass. Table 1.1 shows the movement of employment in metalliferous mining as a whole.

TABLE 1.1[8]

Year	*Underground* *Males*	*Above ground* *Males*	*Females*	*Total*
1873	37,378	20,271	5,034	62,683
1883	30,492	17,773	1,970	50,235
1893	21,240	13,464	1,035	35,739
1903	17,571	11,984	268	29,823

Quarrymen, for most of the nineteenth century, may be said to be statistically invisible, as regards the majority of them. The census records 60,000 of them in 1891.[9] Yet within four years of the passage of the Quarries Act in 1894, no fewer than 134,478 persons were registered as being in quarrying employment[10] – a figure which, on the evidence of the mines inspectors themselves, must have represented only about a half of the total number of those following some kind of quarry-based occupation (the Act was confined to pits of over 20 feet in depth).[11] One possible reason for the failure to number them is that quarrymen were often something else besides. As a divisional inspector of mines put it in 1912: 'You may have a perfectly good quarryman working three weeks or a month in a quarry, and another time he is a farm labourer or working on some other work altogether.'[12] Brickfield workers must also have been underestimated by the census, if only because of the time in which it was taken, in March.[13] Most yards in mid-Victorian times were 'summer yards' with only a nucleus of workers who stayed all the year round. There was a large influx of seasonal migrants when the making season began (in April); and, in the southern brickfields, this was also the chief time of the year when women and children were employed (things were different in the Black Country, and on the Welsh coalfield, where brickmaking was largely in the hands of women). An extreme case of the possible variation in numbers comes from the evidence given at the Children's Employment Commission in 1866 about Messrs Rutter's brickyard at Crayford, Kent – only thirty-three persons were employed there in winter, though the number employed at the height of the season was 382.[14] There were some for whom brickmaking was a casual employment, and many for whom it was a secondary occupation. In Hertford it alternated seasonally with the town's main trade of malting;[15] in the Medway district of Kent with winter work in the cement works; many brickmakers (as Eric Hobsbawm pointed out long ago) were gas-workers in their season off.[16]

The progress of capitalism in the nineteenth century was intimately bound up with the exploitation of mineral wealth, not only of coal (which because of its importance has riveted the attention of posterity) but also of tin and copper, iron ore and lead, sand (the basic raw material in glassmaking), clay (the material of pottery and bricks), mud (mixed with chalk for cement), salt (an alkali for the chemical industry as well as a primary item of household use), and all the different varieties of stone. The Newcastle glass bottle makers quarried their raw material from the flats at Jarrow Slake,[17] the Dublin glass bottle makers from the shore deposits at Sandymount Strand.[18] Sands lined the furnace bottoms in the ironworks; limestone filled them as well as iron ore and coal. In the Black Country all the materials for ironmaking were quarried or mined in the locality – the sand at the numerous cuttings on the sides of the Wolverhampton and Stafford canal,[19] coal and ironstone in bellpits, quarries and shallow shafts, limestone in the Silurian rocks at Dudley Castle Hill and Wren's Nest, the stone pits at Sedgley Beacon, and the limestone quarries at Walsall and Daw End.[20] The whole district was honeycombed with diggings and craters and tips: 'while a coal-pit appears in full work on one side of the road, on the other are pits of ironstone and limestone, equally busy'.[21] The 'heads of the valleys' ironworks in Monmouthshire drew their limestone from the carboniferous outcrops on the northern rim of the coalfield: the Llangattock quarries were one major source of supply; the Trevil mountains another.[22] In the earliest days of the industry the limestone was brought down by pack horse and mule, but by the 1800s an extensive network of iron tramroads connected the Trevil quarries with the works at Tredegar, Sirhowy and Ebbw Vale.[23] The Sheffield hinterland was also thickly quarried, quite apart from the coal mines which, in the 1840s, were still being worked on the city fringes.[24] There was an 'immense excavation' in the sandstone quarry at the foot of Wincobank Hill, near Grimesthorpe, 'very conspicuous . . . from the scar produced by the removal of the green sod';[25] there were many more at Deepcar, where limestone was quarried and burnt.[26] The cutlery trades were served by some dozens of grindstone quarries in the nearby Peak district and in the immediately adjacent parts of South Yorkshire. At Wickersley, near Rotherham, which specialized in the finer class of grindstone, no fewer than twenty-one different owners of grindstone quarries are recorded in *White's Directory* for 1861.[27] The steel industry at Sheffield was equally well supplied with ganister, a refractory stone which was manufactured into steel-making crucibles, and was also used for lining cupolas and furnaces. In the earlier part of the century the ganister had been quarried opencast for roadmaking[28] but with the development of the Bessemer ('open hearth') process of steelmaking, after 1860, it became of 'inestimable value'[29] to the steelmakers (the Sheffield steel industry was using some 14,000 crucibles a week in the 1890s).[30]

Limestone was a basic ingredient in the making of pig iron, the normal charge of the furnace being one barrowload of lime to two of ironstone and three of coal, though the proportions varied according to the quality of iron required.[31] (At Dowlais in 1850 the 'limestone girls' who worked near the tunnel heads of the furnaces, breaking the limestone for use in the smelting, told the *Morning Chronicle* Commissioner that their work was very hard and trying, 'owing to the heavy weights they carry, and the alternations of heat and cold they have to endure'.[32]) Many of the early ironworks maintained their own quarries, but as output expanded these often had to be supplemented (or replaced) with outside sources of supply. The Black Country ironworks were consuming 3,500,000 tons of limestone a year, according to Robert Hunt's mineral statistics for 1856,[33] and local supplies were supplemented by canal-borne imports from the Caldon Low quarries on the Trent and Mersey Canal and the Trevor mountains in North Wales.[34] The mighty Carron ironworks in Scotland had turned to outside sources of supply at a much earlier date, and by the time of the Napoleonic Wars they were relying on shipments of limestone from the quarries in Fife.[35] The Lillieshall Company in Shropshire – the country's leading ironworks in the 1840s – had opened up quarries at Wenlock Edge as well as those which formed part of its original workings, while many other Shropshire ironmasters were by this time dependent on supplies from North Wales.[36]

Limestone was used as a raw material in the manufacture of glass as well as iron.[37] It also provided the 'improving' agriculturist with his fertilizer,[38] the builder with his stuccoes and mortar,[39] the Boards of Health with their disinfectant. (Limewashing was a sovereign sanitarian precaution against cholera and typhus, and earnestly canvassed in the crusade against dirt.)[40] The prime mover in the rise of Buxton lime industry, whose hideously smoking chimneys distracted the traveller's attention from the picturesque,[41] was not the local quarry owners, but Samuel Oldknow, the cotton manufacturer, who needed lime for his bleaching works at Stockport: the Peak Forest Canal was opened (in 1794) to facilitate the traffic.[42] In agriculture the use of lime to dress the land was a development well under way in the eighteenth century, and it was enormously extended by the coming of turnpike roads and canals, which brought lime-burning, as an industry, to districts hitherto lacking their own sources of supply.[43] (The tolls on the lime traffic were the proximate causes of the Rebecca riots in 1843, when the upland farmers of Carmarthenshire came down and smashed the turnpike gates.)[44] The chemical industry, which grew so rapidly in the second half of the nineteenth century, was another major consumer of lime. William Gossage, father of the modern soap industry, maintained his own lime plant at his alkali works in Widnes,[45] so did Brunner Mond, when they pioneered the Solvay process of soda making at Winnington (according to Tom Mann, who worked there briefly as a trade union spy, the lime picking department was the hell hole of the works: 'nobody held out more than two or three months at this job').[46]

The uses to which sand was put – though it was less extensively quarried than limestone (some $1\frac{1}{2}$m tons a year against 10m in the 1890s)[47] – were, if anything, even more various. It was a primary raw material in glassmaking, and in several varieties of soap (Gossage's first soap-making patent, taken out in 1854, specified the mixing of 'about nine parts of soda ash . . . with eleven parts of clean sand').[48] Brickmakers used it to dust the inside of their moulds,[49], and, in the case of silica sands, to make their bricks. As a refractory it served for moulding-beds in malleable iron castings, for repairing kilns and ovens in the potteries, and for lining the saggars and floors.[50] As a filter it was used in the waterworks along with gravel or shells (the Southwark and Vauxhall Waterworks Company used 3ft of Harwich sand to 1ft of hoggin, 9in of fine gravel, and 9in of coarse).[51] The steel industry was another large consumer, taking some 200,000 tons a year in the early 1900s:[52] the ample beds of the Don Valley, where sand was mined as well as quarried, provided the Sheffield works with a conveniently accessible source of supply.[53] The Bunter sands of Belfast Lough were extensively used in the local shipyards as well as being exported to England for floor sand and foundry work in brass.[54] Mansfield sands were considered particularly favourable to casting,[55] both on ground of texture ('which yields the soft, velvety feel so much desired by the metal founder'), and because of the regularity of the bed, which enabled consignments to be kept to sample.[56] Birmingham sands were also highly prized for foundry work, and there were two important quarries in the city itself: the one near Hockley cemetery was worked for more than a hundred years.[57] Charlton played a somewhat similar part for the London metal-working trades, and indeed according to Whittaker (*Geology of London*, 1889) it was the excellence of the moulding sands in this particular corner of south-east London that determined the site of Woolwich Arsenal.[58] The white sands of Headon Hill at Alum Bay, in the Isle of Wight, enjoyed a comparably high reputation with the London glassmakers, and exports (principally to London and Bristol) were running at a rate of 20,000 tons a year in the 1850s.[59]

Sand also made a substantial contribution to mid-Victorian domestic comfort, being used for scouring and scrubbing jobs of every kind. Floors, in the well-kept proletarian household, were sprinkled with sand rather than covered with linoleum or carpets, while a heap of sand inside the door ('with which you were expected to clean your boots') served as a precursor to the door-mat.[60] Sand, as an abrasive, furnished a cheap alternative to soda or soap in washing up dishes; while knives, in the days before stainless steel, were kept clean and polished with Bath brick, 'a peculiar kind of sand . . . obtained from the river bed of the Parret, in Somersetshire'.[61] The trade in house sand was a poor person's enterprise, which ranked with chip chopping and hearthstone selling in the lowest class of back street trade.[62] The sand was hawked from door to door in baskets or on a barrow, and often it was dressed at home. In London there was a whole class of river barges, known as 'Sandies', engaged in dredging it from the Thames ('hard work for very little profit'),[63] as well as those (Mayhew describes some of them) who went out digging for sand at places like Kensington gravel pits and Hampstead Heath.[64] At Todmorden, according to Samuel Fielden, who was born there in 1847, the sand was produced from the refuse piles at the local quarries: in his mother's family the children were sent out there with baskets, 'picking

. . . the whitest scraps, then taking them home, and with a large white stone beating them up into fine sand'; the sand was sold at a rate of one halfpenny a quart 'to the poor people to sprinkle upon their stone-flag floors'.[65]

Canals were largely given over to the mineral traffic,[66] more particularly in the second half of the nineteenth century, when railways captured most of the trade in merchandise and perishable goods. Coal was of course the most frequent cargo, but the Peak Forest Canal was by no means the only one to be promoted specifically for the sake of lime. (One way in which canal promoters canvassed for local support was by offering 'indulgences' to farmers' lime and manure). The carriage of sea sand to the inland farmers – a lime-bearing deposit dug from the beaches at Bude – was the chief traffic in the Bude and Launceston Canal.[67] (Farmers nearer to hand collected it themselves with horse or donkey and cart: as many as four thousand horse loads a day are said to have been taken off.)[68] On the Dudley Canal in the Black Country there were complaints in 1797 that limestone was blocking up every other species of traffic. Passage was impeded, a committee minute records, 'by the loading of limestone into Boats at Charles Starkey's Quarry in the Tunnel, by empty Limestone Boats being left afloat in the Tunnel and By Boats Loaded with Limestone being left in the Canal near the Stop Lock at Tipton'.[69] The rise of the Haytor granite quarries, Dartmoor, which provided the stone for the western face of London Bridge, was directly attributed to the building of the tramway which linked the quarries to the Stover Canal,[70] and the same seems to have been true earlier of the promotion of such large-scale quarries as those at Chapel-en-le-Frith in Derbyshire, and Caldon Low in Staffordshire.[71] The Cheshire salt works was totally bound up with the success of the Weaver and Sankey navigations,[72] and there was sometimes as close a relationship between brickworks and the canals. The canalside was as popular a site for a brickworks as it was for lime kilns.[73] When the boatmen and steersmen employed at the principal brickworks in Oldbury, Staffordshire, went out on strike in June 1876 ('their masters having refused an advance of sixpence per day') some fifteen of the works were stopped, and about 1,000 hands thrown out of employment.[74] The rise of the chemical industry at Widnes was also closely related to the canals, since the Sankey navigation placed them in immediate communication with Cheshire salt, one of their principal raw materials, and with the lead and copper ores which the sailing flats carried inland from Liverpool.[75]

Minerals were also extensively carried by sea,[76] and the early and middle years of the nineteenth century saw the rise of a whole crop of ports devoted chiefly to the mineral traffic. One of the most strategically placed of them was Runcorn on the Mersey, which kept the Staffordshire potteries supplied with southern flints and clay, while at the same time acting as a primary outlet for Cheshire salt.[77] Of the total tonnage which passed through Runcorn Dock in 1884 salt accounted for 130,035 tons, pottery clay for 118,102, coals for 71,127 and earthenware exported for 36,759, these cargoes thus making up 356,023 tons out of a total 518,168.[78] Another such port was Bridgwater, where hundreds of boats were engaged in a largely mineral traffic – carrying bricks from the local brickworks to Liverpool, Antwerp and the Elbe, and trading too in Welsh and Cornish granite and anthracite from Swansea.[79] Some hundreds of sailing barges were engaged in the Kent and Essex brick trade, taking manufactured bricks into London and carrying a return cargo of 'breeze', i.e. ashes and clinker, for the works.[80] 'An inexhaustible fleet of schooners'[81] loaded clay at the china clay ports in Cornwall, and their trade to the north was quite unaffected by the arrival of the railways. In 1885 only 3,683 tons of china clay was sent out of Cornwall by rail: all the rest was shipped at the china clay ports – 114,403 tons from Fowey, a port which had been linked to the industry some ten years earlier; 86,325 tons from Par, a port which had been built in the 1830s to accommodate the increase of the trade; and a further 84,650 from Charlestown and Pentewan, the two original ports of the trade.[82] Poole Harbour was not less important as an outlet for the Dorset clay trade.[83] Teignmouth for that of Devon,[84] and Newhaven and Gravesend for the supply of pottery flints. (On the Trent-Mersey canal in 1836 there were 30,000 tons of flints from Gravesend and Newhaven and 70,000 tons of pipe and china clay).[85] In the spacious yards and private wharves of the Don Valley pottery, when John Tomlinson visited it in 1879, all of the materials had come by boat, except for the coal which was local – 'the flint had been brought from the south coast, the stone from Guernsey, and the clay from Cornwall'.[86] The longer-distance stone trade was also very largely seaborne, a great deal of the stone being carried in two-man barges, or hackers, or flats. The Purbeck stone trade (running at about 100,000 tons a year at the turn of the century)[87] remained largely maritime down to 1914;[88] so did the trade in Aberdeen granite and the Greywacke quarries at Penmaenmawr. St Sampson's, Guernsey was entirely a stone port (shipments increased from 35,000 tons in 1840 to 142,866 in 1861);[89] Portmadoc and Port Penrhyn were built for slate, though Portmadoc schooners later developed an independent carrying trade of their own.[90]

Clay digging and quarrying, though scarcely noticed by economic historians, were crucial to urban growth. They shaped the physical environment of industrial Britain, as well as providing it with raw materials and plant. Nineteenth-century London was surrounded by their activity on all sides, more especially in the no-man's-land on the city's outer edge, 'half torn up for brickfield clay, half consisting of fields laid waste in expectation of the housebuilder'.[91] Sutton, a local historian tells us, was 'canopied . . . with a nauseating stench of burning clay':[92]

> Sutton lies between the gravel of Mitcham and a belt of loam south of Banstead. It has almost equal parts of clay and chalk, with a narrow vein of sand between. In bygone days two industries profited by these deposits, lime-burners and brick-makers. In Victorian times a line of brick-yards lay across the Benhilton district of Benhill Road, Oakhill Road, Balaam Lane, Gander Green Lane. The demand for bricks 1860–90 was enormous and they were working to capacity.

On the northern heights of the city there were the extensive gravel beds at Finchley which Telford had used in building Archway Road;[93] in the south-west there were the ballast pits at Battersea which the Westminster Board of Works was trying to rent in 1867,[94] and 'huge' gravel pits on Wandsworth Common – 'many of them full of stagnant water';[95] in the south-east there were the extensive diggings for loam and sand and chalk;[96] and at Woolwich the sand was dredged from the river 'at great expense', laboriously puddled and laid out to dry, and then sold to the London brickmakers.[97] Further out, the Surrey hills were extensively quarried for lime ('some of the largest and best pits of the country. . . . The . . . lime is . . . sought after by every mason and bricklayer in London'[98]) and the alluvial flats at the mouth of the Thames for cement (the cement industry had been pioneered on the Isle of Sheppey,[99] and was said to have settled at Northfleet and Grays 'not so much from the superiority of its clays to all other clays, as because there was so much of it and so easily got at').[100] The most extensive diggings of all were on the brickfields which had grown up to the west of London, from Southall to Slough, on the low-lying land by the Grand Union Canal and the Great Western Railway – 'the most unpicturesque and uninteresting eighteen miles . . . in our environs':[101] 150,000,000 bricks a year were being made there in the 1890s.[102]

Clay-getting would have been a conspicuous sight in other towns besides London, since there were few parts of the country without some kind of material serviceable for making bricks (Humber silt was used at Grimsby)[103] and few towns without their local brickfields. They took up a great deal of space, since the clay digging was usually done in very shallow pits (sometimes referred to as 'clay holes'), and the beds, in consequence, were quickly worked out. Clay digging on the brickfields had much in common with 'swidden' (slash and burn) agriculture, forever moving to fresh terrain. At the Putton Lane Works, Weymouth, for instance, which were in operation for some seven years only, from 1859 to 1865, the clay pits covered no less than seventeen acres – they were shallow because of the prevalence underneath of stone.[104] In Kent, the brickmaking district of Faversham and Sittingbourne was enormous; in the 1870s the industry employed some two or three thousand persons, and the diggings extended for miles 'as far down as the mouth of the Medway, and again up that river to Maidstone.' The stools where bricks were made by the moulder and his gang appeared as little islands in the midst of desolation, 'dotted about some great tawny flat, which stretches for miles down to the mist waters of the Swale or Medway'.[105]

Clay works were also very much in evidence on the coalfields, where fireclay was associated with many of the coal measures, and both mined and quarried opencast. It was exploited not only for bricks, but also for the manufacture of sanitary ware, drain pipes, sewer pipes, and a range of refractory materials and plant – crucibles for the steel industry, retorts for the gas works, pots for glassmaking, firebricks to line the furnaces and the factory flues.[106] Brickworks were 'very numerous' at Buckley, on the edge of the Denbighshire coalfields, 'great open excavations' met the visitor 'at every turn'.[107] In the Rhondda, E. D. Lewis tells us, there were a number of flourishing brickworks, which in the 1860s and 1870s – when the Rhondda was being built – 'gave employment to a considerable number of girls. . . . The original plant at these brickworks had been erected to manufacture sufficient bricks for the walling of neighbouring colliery shafts and the "fire-clay" was obtained from recently opened levels. So great was the demand for bricks for colliery walling and house building, however, that additional kilns were erected and subsequently bricks from Llwynypia, Bordingallt and Treherbert were sent to all parts of South Wales.'[108] Here is a description of the manufacture of fire-bricks at Dowlais, Merthyr Tydfil, where the work was done exclusively by women: it comes from the *Morning Chronicle* Commissioner's account of 1850:[109]

> I found the girls at work making bricks in a low shed having no windows or opening for the admission of light, except for the doorway through which I entered. Underneath the floors are flues for the passage of heated air, to dry and prepare the bricks for the kilns, which are built adjacent. The clay is ground in mills by steam power; and the women then saturate it with cold water, in a smaller shed opening by a door from the main building. They next temper it with their bare feet, moving rapidly about, with the clay and water reaching to the calf of the leg. This operation completed, they grasp with both arms a lump of clay weighing about 35 pounds, and, supporting it upon their bosoms, they carry this load to the moulding table, where other girls, with a plentiful use of cold water, mould it into bricks. They have to feed and attend to the furnaces used for heating the floors in the open air, exposed to the vicissitudes of the weather and the changes of temperature, alternating between the heat of the drying room and the cold winds outside. They told me that on average they earned six shillings a week when at work, but there are many weeks in the course of the year when, unable to get clay, they are compelled to be idle.

In the stone districts of the country, quarries took the place of brickfields. 'The supply of building stone in the vicinity of Leeds is no less abundant than that of coal,' wrote Sir George Head, when his tour of the manufacturing districts in 1835 took him to the quarries at Woodhead and Bramley Fell. 'The banks of the Liverpool Canal . . . are continually covered with the material in all various sizes and dimensions, such as large blocks, slabs for paving, as well as . . . thinner dimensions termed "grey slate" for roofing dwellings.'[110] Nineteenth-century Bradford was a town built of stone – so late as 1900 a brick building was still 'an anomaly'[111] – and also the centre of an extensive stone trade, conducted by barges on Bradford Beck.[112] Elland Flagstone . . . caps all the hills in the neighbourhood . . . to the east, north and west. It is extensively quarried wherever it occurs'[113] (between forty and fifty quarries were active in the district in 1900, with between 1,200 and 1,500 men and boys employed at them).[114] Nineteenth-century Halifax, too, was 'nearly wholly built of stone', and so were many of the lesser mill towns of the West Riding: 'stone quarries of great extent have existed for a long period; as one bed has been worked out another has been discovered, and the West Riding has not only been able to supply itself with all the stone it has required, but has likewise been able to contribute large quantities to other portions of the kingdom, as well as to the Continent and America'.[115] Building in Aberdeenshire in the nineteenth century was in granite, in Glasgow and Edinburgh in sandstone. In the South Wales mining valleys most of the nineteenth-century cottages were built of Penant sandstone,[116] and the same was true of the ironworking towns at the heads of the valleys. The early ironworkers' cottages at Ebbw Vale, built by the company in the 1800s, had stone walls, stone-tiled roofs and floors paved with stone. Even the spiral staircases were built in stone: 'timber was only used for the doors, ceilings and rafters'.[117]

Urban growth in the nineteenth century produced a multiplication of quarryings, quite apart from the increase of building in the stone districts themselves. Public buildings, such as the new post offices, museums and town halls, were distinguished by the heaviness and abundance of their masonry.[118] Improvement schemes, such as the building of the Thames Embankment,[119] consumed a prodigious amount of stone; so did church extension work, railway stations, esplanades and piers. Bourgeois graveyards – such as Kensal New Town and the South London Necropolis – were filled with polished granite. (The polished granite industry was pioneered in Aberdeen in the early nineteenth century;[120] as well as providing gravestones and mausoleums it helped to promote the rash of civic and patriotic statuary of mid- and late-Victorian years.)[121] Banks and business houses proclaimed their importance on columns of fluted stone, Doric, Corinthian or Ionic, according to the architect's whim. Ambitious drapers followed suit. When, for instance, Marshall and Snelgrove, the silk mercers, opened their new premises in Oxford Street in 1878, the frontage was made up of a promiscuous variety of stone:[122]

> The façade . . . is carried out in yellow malms and Corsham Down stone, all the cornices, string-courses, and weatherings being in Portland stone. The lower portion is divided into bays by pilasters of Portland stone, below which are Shap (Westmoreland) granite pillars on grey Aberdeen moulded bases, the Shap and Portland being finished at their bases with ornamental bronze bands.

The most widespread change of any in mid-Victorian building practice was connected with the rise of the Welsh slate industry. Slate quarried at Penrhyn and Dinorwic replaced locally manufactured tiles as a roofing material: the building of the Chester and Holyhead railway (opened in

1859) and the extension of the coastal trade from the North Wales ports brought them within the reach of every builder. A measure of the change, as well as of the vast increase in building during the 1860s and 1870s, may be seen in Table 1.2, showing slate shipments from Portmadoc.

TABLE 1.2[123]

Years	*Tons*	*Years*	*Tons*	*Years*	*Tons*
1825	11,396	1842	22,190	1859	58,466
1826	13,136	1843	24,716	1860	65,742
1827	10,290	1844	36,344	1861	59,696
1828	9,940	1845	43,858	1862	66,860
1829	10,464	1846	43,472	1863	76,594
1830	11,232	1847	39,601	1864	81,221
1831	12,211	1848	36,503	1865	89,293
1832	14,561	1849	32,467	1866	96,876
1833	13,975	1850	44,874	1867	113,838
1834	15,330	1851	46,338	1868	116,487
1835	18,113	1852	46,224	1869	125,574
1836	20,749	1853	48,858	1870	108,882
1837	23,966	1854	51,109	1871	121,838
1838	25,107	1855	48,279	1872	132,980
1839	27,935	1856	52,463	1873	144,880
1840	32,922	1857	52,697	1874	142,080
1841	29,067	1858	56,314		

Road-making was another prolific source of nineteenth-century quarrying, with thousands of new miles of town and city streets to be attended to, and the revolution in setting and surfacing associated with the names of Telford and Macadam. Closely fitted surfaces replaced the loose scattering of pebbles or flints, and they were bedded in generous layers of gravel, shingle and chips.[124] In the towns there was the further elaboration of kerbs, channels and pavements; causeways, instead of being macadamized, were flagged with hand-made blocks of stone, 'particularly those kinds . . . which do not . . . acquire a polished surface'.[125] Granite setts were particularly favoured by the paving commissioners: they were much more expensive to lay than a macadamized surface (and also more noisy under the horses' hooves), but unlike macadam they did not need to be renewed or repaired so often.[126] In Liverpool it was found that the cost of maintaining macadam was about four times as much as that of setts, and in Manchester macadam was altogether abandoned 'as being too expensive'.[127] 'It has been estimated that granite cubes on a concrete foundation will last from 25 to 30 years, probably much longer, and that the roadways would even then be in good order, or at all events of considerable value for breaking up.'[128] In country districts 'improved' road-making led to the opening up of hundreds of local quarries – parish pits on Poor's allotments and wastes,[129] or land rented by highway surveyors or turnpike trustees[130] ('quarrying' did not always mean hewing out: stone or flint might be picked or dug in the local fields).[131] In the big towns, however, local resources were insufficient, and the authorities came to rely increasingly upon outside sources of supply. Manchester, for instance, paved its first-class roads with setts from Penmaenmawr and Glynnog in Caernarvonshire, and from the Clee Hills in Shropshire, while its second-class roads were formed of millstone grit, quarried in the hills behind Oldham and Rochdale.[132] Plymouth's setts were made on the slopes of Staple Tor, Dartmoor,[133] Stoke-on-Trent's in the Derbyshire Peak.[134]

Sett making was established as a major industry in Aberdeenshire at a very early date. The first paving contract with the City of London was made in 1766.[135] By the 1830s London was established as the major outlet for the district quarries. Of 36,352 tons of stone exported from Aberdeen Harbour in 1831 only 143 tons were of building stone; 3,137 were 'pavement and kerb'; 33,072 tons were 'carriage way'.[136] The carriage way stones were made in six different classes. Those of least dimension were known as 'Common Sixes' and had a pavement depth of 6 inches and a superficial extent of 10 inches by 6; 'Half Sovereigns' had the same extent of surface, but were an inch deeper; 'Sovereigns' were about 10 inches long by 8 wide, and from 7 to 8 inches in depth; 'Cubes' – the variety delivered to London – were not less than 10 inches long and 6 inches wide, with a full depth of 9 inches; while a larger description of stone were known as 'Imperials'. ('Some parcels of these were manufactured between 1824 and 1827, but they have since been discontinued on account of their expense.') Aberdeen itself was paved with a class of pavement termed 'Common Nines', which were a little cheaper than those exported to London.[137] By the 1840s Aberdeen was facing increased competition in London from Guernsey stone and Mountsorrel syenite, which, though more slippery, were accounted harder.[138] London Bridge, 'the busiest thoroughfare in the world', was laid with Aberdeen setts in 1842, so were Cheapside, Poultry, Old Broad Street and Moorgate – the leading City thoroughfares.[139] The Euston Pavement, on the other hand (1843), which exercised a big influence on metropolitan road-making, was laid in Mountsorrel stone,[140] and in later years it seems to have been preferred for roads with particularly heavy traffic.[141]

II

Mining and quarrying activities reached out in many different directions during the early and middle years of the nineteenth century, and it was a long time before they acquired anything approaching a settled character. The scale of enterprise was extraordinarily uneven. Quarries varied in size from an irregular chasm in a field ('an area of a few square yards . . . where one or two persons are occasionally employed')[1] to giant faces of rock. The granite quarries at Mountsorrel, which grew so rapidly in the middle years of the century, with a single company in charge of them, had extended to half a mile of working face by 1879, and were employing about 600 men and boys.[2] The Penrhyn slate quarry was even larger, a gigantic amphitheatre, worked in nineteen separate galleries – the highest 1,000 feet above the lowest – and employing (in the 1890s) 3,000 men and boys.[3] On the Isle of Portland, by contrast, there were no fewer than fifty-six different working quarries at the time of the Parliamentary Inquiry into building stone in 1839, though only about 240 quarrymen on the island were regularly employed.[4] In ironstone mining the spectrum was equally wide, with, on the one hand, mighty industrial complexes, like Sir Josiah Guest's works at Dowlais or William Crawshay's at Cyfartha; on the other scattered bellpits or shallow shafts, worked by small groups or partnerships. As late as the 1870s iron was still being mined at Rotherham by open cuttings in a long line of irregular holes and pits.[5]

Coal mining at mid-century was equally diverse. In Monmouthshire there were the big pits attached to the great ironworks: in the Rhondda, mining was in its infancy, with coal being drawn from levels in the hills. In the nearby coalfield of the Forest of Dean numbers of pits were still being run by the colliers themselves – 'free' miners who qualified for their 'gales' by ancient privilege and right: by being born in the Hundred of St Briavel's, by apprenticeship, or by having worked a year and a day at the pit[6] (a new class of capitalist coalowner was beginning to appear in the Forest: in 1856 the number of coalworks was estimated at 221, but the two largest firms accounted for more than a third of the Forest's output of 460,000 tons of coal).[7] The most advanced pits, from the point of view of mine management and technology, were the 'seasale' collieries of Northumberland and Durham. They were in the ownership of large proprietors, and had a body of trained viewers in charge of them. In 1843 there was an average of 130 miners to each pit.[8] In Yorkshire, on the other hand, there were many small collieries whose owners were said to be no more 'instructed or civilized' than their employees;[9] the pits, 'though numerous' were also shallow, 'many of them being less than 100 yards in depth and few exceeding 250 yards'.[10] In South Staffordshire, too, the pits were shallow, and because of the irregularities in the strata ('one mass of faults')[11] they were quickly worked out. There were twice as many pits in Staffordshire as in Northumberland and Durham – 548 against 270 according to the *Mineral Statistics* for 1856 – but they produced less than half the tonnage of coal.[12] Instead of being run by the owners the South Staffordshire pits were contracted out to butties, a class of gombeen men, or half-proletarian exploiters, who took on the running of a pit for a month at a time, and engaged the men and tools for the job. 'There is no system at all', a Government Mines Inspector complained to the Select Committee on Mining Accidents in 1854, 'the pits are turned over as soon as they are sunk, to the butties. . . . the viewer, as he would be termed in the North, or the ground bailiff, visits the colliery once a week or once a fortnight, by himself or with his deputy, measures the dead work, and enters it in the book. The management is entirely in the hands of the butties.'[13] The *Morning Chronicle* Commissioner, who visited Staffordshire in 1850, was equally dismayed:[14]

> The change from the collieries of Northumberland and Durham to those of Staffordshire seems like going back at least half a century in the art of mine engineering. On the banks of the Tyne and Wear, science the most profound, and practical skill the most trained and enlightened, are brought to bear upon the excavation of coal. The pits are worked under the constant superintendence of regularly educated viewers, each of whom has a staff of assistants, more or less scientific and with practical skill, to carry his directions into execution. In the Staffordshire coal district, on the contrary, everything seems to be done by the roughest rule of thumb. The pits, as regards depth, are mere scratches, compared with those of the North; and, except in the

> case of a few of the thick seam-mines, they are ventilated solely by the agency of the vast number of shafts with which the whole coalfield is honeycombed – anything like artificial means for creating a current of air being seldom or never thought of. The workings in such excavations are, of course, very limited. The labourers could not breathe at any considerable distance from one of the shafts; and the consequence of the whole system is, that the coal is worked in the slowest, most dangerous, and least economical fashion.

Nothing like a class of regular coalowners had yet appeared. They ranged all the way from great nobles, like the Marquess of Londonderry and the Earl of Dudley, to the farmer-colliers, who in the 1850s were still familiar figures in the Wigan coalfield.[15] A Yorkshire owner at Heckmondwike around 1850 was George Brearley, who rented a few fields for coals and covered them with 'day-hole' pits. He was a carman and proprietor of one of the local carrier's carts, 'The New Delight'; among other trades he was also a farmer, a beerseller, a grocer, a blanket weaver and a joiner.[16] Many coalowners at this time were not coalowners in the first place, but landed proprietors, local speculators or ironmasters, whose coal-getting enterprises were subsidiary to their other activities. The Earl of Dudley, for instance, profited from a great complex of money-making activities in the Black Country and in 1866 most of his numerous coal mines were still being worked by butties.[17] In the Monmouthshire valleys, on the south-east border of the Welsh coalfield, there was a whole class of proprietors and lessees who derived a handsome profit from their collieries 'without inter-meddling with the coal trade':[18]

> A man of wealth and substance, or, at all events, having a reputation as such, obtains the lease of a tract of minerals for a term of years, or possibly buys a colliery. In either case he does not work it himself, but takes the biddings of *contractors* for raising and supplying a given quantity of coal in a certain time at so much a ton. These contractors are in most cases men of no capital; generally they have themselves been miners, who, possessing more enterprise than their fellows, and having nothing to lose, take the chance of bettering their condition by entering upon a speculation of this nature. The lessee of the colliery plays off one of these men against another in bidding for the contract, taking such offer as appears most eligible, which, in the usual run of cases, is the lowest. . . . These contractors rarely do well. They live by

> shifts and expedients – by extending credits, and by bills – for a while, till the evil day comes, when they sink, and are succeeded by adventurers of the like character.

Ironmasters, who at mid-century had the ownership of many of the larger coal pits, were interested in them as a source of cheap and accessible fuel for their works, rather than as a means of profit. They were generally worked by sub-contract,[19] not only in Staffordshire, but also in the ironworking districts of Shropshire[20] and Derbyshire,[21] Scotland[22] and South Wales. In the 'heads of the valley' ironworks in Monmouthshire, 'master miners' were appointed to raise both iron ore and coal,[23] and in opencast mining there was a local form of sub-contract known as patching, which in Ebbw Vale survived to the 1870s. 'A patch of hillside where coal or iron ore outcropped would be leased to a sub-contractor employing two or three men: they dug away the overlying soil and shale which was wheeled by women to a "tip" or bank of waste, and the underlying coal and ore conveyed to the ironworks.'[24]

The Cornish mines in the 1840s and 1850s – tin and copper – were more advanced than many of the collieries, from the point of view both of technology and of capitalist finance. Along with the railways they served as a veritable forcing ground of speculative investment. Their ownership was in the hands of companies rather than individuals and their shares were quoted on the London Stock Exchange. The whole pattern of development was in the direction of large-scale works. At Consolidated Mines, Gwennap – the largest of them – 1,730 men, 869 women and 597 children were employed in 1842; the total length of the horizontal galleries was forty miles.[25] The rise of Devon Great Consols, which paid out £763,910 in dividends to shareholders between 1845 and 1861 (within a year of its establishment £1 shares had reached £800)[26] was one of the capitalist success stories of the age. The fortunes of individual mines, however, fluctuated considerably, with sudden jumps and equally precipitous descents. Yields were notoriously treacherous – 'the lode is cut rich to-day and tomorrow poor'[27] – and demand was fickle, depending as it did on the international state of supply: the price of tin could rise or fall as much as 30 per cent or 40 per cent in a year.[28] The custom of the annual share-out of profits weakened capital structure,[29] and companies of 'adventurers' (as the stockholders were called) were constantly being formed and re-formed.[30] Stops were frequent (sometimes temporary,[31] sometimes for good); and the mean average life of a mine was barely five years, as one may see from Table

1.3 drawn up in 1846. (It will be seen that the young mines were much more productive than the old, and this, together with the cost of deep mining, may explain the fact that the life span of a Cornish mine was typically short.)

TABLE 1.3

Copper mines in Cornwall, 1815–1845: Age and productivity of Cornish mines selling upwards of 5 tons of copper ores at public ticketings

	20 years and upwards	*10–19 years*	*5–9 years*	*less than 5 years*
Age	35	40	31	114
Highest average of a single mine	$13\frac{5}{8}$	14	$14\frac{1}{8}$	$26\frac{4}{8}$
Mines producing above 10% copper	2	4	2	13
Mines producing above 5% copper	32	34	29	70

(Source: *Memoirs of the Geological Survey*, 1846, Vol. 1, p. 513.)

Lead mining enterprise was even more unstable than copper and tin, since it was subject to the same geological uncertainties – a vein that would be very productive at one point might prove barren at the next – but enjoyed far less in the way of capital resources to meet them. The lead was worked, for the most part, in narrow, overhand stopes, on levels entered from a hillside (many of the works were in high moorland) with little of the expensive pumping machinery which characterized the profitable Cornish mine. Works were constantly having to be abandoned, as water levels rose, or good veins ran out. But old levels were often opened up again (sometimes by parties of miners working on their own account) and 'trials' were constantly made in the hope of finding undiscovered ores which the 'old men' (the miners of former times) had failed to exploit. The table of lead mines in Cardigan and Montgomeryshire, drawn up by Warrington Smyth in 1847, is peppered with references to workings 'lately abandoned' or 'lately resumed'.[32] Here are some examples from Cardiganshire:

Crown Mines, E. of Llanfair	Lately abandoned.
Rhysgog East	One of Sir H. Myddleton's mines; of late unsuccessfully worked.
Bron y Gar-llan	Working on a small scale.
Esgair Mwyn	Workings lately resumed.
Llwyn Malys	This mine is again raising ore.
Grogwynion, two lodes	Very ancient mine, on which workings are recommenced.
Gelli'r Eirin	Lately abandoned after raising 60 tons of ore.
Llanerch	Small workings lately commenced.
Llancynfelyn, six lodes	This mine has lately been re-opened on a large scale, but with little success.
Cwm Einon	Lately worked on a small scale.

Lead mining enterprise was dispersed over a large number of small workings – some 400 of them raised ore in 1856,[33] more than the total number of tin and copper mines in the country, though employing many fewer workers. In Montgomeryshire, with the exception of the famous Van mine, Llanidloes (itself an on and off affair),[34] all of the workings seem to have been comparatively small; in North Wales there were large mines like Talargoch, 'opened upon to great depth, and affording employment to large numbers of people'. There were also smaller setts, known as 'Poor Men's Ventures', particularly around Holywell and Llanrwst.[35] 'These adventurers work the mines themselves, in many instances after leaving their work at the larger mines, thus prolonging the hours of their underground labour.'[36] The 'Poor Man's Venture' was a small miners' collective, which undertook to work a mine, and divided it on a system of 'ounces' or shares. 'The reason why the men resort to "ventures" and work so much on their own . . . is this' a lead miner told the *Morning Chronicle* Commissioner in 1851: 'wages some time ago were so much reduced by the masters, when the men were working by fathom, that they could not live, and were therefore bound to adventure for themselves.'[37]

Lead mining at Alston Moor, Cumberland, was dominated by the extensive works of the London Lead Company, but there were also 'numerous' small mines. The *Morning Chronicle* Commissioner, writing in 1850, claimed that there were sixty in the area[38] but Thomas Sopwith, a better authority though speaking later, put the number down as thirty-eight.[39] Here is Sopwith's description of the way in which some of them started:[40]

> The practice . . . is, that when any persons discover a vein, or imagine that there is a mineral vein, they apply to the agent of Greenwich Hospital for permission to make a trial. That permission is given for a certain length of time, and they . . . commence to make a trial. . . . If they are successful, then they probably take a lease of the mine, and continue to work it according to the productiveness of the mine. But it very frequently happens that those adventures are failures altogether. There are a very great number of cases in which poor men, having little or no capital, but depending chiefly on their labour, get two or three partners with a little money, perhaps, to assist them, and they continue the trials in that manner.

The real stronghold of the 'Poor Man's Venture' was in the ancient lead working district of North Derbyshire. As late as 1872 there were nearly 200 separate lead mining concerns in the county, 138 of which produced less than five tons of ore during the year.[41] In 1856 there were twenty-six little mines in Cromford, worked by an average of four to six men each.[42] In Derbyshire, as in other lead-working parts of the country, there were a number of attempts to promote larger mines during the mineral boom of the 1840s and 1850s; expensive machinery was installed at several places 'but the mines proving unproductive much of it has been removed'.[43] One of them, Water Groove Company Mine, was employing only six miners in 1856, after a capital outlay which one of the promoters estimated at £30,000; another, High Rake, was divided into six or seven setts 'for poor men' (the miners took them on not for profit but as a winter employment; in the summer they went out to the harvest).[44] 'With a few exceptions', Smyth wrote of Derbyshire in 1876, 'the mining of the present day, although it brings out the fair result of 4,442 tons of ore (in 1873) is limited to "poor men's mines", or small adventures, in which with none but the simplest mechanical aid old works are explored again which scarcely invited the attention of a company.'[45]

Whitby jet digging had close affinities to 'Poor Men's Ventures' in lead. The best stones would realize (in 1864) from 12*s*. to 14*s*., but digging for them was, according to Walter White, 'little, if any, less precarious than gold-digging'.[46] The industry developed slowly during the first half of the nineteenth century, but it grew rapidly after the Great

Exhibition of 1851, helped by the personal preference of Queen Victoria for jet ornaments, and by great funeral occasions, like the death of the Duke of Wellington in 1852.[47] In 1864 some 250 men and boys were engaged in the digging and rather more than a thousand women, children and men in the manufacture.[48] The jet was found among the cliffs in very narrow seams 'not more than two or three inches deep' and the worker had often to be lowered by rope over the cliff until he reached a ledge. He worked in a very narrow space 'almost in a recumbent posture', dragging the rubble and earth outwards on his hands and knees.[49] The diggers whom George Head records in 1835 ('one with a narrow sack of jet on his shoulder') worked at a pitch inland: 'They said they were miners by trade; that they rented the ground where they worked on speculation; that the tradesmen in Whitby gave them a fair price for all the jet they could furnish, and manufactured it into ladies' ornaments; that the price varied considerably – from 3s. 6d. to 10s. per stone.'[50]

In Cornwall the typical evolution in tin and copper had been from small enterprise to large. In china clay – which emerged in the nineteenth century as one of the duchy's leading mineral industries, the tendencies were the other way round. The industry had been started by wealthy promoters from outside. William Cookworthy, who took out the first patent in 1773, was a Plymouth merchant who conducted a porcelain factory in conjunction with Lord Camelford.[51] At St Austell the first extensive works had been started by Josiah Wedgwood, who, in 1791, was toying with the idea of establishing porcelain as a manufacture in Cornwall instead of waiting for the china clay to come to North Staffordshire.[52] But by the 1820s the potters who had played such a large part in the early days of the industry were moving out, and their place was being taken by locally born 'adventurers'. 'A characteristic of the industry during these early years', writes Mrs Barton, 'was the frequent subdivision of much of the clay-bearing land into a multiplicity of small setts. This was done by the mineral lords in order to meet the modest means of those taking out leases. In most cases the setts were far too small to allow of proper development and the leases too short to encourage this.'[53] The small scale of production nevertheless persisted. Of the eighty-nine pits open in 1858 only two were producing more than 2,500 tons of clay a year and there were none as yet which had blossomed into firms.[54] Pumping engines did not come into general use until the 1870s, when they were bought second-hand from the bankrupt tin and copper mines, so that workings remained for a long time restricted in depth, and the industry (which doubled its production in each of the later nineteenth-century decades) grew in size by taking up more and more ground. Competition among the small producers, and the comparative ease of entry to the industry, kept prices low, but the attempts to bring about an amalgamation did not succeed until the formation of China Clay Firms in 1921.

There was a remarkably similar cycle of development in the Cheshire salt industry, which Brian Didsbury writes about. Here too was an industry which was easy to enter, and where the expansion of production came about through an increase in the number of small producers rather than a growth in the size of the firm. The Liverpool merchants, who played such a big part in the industry's eighteenth-century development, had, by the nineteenth, disappeared. Small proprietors, on the other hand, were continually springing up, encouraged by rising demand and the small capital needed to start a works, or rent a family pan. Attempts to limit competition amongst the producers – it was even more ruinous and fierce than in the case of china clay – broke down because of the ease with which newcomers could enter the trade: some of them were the boatmen who carried the salt down to Liverpool on the river Weaver; some started off as lumpmen sub-contracting for a pan.

In brickmaking, too, growth in the early and middle years of the nineteenth century came about primarily through a multiplication of small producers. In one of the largest yards in Bristol (according to an account of 1866) the whole number of persons in employment, including children, did not exceed thirty; others in the neighbourhood were much smaller, 'one or two moulders, with a few helpers to each, being all that are employed'.[55] According to a companion account there were about eight large brick and tile yards in the Staffordshire Potteries at this time and then 'a great number of small brickyards in which bricks are principally made in the summer season'.[56] The successive reductions in the brick tax removed the burdens on the small producer and made it easier to enter the industry (the removal of the salt tax in 1824 had a similar effect on saltmaking), while the diffusion of building activity, and the small-scale nature of the building industry, put a premium on local suppliers and cheap freights (as in the case of road-making materials, haulage could be a high proportion of total cost). Despite the vast increase in production (in 1855 no fewer than 130,000,000 bricks are said to have been produced in the Manchester region alone),[57] the industry remained essentially local in character, each town being served by a perimeter of brickworks (there were twenty-four of them at Southampton in 1856, according to an interesting and detailed account drawn up by Robert Hunt).[58] In the countryside the industry was very widely dispersed. 'Almost every parish in Staffordshire . . . has its brickfield', a factory inspector complained in 1872, 'seldom employing more than one boy, but occupying perhaps nearly half a day in visitation.'[59] A 'profusion' of such small country yards has been identified near Dorchester, though only one, Broadmayne, is mentioned in the trade directories of the time: others were at Tolpuddle, Puddletown, Bryantspuddle, Warmwell, Owermoigne, and (a little further off) Coombe Keynes and Bere Regis.[60] Tileries spread in much the same way, as the historian of Westmorland agriculture records.[61]

> Consequent upon the inclosure of the commons a new industry sprang up in the northern part of the country in the shape of tile works. Tile making was introduced into East Cumberland in 1821 on the Netherby estate and thirty years after there were no fewer than 42 tileries in the district.

A proprietor needed little in the way of capital to start a brickworks, and he was often a man with other interests besides – a farmer (or clergyman) wanting to profit from the claybeds on his land, a builder (or a grocer) with a sideline in bricks.[62] In Birmingham the 'old school' of brickmakers, according to their chronicler, were brickmasters in the summer and maltsters and brewers in the winter. In the Manchester district (according to Meade-King, the factory inspector in 1876), the occupiers of brickworks 'often farm to some extent: if they find that they are not allowed to employ children in the brickyards they set them to work on their farms'.[63] Instead of undertaking the management of their works, brickmasters would often sub-contract them, either to a middleman, who undertook responsibility for the season's operations (and perhaps contracted out the work himself), or to a moulder, or by a series of separate agreements to cover the different components of the work – clay getting, tempering, wheeling off, moulding, kiln-burning, and cartage. A variety of arrangements were possible: sometimes the moulder contracted for the whole job and then sub-contracted in turn to the temperer, sometimes he employed him, and sometimes both moulder and temperer were contractors on equal terms.[64] Usually it seems to have been the moulder who was the dominant figure on the brickfield, with the kiln-burners as a little sect on their own.

In quarrying, the nineteenth century saw the simultaneous growth of both large works and small, but it was the smaller works which predominated. Out of a total of 7,132 quarries recorded by the Mines Inspectors in 1912, the number employing more than ten 'inside' workers was only 1,017, and the number employing more than thirty was 334.[65] Some of the largest works – such as the Mountsorrel Granite Company and the Darbyshire's works at Penmaenmawr – developed around sett making and the quarrying of road stone. But road-stone quarrying was also the occasion for a multiplicity of small and often short-term enterprises, and sett-making as an industry (like its cousin, dry stone walling) was widely diffused.[66] In a smaller quarry it was often undertaken by an individual workman, who agreed to hew and dress the stone at so much a cubic yard.[67] He might contract for the work directly with the highway authority, or indirectly, through one of the local builders. Father and son might do the work together – or a pair of brothers – with one or two outside hands to help them. The writer of *Reminiscences of a Stonemason* took on such temporary jobs in rough times, when other kinds of work were difficult to get, and he gives us a number of examples of them.[68] In such a situation the quarryman was both a rock man and a sett maker – 'he is entirely on his own responsibility as far as the whole thing goes'.[69] On the Wiltshire Downs flint digging for the road works was undertaken on a similar basis.[70]

> Generally each man has his own station and tools, and works singly. His apparatus consists of a pickaxe, a stout iron drag, a steel fork with eight grains, a shovel, a riddle, or 'ruddle' (sieve) and a wheelbarrow. Along the winter and spring he has a movable shelter made of a couple of close hurdles covered with a rough canvas sheet; whichever way the wind is setting he adjusts this against a stout prop, and works behind it.

'Purbeck kerb' (kerbstones, pavement setts, and gutterstones), which in the 1880s (according to one account) was to be found 'in almost every town' in south-east England ('Brighton, Portsmouth, Winchester, Salisbury . . . have made use of it in almost every street'),[71] was made by a tribe of quarrying individualists, the Purbeck marblers, who worked the cliffs and downs in the neighbourhood of Swanage and Corfe Castle. This was an ancient race of quarriers, the descendants, according to one conjecture, of a colony of Norman stone workers. They were organized into a company ('a kind of trades union . . . at the meetings of which the more intelligent men are generally outvoted, as in other and more important assemblies')[72] and held their own court to enforce their rules. They traced their privileges back to the fourteenth century, and claimed a hereditary right to dig stone at will – and in perpetuity – within the boundaries of the 'isle'.[73] Employment was a matter not of labour hiring but of double descent and the serving of a seven years' apprenticeship: none was allowed to work in the quarries 'unless the legitimate offspring of a quarryman and the daughter of a quarryman, and properly apprenticed to the craft'.[74] Apprenticeship was from father to son, and many of the quarries seem to have served as family berths, though for the larger ones a small company of quarriers would combine.[75] The marblers were stonemasons as well as quarriers, manufacturing their wares in little stone-built sheds with lean-to roofs, open at one end, placed near the top of their diggings; sinks, horse troughs and gateposts were apparently (along with kerbs) the staple products.[76]

The typical quarry of the nineteenth century – if it is possible to generalize about so various a phenomenon – was rented rather than owned, and passed through a succession of different hands. It might be intensively worked at one time, when there was a particular order to fulfil, but at other times lie unrented and fallow. This was particularly the case with building-stone quarries, where demand, even at the larger ones, was very up and down, and where a lease might run only for the duration of a building contract. Leases were often taken for short periods of time, from year to year in many cases, so that there was no inducement to cut drains, or build approach roads, or install plant.[77] The following advertisement, which appeared in the *Oxford Chronicle* in 1847, is expressive:[78]

> HEADINGTON QUARRY, NEAR OXFORD.
> To be LET, either from year to year, or for a term of years, as may be agreed upon, and with immediate possession – a large and productive STONE QUARRY, with most excellent CLAY for BRICKMAKING. The Quarry produces freestone, wall, lime, block, and hard stone. Apply to Mr. Matthews, Solicitor, Oxford.

With a farmer's quarry the use to which it was put might depend on the state of family labour: if there were sons of the right age – or a big family of brothers – it could be worked in conjunction with the farm; if not, it might be rented out to a local mason, publican or contractor. Roadstone quarries were often shared by different users. An example is given by the

Cumberland quarryman who appeared before a Royal Commission in 1912 and described his place of work:[79]

> Part of it is a parish quarry and part under royalty. There are four or five different classes of men going to work in that quarry a year under different headings. There are road men. They will quarry a bit and then the foreman of another section will go with his men and quarry a bit. There are contracts let for widening district roads. At the present time there is a contract to a coach driver who is not a practical man, although he has the contract, he is quarrying, and sometimes farmers and such like men.

In Clydach Vale, Breconshire, where limestone was extensively quarried, the whole range of enterprises already discussed co-existed in one place. The following account of their activity in the 1900s is based on the memories of people who were alive and connected with quarrymen at the time.[80] The largest quarries (about forty men were employed on them) were the two owned by the Clydach and Abergavenny Limestone Company, the 'Cuckoo' quarry by the railway station, and the quarry at Llanelli hill. At one time the company had a contract to supply the Ebbw Vale steelworks with lime ('it had to be broken down to a 6 inch gauge for the furnaces'), and later on turned to making roadstone. Pwlldu quarry, on the mountain top dividing Clydach Vale from Blaenavon, was worked by the Blaenavon Iron Company for its blast furnaces, the lime being taken to the works on an iron tram-road. The vast Llangattock quarries on the other side of the valley – the face is some two miles in length – had at one time been worked for the ironworks at Nantyglo and Beaufort; at a later stage it seems to have been quarried by or on behalf of the Brecon and Abergavenny Canal Company;[81] by the 1900s it was in the possession of a family of Thomases, who rented it from the Duke of Beaufort and worked it as a lime burning enterprise. The lime, which was sold to farmers over the whole district from Abergavenny to Crickhowell, was carried about the country by twelve pack mules, with two riding ponies and a driver, Aneurin Hopkins, to lead them. Each would carry two one-ton sacks, and they travelled together as a caravan, making one or two journeys a day. John Thomas, who started the lime business, was a small farmer, with a holding of ten acres, which he built up in the course of a short lifetime to forty. He died leaving three small farms to the family, and the quarrying was taken on by Thomas Thomas ('Thomas the Mules') one of his sons. Three of his brothers worked with him – Reg, Dai and George; three sons – Arthur, Billy and Albert; a man called Low Morgan ('a regular rockman'); Howell Booth (a labourer who lived at one of the family's farms and was chiefly employed at the Kilns); Benny Edmonds, and Aneurin Hopkins (a cousin) who drove the mules. The three sons were chiefly engaged at farm work, but came to help at the quarry during spring and autumn, the peak seasons of the year. 'It was a family affair . . . the sons was all working through and through, on the farm and on the kilns.' Blaenonneu quarry, on the other side of Llangattock mountain, was worked by Jack Hopkins, cousin to the Thomases and father to Aneurin who drove the mules. (When Jack died his son Emlyn took over the working of the quarry; he was club-footed and Aneurin left the Thomases to go to help him out). Jack Hopkins is remembered as 'the typical farmer with his quarry . . . 1 or 2 kilns which would produce 4 or 5 tons a day'. He farmed 'when it was right for farming' (his farm was a mile and a half away from the quarry), and when it was not, he went quarrying. Another family who ran their own quarry were the Watkins, who had the Wenelt quarry at Gilwern, by the canal wharf – which they rented from a small farmer – and a building stone quarry at Govilon, further down the valley. There were five brothers in the family. One of them, Tom, was a railwayman, the others all worked as quarrymen, sometimes on their own account, at Gilwern and Govilon, sometimes as employees. Their stone was taken away by canal, with one brother ('Ben the Boat') in charge of it. (At Whitsun the boat was used to take the Sunday School children 'pleasuring' at Llangattock). The Wenelt quarry was an up and down affair. It had a good trade in 1912, when the Great Western Railway was extended from Nantyglo to Bryn Mawr, and the quarry supplied the stone for building the new station. But the work was never regular, and when orders were slack the brothers took up employment as rockmen at the other local quarries: 'Ben the Boat', died at the face (from a rock fall according to some accounts), working for the Clydach and Abergavenny Company in the quarry at Llanelli Hill. The Black Rock quarry, the one at work today, was, in the 1900s, disused, though earlier it had sent limestone down to the canal wharves at Gilwern. In the 1920s it was taken up by A. L. Watkins, a cattle haulier from the Usk valley who 'did a bit of farming' at Llangattock, and later by a Mr Weaver of Llanfoist, 'more or less a timber merchant as well as a quarryman'. Today it is worked by a roadstone company, which also supplies the Ebbw Vale steelworks with lime.

In the early and middle years of the nineteenth century a mineral enterprise was distinguished from a factory by the comparative paucity of plant. In the simplest kind of mine – the day-hole, bell-pit or drift – there might be nothing more than a few barrows, some iron rails, and a hand-winch or gin to draw the minerals to surface: 'scores of coal-pits' in the Wolverhampton district were said to bear this character in 1850, 'usually' with about ten men underground 'and a couple of men and as many women at bank'.[82] The 'Balance pit' – a very common 'improvement of the 1820s and 1830s in the South Wales coalfield'[83] – was hardly more elaborate. The winding gear was worked not by water but, in Dowlais at least, by girls:[84]

> At the mouth of the coal-pits . . . they alternately raise the trams loaded with iron-stone and coal, and let down the empty ones to be filled below. These girls work at what are called 'balance-pits' – the principle being to raise the full tram by the descent of the empty one, assisted by such a weight of water as will produce an equipoise. As soon as the loaded tram reaches the mouth of the pit these girls drag it away; two of them then step on the platform which supported the tram, and haul at a line passing a pulley overhead, which by a valve lets off the water from the tram at the bottom of the pit. In doing this, one foot of the girl on the open side of the pit's mouth is often suspended over the abyss. One of these girls sets the drum in action, regulating the velocity of the ascending and descending trams by a 'break' acted upon by a pulley.

There was little in the way of plant in a brickfield; the moulder worked under a rude thatch covering ('a protection from the sun . . . not from the weather'),[85] drying was in the open air, and if burning was by open clamp – quite a common practice on the southern brickfields in the 1860s[86] – there might not even be a kiln. In quarrying at mid-century everything was done by hand, even a great part of the haulage. 'All the means and appliances of labour about the quarries are of the rudest description', the *Morning Chronicle* Commissioner wrote of the Isle of Purbeck. 'Main force is the element principally relied upon. . . . Long as the district about Swanage has been quarried and immense as has been the quantity of stone shipped from it, it does not, even to this day, possess a pier or jetty of any description. . . . The stone is dragged from the shore by very tall horses, in carts with very high wheels, as far into the sea as such an apparatus can venture with safety. . . .'[87] Cornish mines were equipped with engine houses at an early date, and their methods of drainage and ventilation were the envy of mining engineers in more backward places; but men went up and down the mines by ladders and their work itself was untouched by machines. In Derbyshire lead mining, drainage was still characteristically by water-wheel, and by the seventeenth-century innovation of 'soughs' (an underground tunnel dug beneath the level of the mine to draw off surplus waters).[88]

A mineral enterprise was also, by comparison with the factory, short-lived. This was partly because of the shallowness of the workings, and the frequency with which they were exhausted, and partly because of the element of speculative risk. 'In my young days', wrote Osborne O'Hagan, an active participant of the 'new' mining of the 1870s and 1880s, 'deep sinkings involving much capital had not been favourably considered.'[89] In some cases the subsidiary character of a mineral enterprise acted as a bar to capital investment, in others the bar was the diffusion of ownership and control. In South Staffordshire coal-pits were short-lived for reasons of geology alone, because of the frequency of faults. 'There are no large pits', wrote Rowe in 1927, 'a small pit is sunk and worked out, and a new pit sunk the other side of the fault. The district is dotted all over with disused shafts.'[90] Cornish mines were literally 'Adventures' – with an element of gambling speculation which meant that the company was put at risk when yields proved disappointingly low. A brickworks at mid-century was not expected to last long: 'The proprietor . . . usually rents the necessary land at a price per acre, and in addition pays for all clays removed at a set price . . . as the brick earth is exhausted, or the workings reach an inconvenient depth, the ground is levelled and again thrown into cultivation.'[91] In the case of quarrying it was demand rather than supply that was liable to give out. Even a larger quarry was apt to be worked by fits and starts, depending on the state of orders. Some 450 workers were employed at the Kirkmabreck granite quarries, Dumfries in the 1830s, to supply the dock extension works at Liverpool, and no fewer than eleven schooners, specially engaged by the Dock Trustees, were used to carry off their stone; in 1844 the works came to an end, and the number of quarrymen employed at Kirkambreck fell by two-thirds.[92] A smaller quarry was likely to be worked 'at some time or other' rather than continuously.[93] In the case of a family quarry – like the Thomas Thomas's limeburning enterprise at Llangattock, or the Purbeck marblers' pits – a lot depended upon the balance in the family between able-bodied members and dependants: there had to be enough grown-up brothers or sons to go round. The same was true of family berths at the saltworks, though in this case it was often a matter of daughters as much as sons.

Another feature of mineral enterprise, especially pronounced in the early and middle years of the nineteenth century, was the very wide use of sub-contract. In some cases the proprietor sub-let the entire working of his premises and plant; in others it was a matter of hiving off particular classes of work – limestone getting, in the case of an ironworks,[94] for instance; haulage in that of a lead mine, coal pit or quarry.[95] (The London Lead Company was paying out £2,000 a year to local farmers in the 1850s for horses, carts and teamsters).[96] In some cases sub-contracts were made by the proprietor with a labour master: the 'driftmaster' in the Cheshire salt mines was responsible for engaging the rock getters, and he paid them out of the salt delivered on the wharves or at bank;[97] the 'washing master' at a lead works recruited gangs of children for dressing the ores (in a remote place like Patterdale he might have to lodge them as well as paying them for the week's work);[98] the 'charter masters' and butties in the coal and iron mines of Staffordshire and Shropshire had the entire underground working of their pits, though they in turn sub-let part of the work to the facemen (the holers and pikers), and surface workers might be in the charge of the separately engaged 'butty banksman'.[99] In metalliferous mining and quarrying, on the other hand, under the 'bargain' system, contracts were usually made with a workers' collective, a company of men who would have a large degree of autonomy; in 'Poor Ventures' (such as those in the lead mining districts of North Wales) they might have complete control. There was also the disguised form of sub-contract by which a man could only make fair money if he had help from members of his family. This was very general in brickfields and saltworks, and in the more limited form of father and son partnerships it was also a common arrangement in mines and quarries.

The prevalence of sub-contract was closely linked to the often provisional character of mineral enterprises. It encouraged the lateral extension of activity rather than vertical integration in depth. It was a system which made it easy to open up new or temporary workings, because it limited the proprietor's responsibilities and reduced the demands on his purse. It enabled him to share his risks with the middleman and to some extent with the worker too. It flourished under owners who were absentee or otherwise preoccupied, allowing the landed gentleman to start up a coal pit, or the farmer a brickworks, without limiting the development of his assets, or laying too much claim on his time. These systems of indirect employment survived, however, even when a proprietor undertook the management of an enterprise himself; in metalliferous mining, and in certain classes of coal mining and quarrying, they continued to govern the form of labour hiring, if not always the content, down to the end of the century.

III

Mining and quarrying were distinguished from factory labour by the fact that they were, first and foremost, sweat and muscle jobs; and little that happened in the nineteenth century impaired their labour-intensive character.

The Cornish miner inched his way forward by painful degrees. The tin and copper were thinly disseminated in very hard rock, and progress in boring them was slow: 'one, two or three feet in a week, or a few inches daily, is often the whole amount of the united operations of twenty or thirty men', wrote Joseph Watson in 1843.[1] The spread of deep mining, which followed the introduction of pumping engines in the later eighteenth century, and the vast increase in mining for copper, multiplied the need for hands, not only because of the increased volume of production, but also because the deeper veins were more difficult to excavate. The dressing operations too demanded much more work, since the deep-mined ores were inferior to those which in earlier years had come from shallow shafts or open streams. 'Where ten persons previously would have . . . broken up the quantity of ores to produce a ton of copper, it now required the labour of twelve or fifteen.'[2] In the larger mines there might be as many hands engaged in preparing the ores for market as there were in ore-getting underground – in 1836 there were some 1,400 of them at Consolidated and United Mines, Gwennap. Most of this work was carried out by women and children,[3] and until the introduction of stamping and dressing machinery, in the second half of the nineteenth century,[4] the work was done with the hands; 'spalling' – breaking the lumps with a small sledge-hammer; 'cobbing', 'in which the blow is directed with the object of knocking off a piece of poor rock from a lump of mixed ore and refuse'; and 'bucking' – breaking the ore with a very broad flat hammer in order to reduce it to powder ('ragging', the breaking up of the very large lumps of rock as they came from the mine, was a work mostly done by men, using a large sledge-hammer weighing about 10 to 12 lbs).[5]

The intensity of labour was increased as well as its amount. In the deeper levels the air was so thin that the miner's candle could only burn with difficulty.[6] Dust was another hazard which grew more deadly in its effects from the increased use of gunpowder and the metallic dust of drilling. In a copper mine the worker had also to contend with great heat. The 'fast ends' which the tutworker encountered at the blind extremities of a shaft were possibly the most murderous mining conditions in the world, with so little oxygen in the air and so much heat that it was impossible to work for more than twenty minutes at a stretch (to recover, the worker had to take an hour's rest and bathe himself repeatedly before he was fit to resume his stint).[7] The temperature in the United Consols Mine at Gwennap was said to be as high as 125°: 'the men can only work by short spells, and are constantly supplied with cold water for drinking, which soon becoming hot in the warm atmosphere of the mine, is sent down from above at very brief intervals'.[8] Another health hazard was the lengthened journey to work, which involved long hours of toiling up and down perpendicular ladders. There were mines so deep that 'not less than three hours' were said to be expended on them, and an hour's journey each way (according to the Medical Officer of the Privy Council, writing in 1861) was typical.[9] The first symptoms of failing health among miners – and the onset of what was known as 'the miner's disease' – were dizziness and exhaustion on the ladders. (Older miners, unable to manage the long climb, might be put to work at bank or in the shallow workings.)[10]

Cornish engineers pioneered most of the major advances in mining technology, and in the nineteenth century they carried the Cornish pumping engine (and often Cornish miners too) to every part of the world. But in Cornwall itself their achievement was extraordinarily one-sided. They made it possible to discover and bring up ore from the deepest levels; but for all their winding engines and pumps and their maze-like ventilation systems, they invented little or nothing to lighten the miner's work load. When it came to transport, the crooked shafts and slanting galleries seem to have defeated them. The primitive 'man engine' (a rudimentary and perilous version of the 'cage') was never generally installed, despite the faith expressed in it by witnesses before Lord Kinnaird's Commission in 1864: the miner was left to climb it. Some ideas of the poverty of provision for carrying the men up and down the shafts, or for assisting them at the face, can be obtained from an interesting inventory drawn up by the Mines Inspectors in 1890.[11] (See Table 1.4.)

In coal mining, too, work got harder in the nineteenth century, and

TABLE 1.4

RETURN showing the Names of the various METALLIFEROUS MINES within the Stannaries of *Devon* and *Cornwall* now Working, and the Number of MEN working Underground in each Case; which of them have Machinery other than Ladders for raising and lowering the Men; which of them employ Boring Machinery with Compressed Air; and how many of such Boring Machines are in use in each Mine.

CORNWALL.

NAME OF MINE.	Number of Persons employed Under-ground.	Machinery other than Ladders for Raising and Lowering the Men.	Number of Boring Machines worked by Com-pressed Air.	REMARKS.
Agar, Wheal – – –	129	1 gig –	4	
Basset, Wheal – – –	156	1 gig –	3	
Basset, West – – –	172	1 gig –	None –	Flat lode.
Blue Hills – – –	75	None –	None.	
Blue Hills, East – –	23	None –	None.	
Botallock – – –	237	None –	2	
Callington United – –	169	Gigs –	3	
Cambourne Consols –	46	None –	None –	Working on small scale at present.
Carn Brea – – –	437	Gigs –	3	
Carsize, West – – –	8	None –	None –	Shallow mine.
Condurrow, South – –	116	None –	None –	Preparing for cage.
Cook's Kitchen – –	116	Man-engine and gigs.	3	Power sufficient for five.
Cook's Kitchen, New –	22	None –	3	
Crofty, South Wheal –	65	Gig –	3	
Danescombe Valley –	16	None –	None –	Entered from adit.
Dolcoath – – – –	605	Man-engine and gig.	10	
Drakewalls – – –	46	None –	None.	
Eliza Consols, Wheal	166	None –	None.	
Fortune, Great Wheal –	45	None –	None –	Shallow mine.
Frances, South Wheal –	169	Gig –	2	
Frances, West Wheal –	135	None –	1	Flat lode.
Friendly, Wheal – –	6	None –	None –	60 fathoms from surface.
Grenville, West Wheal –	28	None –	2	Flat lode.
Grenville, Wheal – –	163	Cage –	2	
Gwin and Singer – –	37	None –	None –	Shallow mine.
Killifresh – – –	108	None –	None.	
Kitty, Wheal – – –	98	None –	None –	Flat lode.
Kitty, West – – –	62	None –	None –	Flat lode, 130 fathoms from surface.
Levant – – – –	350	Man-engine	4	Ladders from 230 to 302.
Levant, North – – –	56	None –	None.	
Lovell, The – – –	10	None –	None –	Shallow mine.
Lovell, East Wheal –	7	None –	None –	Shallow mine.
Metal and Flow – –	8	None –	None.	
Owles, Wheal – –	58	None –	None.	

advances in output were achieved not by mechanization but by changes in hoisting and haulage, by improved hand tools, and by adding to the labour force in ever greater numbers. Boy labour was freely recruited from the workhouses and reformatories, agricultural labourers from the land, whilst within the mining community fathers were followed into the pit by their sons in quasi-automatic progression. (At Worsley the Duke of Bridgewater's Trustees tried to put a stop to this in 1849, but two years later the general manager complained that there were not enough colliers to meet the requirements of the trade, and as a result of his remonstrances a further forty-three boys were admitted to the collieries).[12] In times of boom – during the Coal Famine of 1873 for instance – the increase in numbers was phenomenal, and the Census figures (Table 1.5) show that it was maintained from decade to decade right down to 1911.[13]

TABLE 1.5
The increase of miners, 1871–1921

Year	*No. of miners*
1871	370,900
1881	495,500
1891	668,000
1901	806,700
1911	1,067,200
1921	1,131,600

(Adapted from B. R. Mitchell and Phyllis Deane, *Abstract of British Historical Statistics*, Cambridge University Press, 1971, pp. 118–19.)

As in the case of Cornish mining, technical innovation (the spread of deep mining in the 1860s and 1870s) multiplied the need for hands. Steam-driven machinery, in the form of the pumping engines, ventilating fans, and steam-winding gear, which were being extensively introduced in the third quarter of the nineteenth century, made it possible to open up workings in new and previously inaccessible levels. Improved ventilation allowed working places to proliferate: they were no longer tied to the foot of the shaft, but spread in all directions, and in a large mine there might be as many as twenty miles of tunnelling underground. The shift (admittedly an uneven one) from 'pillar and stall' to longwall meant that more hewers could be accommodated on a face, and the greater use of explosives, especially after the introduction of dynamite, meant that there was more material for them to fill. The number of putters also increased: there was more coal for them to handle, and longer galleries to work. Longer galleries also meant more roads to keep up, more rails to lay down, more roofs to attend to, and led to the formation of a whole new class of underground worker – the stonemen or 'rippers' – who had the job of extending the levels. Timbermen were another new class of underground worker who appeared in the second half of the nineteenth century, partly as a result of the greater division of labour which accompanied the growth of the work force, partly because hewers were reluctant to undertake work which lost them earning time.[14] (In earlier years, the hewer had been expected to do his own propping, but by the second half of the nineteenth century this was being vigilantly resisted.)[15]

The only change that took place in the miner's tools during the second half of the nineteenth century was the replacement of iron by steel in picks and borers, though gunpowder – a chemical aid rather than a mechanical one – was more frequently used than in earlier times. The double-headed pick, 'wielded by a strong arm and directed by a sure eye', was still (wrote Smyth in 1875) 'the queen of weapons'; coals were still filled by shovel, shot holes prepared by the hand-hammered borer.[16] Whatever the changes at the pit head, manual undercutting still prevailed at the face:[17]

> In a small corner-like recess, full of floating coal-dust . . . glimmer three or four candles, stuck in clay which adheres to wall and roof; or there may be only a couple of Davy lamps, each of which may be truly styled *lucens a non lucendo*. Close and deliberate scrutiny will discover one hewer nearly naked, lying upon his back, elevating his small sharp pickaxe a little above his nose, another picking into the coal-seam with might and main; another is squatting down and using his pick like a common labourer; a third is cutting a small channel in the seam, and preparing to drive in a wedge. By one or other kind of application the coal is broken down; but if too hardly embedded, gunpowder is employed, and the mineral blasted; the dull, muffled, roof-shaking boom that follows each blast startling the ear of the novice, who commonly concludes that the whole mine has exploded and that his last minute is near at hand.

There was more change in haulage, with the introduction of pit ponies and underground wagonways; except in the smaller pits boys no longer had to crawl on their hands and knees, dragging the trams along behind them. But the haulier's job remained a gruelling one, as Dave Douglass points out. The main change in the later nineteenth century was the replacement of a younger by an older set of boys.[18]

Mechanical coal-cutters were being frequently patented from the 1860s, and high hopes were entertained of them when they were first introduced into the Barnsley coalfield in 1864, at the end of an eighteen weeks' lockout: 'If they are as successful as expected they will . . . diminish the chance of turn-outs in the future.'[19] But the 'great revolution' expected of them did not materialize: in 1901, forty years after the first wave of patents, only $1\frac{1}{2}$ per cent of the total output of coal in Great Britain was cut by machinery (as compared with 25 per cent in the United States).[20] It proved difficult to get the machines a clear run. In Lancashire the face was pronounced too thin.[21] In South Wales poor roofs, dipping seams, and numerous faults were peculiarly obstructive, 'the pipes transmitting the power were liable to be damaged by the numerous falls'.[22] Conditions were not much more favourable in the Lothians, according to the account of the area left by Paul de Rousiers in 1896:[23]

> The most characteristic feature of the mines is that the coal is dispersed in irregular seams, that it is constantly intermixed with rock, clay and other impurities, and that the miner's pick must continually be wielded with discernment. The machine lacks this discernment, it is essentially blind, and therefore it does not seem likely to have an important future in the extraction of coal. In any case its role at the present moment is a negligible quantity.

Even in Northumberland and Durham, where the seams were more regular, the mechanical cutter was slow to make its way. (In 1913 only 6 per cent of production there came from machine.)[24] Beyond this there was the question of comparative costs (the machines were subject to frequent breakdowns and expensive to maintain: labour was comparatively cheap), and finally there was the workers' opposition which affected the mechanical coal-cutter as much as any other innovation in the mines, even if it did not excite as many strikes as the Davy Lamp. As *The Times* Commissioner put it after a visit to the Rhondda miners in 1873:[25]

> To hint at the possibility of a coal-cutting machine is to make them regard you as their mortal enemy. Steam engines to raise the cages, pump the water out, and work the trucks are all regarded as innovations designed to cheat the collier out of the profits and delights of existence.

In ironstone mining, as in copper and tin, the chief work of the miner was boring blast holes. The holes were bored with a long iron chisel 'plied lustily' by a heavy iron hammer until it had pierced the rock:[26] it was a slow, laborious process, and even in the 1870s the average rate of progress was no more than a foot an hour.[27] Steel or steel-tipped borers replaced iron ones in the third quarter of the nineteenth century,[28] but the mechanical rock drill was slow to spread. It had been patented even earlier than the coal-cutting machine, and had been anticipated in 1812 when Richard Trevithick brought out a rotary machine for boring which, with a weight of 500 lbs placed over the drill, bored 1½-inch holes in Plymouth limestone at the rate of one inch per minute.[29] Yet in 1880 when Davies published his treatise on *Metalliferous Mining*, drilling machinery was 'as yet in its infancy' and he thought it 'perhaps too soon to judge of its cost in comparison with hand labour'.[30] In Cleveland, the most advanced ironstone district in the country, rock drills in 1880 accounted for a quarter of a million tons of production, compared with six million tons obtained by hand drilling.[31] Ten years later, in 1890, the proportion of machine-drilled stone in Cleveland had risen to about a sixth.[32] There was rather less incentive to introduce the drill in the haematite iron district of Cumberland and Furness, since the ore in many places was comparatively soft, and could be removed by a miner's pick without recourse to blasting at all.[33] At the Roanhead mine, Furness, within living memory, all the work was done by hand.[34]

> The face of the ore . . . was bored . . . by means of a hammer and steel jumper, and the companies, consisting of a basic unit of two miners to a labourer, therefore relied on the conscientious services of a blacksmith's shop to keep them supplied with sharp tools. . . . When the face . . . had been blasted loose, the labourer and miners shovelled the haematite into a waiting bogie . . . and a further labourer trailed the bogie to the main roadway and a contract team took it to the haulage shaft.

The rock drill did not begin to be introduced into stone quarrying until the 1870s[35] and it was far from revolutionary in its effects. Steam drills of the Ingersoll-Sargent kind (in general use at Mountsorrel in 1896) had a rate of penetration 'going full speed' of three feet an hour, compared with ten inches an hour with a hand-driven borer.[36] Drills like this seem to have been widely used by the 1900s in the larger granite quarries,[37] but nowhere else. Building stone quarries did not use them at all, since the material was too delicate for blasting, and they made very slow progress in slate.[38] In the giant slate quarry of Dinorwic only three were in use in 1902.[39] In the limestone quarries of Clydach Vale – as in many other places – the universal borer was the 'jumper', an iron bar about 5 ft 6 ins in length with a chisel-pointed end, locally known as a 'churn'. The stone was bored by working the 'jumper' up and down, and, according to the old quarryman, about a foot an hour was workmanlike progress: 'You kept on turning and twisting the bar – dampening down. It was like butter making in a churn. You'd start the hole – put water in (there would be a leather strap on the bar to stop the sludge coming over your leg) turning all the time with water to soften the stone.'[40]

In stone dressing the whole tradition of the work was craft rather than industrial in character, with quarrymen who combined a miner's skill and a mason's with their own. As the Leicestershire quarrymen put it in 1892, scorning the notion of machinery: 'skilful men with sharp-edged hammers do all the important work at stone quarries'.[41] The hearthstone raised at Godstone, Surrey was hewn into neat blocks by a peculiar double-headed axe;[42] roofing slate was chopped into rectangular pieces with a large knife and the splitting was done by hammers and wedges, similar in shape to the carpenter's chisel, and requiring the same niceness of judgement.[43] Flint knapping, too, required working to fine limits. The flints were shaped by a quartering hammer; then by 'flaking' ('the most difficult branch of the business. . . . Many knappers are unable to flake, and but few attain great proficiency in the art'); and finally by 'knapping', another hand-hammered work which formed the flakes into gun-flints.[44] Paving stones were fashioned entirely with hammers. At Mountsorrel, one of the great centres of the industry, the sett-makers depended on a variety of them: the 'burster,' 'an immense tool weighing 30 lbs', which was used for breaking up the large blocks and irregular lumps; the 'knob hammer' for knocking off the larger knobs; and the 'squaring' hammer, 'universally employed' for squaring setts.[45] In Aberdeenshire, too, sett making was a work of the hammer, and more substantial work in granite – for example to form the flutings of Doric columns – was executed by axe-dressing.[46] 'The dressing of granite by mechanical means is a problem that has engaged the attention of engineers for many years', wrote Powis Bale in 1884, 'and is still, practically speaking, unfulfilled.'[47]

'Scientific invention has not been able to do very much in the way of assisting man in the art of stone getting', wrote a Leeds newspaperman in 1875, after visiting a large sandstone quarry near Bradford. He noticed two 'Goliath' cranes, and several smaller ones besides, and also lines of rails to every part of the quarry. But for the rest 'the pick, the hammer, the wedge and the charge of gunpowder are still the chief agencies that the delver employs'.[48] Quarries for building-stone (like the softer classes of limestone and sandstone)[49] were particularly conservative: they did not use explosives for fear of bruising the stone; and the discontinuous nature of their operations made any kind of fixed investment costly. As late as 1912 it was 'all hand labour' at the Bath stone mines. No explosives were in use and the blocks were dislodged by the crowbar and wedge; the 'ordinary axe a quarryman uses' was all that was thought necessary for cutting up the stone.[50]

The most widespread measures of 'improvement' were in haulage. Cranes had been introduced into Aberdeenshire as early as 1835,[51] and by the 1900s there were few larger quarries without at least the hand-operated version. In some quarries stone was carried away on overhead rails known as 'Blondins';[52] works tramways were to be seen in large quarries like the Darbishire's works at Penmaenmawr. There were other quarries, however, where stone was still wheeled out of the quarry on barrows, or dragged up by a horse and cart, or raised by a hand winch. At the Llanberis slate quarry in 1912 there was no hoisting machinery at all.[53] On the Isle of Purbeck each individual quarryman was responsible for his own haulage. Fred Bower, whose family came from Langton Matravers, and who worked there briefly in the 1880s, describes the laborious procedure then in use: the quarrier, having cut his stone, loaded it on to a bogie, to which he harnessed himself 'just like a horse', and dragged it to the foot of the incline. From there it was hauled up on a chain by means of a windlass worked by a donkey or mule at the top.[54]

Clay-getting was not a dangerous trade like quarrying, where the accident rate was high, but it was possibly more laborious, and certainly less affected by 'improvement' than any other branch of mineral work.

In the china clay industry of Cornwall and Devon there was nothing in the way of mechanical aids (pumping engines were an innovation of the 1860s and 1870s[55] and kiln-drying came even later). The clay digger worked with a heavy pick (the heaviest in Cornwall, according to one authoritative account, published in 1875),[56] and a long and square-mouthed shovel. The 'balmaidans', who had the job of cleaning the final impurities off the clay – they cleaned two or three tons at the rate of a shilling a day[57] – worked with a scraper described in 1860 as 'resembling a small Dutch hoe'.[58] The clay was puddled by diggers trampling on it with their heavy boots and stirring it with their dubbers, until it was reduced to a bed of slime. It was dried in the wind and sun. It was carried about, from one stage of the sifting process to another, on hand-borne wooden trays. At every stage in the process it was all hand work, and the clayworkers had the reputation of being the strongest men in the country, and engaged in the lowest though not the worst paid class of work in Cornwall. 'They reckoned to shift 20 tons or more . . . in the day, with the help of nothing but their own arms and a shovel.'[59] At Wareham in Dorset the quarries and mines for ball-clay had a rather different character. The clay there was coarser, 'being such as is used in the manufacture of common stoneware', and required less in the way of sifting before it was despatched. 'It is not worked previously to exportation to the potteries, being sent thither . . . just as it is extracted from the pits.' The chief instrument in use in 1850 was a narrow, heavy spade, 'weighing from 17 to 20 lbs' which served both as a digger and as a slicer.[60]

On a brickfield the clay getter's work could be just as hard, but it had less to do with the digging (an autumn job) than with the more protracted processes involved in cleaning the clay and reducing it to plasticity. The most arduous part of this work was the tempering, usually undertaken by a gang, which involved repeatedly turning the clay with spades, mixing it with water, and treading it into a sodden mass. By the 1850s there were pug mills to perform some of this work, but many yards still relied on the naked feet of labourers 'from long practice, sensitive to . . . the smallest stone or roughness which interferes with the uniform texture of the mass'.[61] Stones were picked from the clay by hand, 'a tedious operation, but one which cannot be neglected with impunity as the presence of a pebble in a brick generally causes it to crack in drying, and makes it shaky and unsound when burnt'.[62] Tileries preferred to rely on 'slinging', an equally protracted process which involved cutting the clay into thin strips and then dashing them together, 'during which operation most of the stones fall out'.[63] Haulage, which was done more by boys in the southern brickfields, and more by girls in Staffordshire, was another arduous part of the work; a child might make 2,000 journeys in a day, carrying bricks from the moulder's table to the hacks, each time with a 10 lb weight of clay.[64] The puggers-up, who carried the clay from the clay pit to the moulder's table, had fewer journeys to make, but more to carry. At the Tipton yard recorded in the Factory Inspectors' Reports for 1865, the younger children carried 'on average' about 20 to 40 lbs of clay 'on the head' and about 10 to 20 lbs 'in their arms'; the older ones – girls aged fourteen to sixteen years of age 'and occasionally older' – carried an average weight of 60 lbs each journey.[65] For the heaviest loads of bricks a child would sometimes be harnessed to the barrow 'like a little donkey'; Lakeman, Factory Inspector for East Anglia, recorded a case in 1871 where two of them were drawing the barrow in tandem, 'one in the shafts, the other a leader pulling along with a rope'.[66] Much of this work, including the heaviest parts of it, was performed by women and children, the men being chiefly employed as moulders or at the kilns (in South Staffordshire the women were the moulders too).

The development of machinery in brick making was extraordinarily uneven, both as between different regions, and in different departments of the work. The pug-mill, which was the most widely diffused, still left much of the cleaning and preparatory work untouched, and, if anything, increased labour, at least for the puggers-up. In the 1860s its use was by no means universal. 'Pug mills are not general in Essex and Suffolk', wrote Factory Inspector Lakeman in 1872, 'the earth is trodden by children who are kept at work tempering a heap of clay from morning to night.'[67] In North Wales and Cheshire, too, according to J. H. Bignold, the district Factory Inspector, traditional modes of tempering were untouched. 'I do not think there are any pug mills about here', he told the Factory and Workshops Commission in 1876, 'the clay is prepared by hand; they commence in the early part of the season, what they call the casting of the clay . . . the boys who are carrying in the summer are picking stones out of the clay which is going to be cast in winter.'[68] Moulding machines were being widely canvassed in the 1860s, but they met with varying success, depending partly on the nature of the brick (the machine was generally judged unsuitable for high quality bricks, even when it was adopted for common stocks),[69] partly on the size of enterprise (the machine was too expensive for a smaller yard to maintain) and crucially on the nature of the local clay:[70] in the London clay basin, all kinds of difficulties appeared and the very extensive brickyards of Middlesex and Kent were, in the 1890s, still almost all hand-making 'summer' yards.[71] In the 1870s brickmaking machines were said by the brickmasters to be of 'very little use' in the Black Country,[72] but in Nottingham they were being introduced 'almost universally'.[73] In the Manchester district, where the brickmakers (all male) were very strongly organized, the machine met a bitter, sustained and often violent resistance – the source of the 'Manchester outrages' inquired into by the trade union commission in 1866: ten years later the would-be innovator was still liable to find his premises blown up. (A hand-making industry was still in existence around Manchester in the 1890s,[74] though whether it survived because of the small size of local enterprise, the nature of the local clays or the enduring strength of resistance is a matter which invites historical inquiry.)

Clay digging remained down to the end of the century a matter of pick and shovel or spade.[75] Only after the rise of nationally orientated industry, with the development of brickmaking in the Fletton–Peterborough district, was the 'steam navvy' brought into requisition in the brickfields. Like other innovations in the industry, its fortunes were uneven. 'Steam navvies are used where the material is sufficiently soft and uniform', wrote Searle in 1913, 'but their use is unpracticable in many brickyards on account of the need for selecting certain portions and discarding others from the quarry face.'[76]

Fireclay works, which supplied such modern appliances as furnace linings and gaswork retorts, were paradoxically amongst the most recalcitrant to change. Hand picking, 'one of the most primitive means of removing the coarser impurities from clay',[77] still ranked high among the preparatory processes when Searle was publishing his books in the 1900s. Pug mills were by this time generally in use for firebricks; but in crucible making for the steelworks, and in the manufacture of gas and glassworks appliances, a more ancient form of puddling still held sway – treading underfoot. This is how Searle describes it:[78]

> When the materials are to be mixed by treading, they are spread out on a concrete floor and are sprinkled with water. The mass is turned over repeatedly with spades and, when it becomes too pasty to be worked in this way, it is again spread out and is trodden by men with bare feet, who squeeze the clay between their toes, and so mix it thoroughly. Each portion of the material has to be squeezed between the toes, compressed and then pressed on to the previously worked paste. The treader stands in the middle and, working his toes, goes over the whole surface of the clay in a spiral direction, always working towards the edge. Having reached the edge, he turns round and walks in the opposite direction until he arrives at the starting point. Some treaders prefer to walk in straight lines instead of in a spiral direction. The trodden mass is then made up into balls of 40 to 45

lbs weight, and is afterwards beaten into a dense mass. In some works it is 'pugged' after being trodden.

Firebricks themselves continued to be made by hand, long after machinery had made inroads into the red brick trade.[79] 'The machines that have been tried have proved failures', the Factory and Workshops Commission was told in 1876 (only one employer had them in use at Stourbridge).[80] 'The hand method still holds sway', wrote Searle in his *Refractory Materials* (1908): 'most makers consider that firebricks will never be made successfully in any other way, and few . . . will even consider making them by machinery.'[81]

Technical change in the nineteenth century made mineral work more productive, but did not impair its handicraft character. Explosives were more widely used but they did not put an end to hewing. Moulding machines were introduced but they did not lighten the clay getter's task. The basic tools remained unchanged – in mining the double-headed pick, in quarrying the 'jumper', in the case of blasting work, or the crowbar and wedge; in clay digging, the long-handled, square-mouthed spade.[82] The work was rough and laborious, yet highly skilled. It continued to demand enormous inputs of physical strength, while at the same time leaving everything to the worker's judgement. It was simultaneously open to all comers – freely recruiting labour from all sides – and yet, at the same time, hereditary tendencies were strong. The rockman perilously perched on his ledge, drilling at a high-level face, the tutworker in the airless extremities of a winze, the collier in his stall, all had to combine a labourer's strength with an artisan's ability to work to fine limits. The moulder or temperer on the brickfield had to be really choice in his selection of material. The stone dresser, like the mason, retained his own individual brand.[83]

When machinery was introduced it was extraordinarily lop-sided in its effects, and if it abridged labour in some directions (e.g. in coal mining, by the elimination of trap boys), it greatly increased it in others. Steam power was harnessed to ventilation and drainage, but there was an almost total lack of mechanization at the point of production itself – at the face. Blasting was still a matter of hand-hammered holes and gunpowder, hewing of mandril and pick. So far as the face worker was concerned, the significant changes involved 'improved' hand tools rather than a new technology. In metalliferous mining the introduction of cast steel boring rods – 'very rapidly brought in since 1851'[84] – allowed the rate of drilling to increase, but did not change its essential character. The same was true of the analogous change from iron to steel in hewing: it left the miner still working in a crouch, wielding a short, sharp pickaxe, streaming with perspiration in choking dust. In quarrying, too, the changes were of an intermediate kind, such as the introduction of the so-called 'patent axe', 'an instrument composed of a number of thin slips of steel tightly bound together', which first found its way into Scotland about 1818 and by the end of the century was in use at granite-working centres all over the country;[85] or the spread of the feather and tare method of stone splitting, which replaced the older method of wedge and groove, but still required the quarryman's whole strength. (The 'tares' had to be repeatedly hit with a sledge-hammer until the rock eventually broke.[86])

Technological improvements in mining and quarrying encouraged owners to take risks with safety and health. The miners realized this when they opposed the introduction of the Davy Lamp – it would encourage owners (or their viewers) to experiment with work in much more dangerous levels; and perhaps it was this which the Rhondda miners had in mind when they expressed their hostility to steam pumps. The use of explosives greatly increased the risk of fatalities in both quarries and mines. It none the less spread. As Warrington Smyth, a Mines Inspector, put it in 1875: 'to the objection . . . that it shakes the coal more, and that it is sadly conducive to accidents, there is always returned the answer that without the effective agency thus obtained, it would be difficult for certain collieries to maintain their place in the general competition'.[87] With the spread of deep mining in the third quarter of the nineteenth century, the colliery disaster became a regular and half-expected calamity on the coalfields. In quarrying, too, as often though less dramatically, the increased use of explosives (along with larger and more perilous faces) produced injury and death. Dust levels were another occupational hazard which nineteenth century improvement increased. There was also bad air in the deeper levels, and on the face that had been fired with explosives. Pneumatic, or steam-powered, drills raised a great deal more dust than the 'jumper', and when they were introduced into metalliferous mines and quarries the 'miner's disease' of phthisis soared. Straps and pulleys and unfenced gearing were added to the health hazards of a brickfield when machinery was introduced, and the same may be said of the introduction of cranes to the quarries.[88] Technological advance improved mineral output, but often at the expense of the worker's comfort and health.

IV

Industrial discipline in mining, quarrying and other classes of mineral work was very different from that in the factories and mills. There were no purpose-built premises to wall the worker in, but endless galleries, scattered excavations, or open-fronted sheds. Detailed supervision was difficult because of the open air (or underground) nature of the work – the 'miner's freedom' as Carter Goodrich pointed out long ago, owed a good deal to secluded location;[1] nor did the employer necessarily see a need for it: there was nothing which would go up in flames on a clayfield, nothing which the worker might 'take home', or trade rivals be tempted to pinch; there were no valuable raw materials to husband – no need for elaborate stocktaking or police. A quarry, indeed, might not be marked off as private property at all, and one of the chief tasks which the Mining Inspectors pursued, when the Quarries Act of 1894 extended their jurisdiction, was to try to force the owners to put up fences.[2]

Another sanction often missing was the threat of unemployment. There was no 'reserve army' of labour in places like the Isle of Purbeck or the Forest of Dean, but, on the contrary, auxiliary sources of livelihood or employment which enabled a worker to support himself in multiple ways. On the coalfields shortage of labour rather than excess prevailed: even in bad times owners were anxious to keep men on, and preferred to work them half-time rather than risk being shorthanded when good times returned. Cornwall, too, in the time of its mining prosperity, was hungry for hands: job changing there was endemic and a matter not only of necessity but also of choice: when a pitch proved unrewarding the Cornish miner jacked up, confident that he could find an alternative berth.

'Daywagemen', the class of workers most closely dependent on management, were often a minority in mines and quarries, and relegated to the inferior classes of work. (In the Cornish mines they were disparagingly referred to as 'owner's account men'.)[3] Most of the work was performed by sub-contract, by 'bargain' or by piece – i.e. by payment systems which left the workers a degree of latitude in the performance and timetabling of their tasks. This was partly because of the need to provide incentives for what was difficult and often dangerous work, partly because of the prevalence of indirect systems of employment (an answer to shortage of capital, weakness of management, or both); and in some cases it was because of the seasonal or discontinuous nature of the work. At Beaumont and Blackett's lead mines, Weardale, 'wagemen' were confined to those in subsidiary occupations and formed a small portion of the labour force. The miners themselves worked on 'bargains', some being paid by 'bing' (i.e. according to the value of their get), some by the fathom;[4] in a smaller lead mine there might be no wagemen, the whole work being taken on as a 'bargain' by a collective of miners. In the Merstham lime works Surrey (a major source of lime for the London building trade) wagemen were confined to an undifferentiated class of 'labourers'; the majority of the men worked in what are referred to in the wages books as 'companies', being paid by piece according to the different classes of work – 'Chalk dug', 'Kilns filled', 'Lime removed'.[5] The 'bargain' system which Merfyn Jones writes about in the slate quarries of North Wales was a kind of speculative system of piece-work in which the worker took a lease on a pitch, and was paid in proportion to the minerals he extracted from it. This system was particularly well suited (from a proprietor's point of view) to the vagaries of metal-bearing lodes, and it was very general in ironstone mining, tin, copper and lead. The 'bargain' covered a spectrum of different arrangements, and the terms were subject to revision, depending on the state of the lodes. 'When the veins are exhausted', wrote Richard Heath, after visiting the lead miners in Swaledale, 'the mining companies offer them a large percentage to search out fresh veins; but when they are found, they reduce the amount rapidly, as they know that labourers will then come flocking in'.[6] Terms were also liable to be much tighter where a large firm, like Beaumont and Blackett's, had charge of the work. The speculative element varied, too, from one mineral to another: in ironstone mining the ores were more regular than in copper and lead, and the 'bargain' less liable to fluctuate.

'Bargain' work was generally taken on by collectives. Amongst lead miners these 'co-partnerships', as they were called, usually involved from four to eight men; sometimes they were got up extempore to take on a particular pitch; sometimes they worked together in a semi-permanent collective, recruited on the basis of kin.[7] The Cornish miners' collectives were called 'pares'. 'These, when the lode is wide, will sometimes all work together, but generally they divide themselves into parties, relieving each other at the expiration of eight hours.'[8] In the haematite iron mines at Hodbarrow, Cumberland, the collectives taking up a 'bargain' varied (in 1881) from a company which consisted of only two men to another which involved as many as seventy-five. 'Although much of the work was done by small companies of miners', the historian of Hodbarrow tells us, 'it was not unusual for the bargain to be let initially to a large group of men who then divided the task among themselves.'[9] In the slate quarries of North Wales the 'bargain' usually embraced both the hewing and the dressing of the stone, though in some cases only the rockmen and splitters were in partnership, and dressers were taken on by them as wage labourers. At Dinorwic the bargains were taken by two partners and a journeyman to help them:[10]

> It is the custom for both partners to work inside the quarry for a few days at the beginning of each month, drilling and blasting to procure blocks for the dressing shed, as it is found that one man could not very well keep his partner and a journeyman supplied with sufficient blocks for dressing, unless he is assisted at the beginning of the . . . month, for the purpose of clearing down tops and bottoms and drilling holes. After the first week, generally, one partner will keep the other partner and journeyman going for the rest of the month.

Another form of engagement which had an element of 'bargain' in it was the contract for an individual job. Mine sinkers, an itinerant class of men, half miners, half navvies, contracted for their work at so much a fathom: their earnings depended on the difficulty or otherwise of the ground. In the Wareham clay trade a rather similar system was employed for those who undertook to take the overburden (or 'heading') off the clay:[11]

> A party of 'excavators' . . . engage to remove it at so much for the cubic yard. Sometimes the clay is but a few feet beneath the surface; at others it may be upwards of a hundred. Other things being equal, the deeper it is, the adventure is, of course, the less profitable.

Piecework – the most common system of payment in the collieries, brickfields and saltworks – involved less formal abdication of responsibility by the owner than the 'bargain', but in practice it could leave the worker with a wide latitude in the performance and timetabling of his tasks. It also included an element of sub-contract. The hewer was usually the employer of the drawer (often they were father and son); brickfields and salt pans were commonly worked on a system of family (or semi-family) berths, with one person contracting for the place, and then taking on relatives to boost their earnings.

None of these payment systems encouraged punctuality, and there was no steam-powered machinery to regulate the worker's hours, or keep him up to the mark. At the Clydach limestone quarries timekeeping was definitely contingent. There were regulation hours, but since the rockmen could enter the face at will, with no works' entrance at which to clock in, or foreman's looking glass to spy them out, little was done to make sure they were observed. 'A quarryman was allowed to come in his own time and finish in his own time – his wages were according to what he quarried and filled.'[12] Some Cornish mines had company rules which forbade men to come up early, and threatened them with fines when they did so; but there does not seem to have been an adequate police to enforce them, to judge by the following exchange which took place before Lord Kinnaird's Commission in 1864:[13]

> Who takes your time? – The agent here.
> Is he always here when you come out? – Not always.
> You generally fire a shot, I suppose, before leaving your work? – Sometimes we do, and sometimes we do not.
> Do you not generally . . . do so? – No, we do not always.
> Do you leave your work before the other men come in to relieve you? – Yes, sometimes.
> What is the rule, are you generally out before fresh men come on? – We generally calculate to get up before they go down.

In a coal pit at mid-century there might be no regulation hours at all so far as hewers were concerned, especially in the smaller class of pit, where work might be performed in a series of sub-contracts. Things were 'rough and irregular' at Cramlington when Thomas Burt went to work there in the 1850s. 'There was no recognized starting time or ending time for the day's work to the fore-shift men. The coal-hewers went into the pit and came out when they liked, and some of them apparently liked to go in at ten or eleven o'clock at night and to remain till about noon the next day.'[14] A more common practice in the collieries was for the hewers to go down the pit in the 'owner's time' – i.e. at the fixed starting point for the day – but come out again in their own time, when their stint for the day was completed, or when they felt they had earned enough money with their tubs.[15]

Hours of work in mining and quarrying were definitely shorter than those in indoor occupations, but day-workers – men and women, boys and girls – fared a good deal worse than those who were paid on contract or by the piece. In County Durham, in 1849, the day's work for a hewer was defined as 'not exceeding eight hours'; that of the operative on fixed wages was twelve, 'the period commencing with the hour when the engine begins to draw coals to the surface'.[16] Drawers (often the hewer's son) generally had to work longer hours than hewers, in order to clear the tubs, and this seems to have been the case whether they were employed by a coal-owner, a butty, or the men.[17] The Mining Commissioners report of 1844 gives two examples of how the number of days might vary too. Both are taken from pits in the Lothians. The first is that of Robert Pentland, 'a man of fair character', who had two sons working with him. In the fortnight ending 16 September:[18]

	The father worked	9 days
	The sons	12 days
F/night ending	father	10 days
Sept. 30	sons	10 days
F/night ending	father	9 days
Oct. 14	sons	11 days
F/night ending	father	5 days
Oct. 28	sons	7 days

The second is that of George Reed, 'a man of indifferent character', who worked with his son as follows:

F/night ending	father worked	8 days
Dec. 8	son	9 days
F/night ending	father	10 days
Dec. 23	son	11 days
F/night ending	father	9 days
Jan. 6	son	11 days
F/night ending	father	6 days
Jan. 20	son	10 days
F/night ending	father	5 days
Feb. 3	son	7 days

An eight-hour day was very widely (if unevenly) established in coal in the 1860s, at least for hewers. In the Yorkshire coalfield it seems to have been won by the strike movement of 1858;[19] in Northumberland a little earlier (according to one commentator it was associated with the change from wedging to blasting at the face);[20] in Scotland it was being vigorously agitated for in 1866. The five-day week – in the form of a ten or eleven-day fortnight – was also common, sometimes because of voluntary 'laaking', sometimes as a result of forced stops. William Baxendale, a Wigan miner who appeared before the Mines Commission of 1866, reckoned to work 'about five days a week' or ten a fortnight. 'Sometimes I work 11, sometimes eight or nine days; we are stopped from working by break downs and accidents, and so on, sometimes.'[21] In the wages books of a large Northumberland colliery, cited by the Rev. R. F. Wheeler in *The Northumbrian Pitman* (an unrancorous account published in 1885) the hewers drew an average pay of eight days in a fortnight and worked an average of four days a week. Shifters and stonemen worked, on average, one day a week more.[22] In Ayrshire, according to evidence given before the Royal Commission on Labour in 1892, 'Few miners or colliers care to work more than five days per week. Some districts have a set holiday, without pay weekly; in others each man takes a day as it suits him.' In Blantyre 'the majority of the districts work five days a week, observing Thursday or Saturday as an entire holiday'.[23]

One reason for the short hours in the pits was the miners' 'darg' or self-imposed limitation on production, which was maintained by secret understandings amongst the men irrespective of whether or not they happened to be in union. The 'darg' served a variety of purposes – for instance it could be imposed as an early warning to employers, or as an alternative to strike.[24] But above all it was a way of keeping up piece-rates.[25] As *The Times* Commissioner wrote in 1873 after his visit to the Rhondda:[26]

> they cut out block after block of coal till they make up their day's work of about three tons. They could do more, each man with his boy, but . . . they like to stop at three tons, because they imagine that, by working less, they keep up the price of coal, and by this means maintain the standard of their wages. . . .

In the middle decades of the century there is evidence for the existence of a 'darg' in coalfields all over the country.[27] It was particularly frequent in Scotland, where in later years it was incorporated into the strategy of the miners' unions:[28] in Lanarkshire and Ayrshire, according to

Alexander Macdonald, the 'darg' had the force of what he described in 1866 as 'old custom'.[29]

Absenteeism, or 'laaking' as miners called it themselves, was a very emblem of the miner's freedom, and it was supported by a complex of commitments which the outsider could only dimly discern. 'There are leaders in every pit who regulate these matters', a Mining Commissioner wrote in 1844, 'If they hold a meeting in the mines, where they always hold their meetings, and fine a man, they come up and don't work that day'.[30] Accidents brought miners to the surface, in solidarity with the dead; 'superstition', it was said, kept them from going down. Absenteeism varied with the state of the weather – according to miners' tradition 'laaking' was most likely to occur on a fine summer's day, when a brick might be tossed in the air to decide whether or not to stay up top. It also depended on the state of wages – 'Pay Monday', the Monday after the fortnightly or monthly pay, was usually a holiday: in Merthyr Tydfil, where ironworkers and colliers enjoyed it together, it was observed as a kind of jubilee – *Dydd – Llun – dechra'r – mis* or 'Monday the beginning of the month';[31] 'Cavilling Monday', which Dave Douglass refers to in his chapter, was celebrated by the Durham miners in one of their songs. Sensational accusations of slacking were bandied about during the Coal Famine of 1873, when high wages were said to have reduced the miner to a two or three-day working week. A more qualified picture emerges from the statistics produced, towards the end of the nineteenth century, by the Mines Inspectors and the Labour Department of the Board of Trade. These show steep regional differences in absenteeism. A table preserved in an appendix to the Royal Commission on Labour conveys in remarkable detail how attendance might fluctuate not only from day to day (with 'Saint Monday' by no means always preferred to the more secular days of the week), but also from one fortnightly pay to the next (Table 1.6).[32]

In a coal mine there was very little in the way of supervision so far as face workers were concerned. The pit deputy was not so much an overlooker as a man with specific jobs to do, principally (at mid-century), shot firing, and fixing props and brattices. His travellings underground had to do with firedamp, ventilation shafts and the state of the woodwork rather than the progress of the hewer on his stint (a personal matter in which he was not called upon to interfere).[33] At the coal face his position was ambiguous. He was a representative of higher management, yet his identity and standing depended upon the performance of his craftsmen's skills. He had no time-sheets to fill in, for

TABLE 1.6

Messrs. Jas. Sparrow and Sons
Ffrwd Colliery, near Wrexham, Wales

Employment has been steady for the last 4 years, but the colliers' attendance has been irregular, as follows:

Absentees from Colliery during 1891.

Fortnight ending	April 4.	April 18.	May 2.	May 16.	May 30.	June 13.	June 27.	July 11.	July 25.	Aug. 8.	Aug. 22.	Sept. 5.	Sept. 19.	Oct. 3.	Oct. 17.	Oct. 31.	Nov. 14.	Nov. 28.
Monday	—	19	33	27	—*	25	46	23	35	18	87	35	23	41	36	36	34	22
Tuesday	—	27	19	26	—*	34	31	16	31	16	28	52	26	28	86	24	32	38
Wednesday	—	16	17	46	39	26	18	29	21	14	17	16	24	36	17	24	29	27
Thursday	—	14	18	47	10	20	28	18	16	11	23	15	15	20	23	23	24	31
Friday	—	12	3	18	12	21	23	11	17	16	21	19	20	13	12	21	14	25
Saturday	—	7	11	12	10	22	28	31	30	8	17	15	19	17	15	15	21	12
Total pay week	—	95	101	176	71	148	174	128	150	83	193	152	127	155	189	153	154	155
Average	—	15·83	16·83	29·33	17·75	24·66	29·00	21·33	25·00	13·83	32·16	25·33	21·16	25·83	31·50	25·50	25·66	25·83
Monday	—	32	41	24	36	51	72	—†	43	—‡	56	—§	60	38	25	41	51	—
Tuesday	—	38	24	37	32	44	58	—†	44	—‡	44	—§	34	50	29	38	62	—
Wednesday	48	17	21	19	21	27	40	—†	24	48	28	38	40	25	25	26	57	—
Thursday	32	17	15	17	21	18	63	28	26	17	23	13	25	37	26	28	19	—
Friday	17	5	11	13	27	18	19	28	35	18	15	11	30	9	20	14	32	—
Saturday	19	6	14	14	40	19	26	16	42	24	17	16	28	15	22	14	17	—
Total weeks after pay	116	115	126	129	177	177	278	72	214	107	183	78	217	174	147	161	238	—
Average	29·00	19·16	21·00	21·50	29·50	29·50	46·33	24·00	35·66	26·75	30·50	19·50	36·16	29·00	24·50	26·83	39·66	—

* No work. † Pumps broken down. ‡ Bank Holiday. § Collier's demonstration.
The above averages are all confined to colliers. Surface men's attendance has been regular.

the hewers were paid by the piece, and their earnings were a matter of pit-head wrangling with the checkweighman at fortnightly or monthly 'pays'; nor would he need to keep elaborate accounts of stores – or checks on them – since miners supplied many of their own working tools themselves. There was in fact so little in the way of paper work for the deputy to do that – like the Northumbrian deputy Dave Douglass refers to – he did not need to be a writing man at all. (Three of the four overmen, and four of the five deputy overmen, involved in the Penygraig disaster of 1880 could not write at all; reports when needed were written for them by the timekeeper or the lamp man).[34]

In the days of the pillar and stall system the miner's workplace was peculiarly his own, separated from its neighbours by thick walls of coal, and inaccessible to management's prying eyes: 'an amount of supervision which could be sufficient at the surface is utterly inadequate below ground, because the working places are not within sight from any point, and can only be reached by traversing low and tortuous passages . . . '.[35] The make-up of the work group was also the miners' own; they chose their own mates and they employed, very often, their own drawers to keep them supplied with tubs. In Northumberland and Durham, under the cavilling system, men changed their places quarterly, by lot; elsewhere a pair of men might be given a stall to work until it was finished, and then move further along the tunnel to start a new place of their own. In Durham the stall was used by the alternative shift, but earnings were shared under the system of 'marraship', which Dave Douglass describes. In South Wales, on the other hand, only a single shift was worked and there was bitter resistance to the introduction of the alternative shift: one reason given in 1873 was the miner's dislike of a 'stranger' working in his stall. 'They said four men would be required to work each stall on the double shift, two on and two off; and that four colliers could not work together without two, or one of them, at least, proving dishonest towards the other'.[36] (One way of getting round this problem when double shifts eventually were introduced was for both to be worked by members of one family.)[37]

A good deal of the working equipment in a mine belonged not to the owners but, as far as face work was concerned, to the miners themselves. This was a measure of economy on the owners' part, but it also involved the alienation of an element of supervisory control. The miner possessed his own personal picks and drills – 'at least half a dozen picks' according to the *Morning Chronicle* account of County Durham – and he was personally responsible for keeping them fettled. 'A . . . hewer . . .

generally fits the hafts . . . himself, and a blacksmith, partially paid by the colliery, keeps the iron part in order. The picks have to be sharpened every day, so each hewer when he ascends goes straight with his implement to the blacksmith's shop, and next morning finds it laid out in readiness for him.'[38] The hewer paid *2d.* a fortnight for this service. He was also responsible for finding his own explosives. At some collieries in County Durham the men had established their own powder magazines, 'attended to by one of themselves' in a detached building; at others they bought powder of the owners or the overmen. Cartridges or charges were sometimes manufactured by the miner at home. In Leigh, Lancashire, this is remembered as a dangerous time:[39]

> Some nights the collier would be busy sharpening and putting new shafts in his pick, or making cartridges and filling them with gunpowder; and so heedless of danger were some men that they would make these cartridges on the hearthstone where a spark from the fire, falling into the powder-can, would have blown the house up.

In the days of candles the miner was also responsible for finding his own light ('about two pounds of candles in a fortnight'); 'pit candlesticks' were made with lumps of clay 'generally obtained by the wives and children of the hewers' (in Durham the hewers would threaten to stay off work unless they were kept well supplied). When Davy lamps were adopted, the owners paid for them, but the miner rather than the lamp man was responsible for keeping them in trim: 'The men carry the gauze cylinders home . . . and clean them carefully, as, should the wire net work become painfully clogged with coal dust, the danger of explosion is greatly increased.'[40]

The Cornish miner did not necessarily have to find his own working materials or tools – they might be loaned him by the company and charged on his month's account[41] – but in other respects he enjoyed a wider latitude at work than the collier. Working places were let out on a contract basis and put up to auction each month. The contract was taken by a 'pare', a group which might vary from two men to a dozen or more;[42] while it lasted the pitch was formally their own. (Miners who had been allotted a profitable place would hang on to it for as long as possible, at the same time trying to conceal its profitability from the 'captain' by holding back some of the more valuable ores.) The tributer, who contracted for the ore-getting pitch, was not so much a wage earner as a small-scale prospector renting the temporary use of a pitch; he was paid according to the value of his lodes – in proportion to profits and results; a lucky strike would put him temporarily in clover (but only temporarily, because if a place proved profitable its price at the auction would be raised). The tutworker, who had the job of sinking the shafts and driving forward the levels, also rented his pitch, and had to bid for it on the monthly setting day. But his franchise was narrower than that of the tributer, and he was paid by the fathom rather than in proportion to tin or copper raised. His earnings were more regular, but he had very little scope for speculative gains. Both tutworkers and tributers had their own choice of mates. Father and son, or brothers, worked together on a pitch, and it seems that men sometimes travelled together and took up employment in 'pares'.[43]

The 'tribute' system virtually eliminated the function of management so far as day to day supervision was concerned. The working place of the tributer was visited only when it was a matter of assessing its potential value for one of the monthly bargains, that of the tutworker only to measure up his work. There were no foremen or overlookers in the Cornish mines, but only captains who had a very much wider brief, being charged with superintendence of the works in all their aspects, sampling and smelting, book-keeping and marketing, as well as the underground working of the mine. There were very few of them in relation to the number of workers employed – twenty-eight for instance, at Consolidated Mines, Gwennap in 1836, when 2,387 workers were employed. At Fowey Consols, with 1,706 workers, there were fourteen 'Agents, etc.' and one engineer.[44] At Dolcoath mine, one of the larger and more profitable mines on the Redruth copper belt, there were, in 1864, only two underground agents. 'All the principal parts of the mine are inspected twice a week, but a couple of men may take a pitch in some distant part of the mine, and if that is seen once a fortnight that is as much as we think necessary.'[45] There is no doubt that the miners welcomed this lack of supervision, and indeed tried to increase it by making access to their pitch difficult 'or, at all events troublesome' to strangers. 'They are constantly endeavouring to outwit the agent by fair means or foul, and will candidly confess that "the whole art of mining is fooling the captain".'[46]

In the smaller Cornish mine, or on a tin stream, the function of management might be dispensed with entirely. In such cases, H.M. Johnstone, a Divisional Mines Inspector, told the Royal Commission of 1912–13, the mine was worked wholly by the men, with several independent sets of tributers at work.[47]

> They make their arrangements, usually verbal, direct with the lessee who may have no representative on the mine premises, but may occasionally, perhaps twice a year, pay a visit to the surface of the mine. Special areas are not allotted to each set, but each set of tributers select their own pitch and have the exclusive right to work in it during the currency of their agreement. They have the right to sink shafts and provide their own means of access and transport. They employ all labour and find all materials and tools.

In quarrying there was less need for management than in a mine, since there were no expensive steam engines to maintain, and in the smaller or simpler kind of works there would be fewer different activities to co-ordinate. In the Buckley clay pits, according to a witness who appeared before the Quarry Committee of Inquiry in 1893, there was no one in command at all.[48] The same was true at Waterslip Quarry, Somerset, in the Mendips, 'a large quarry in which 80 or 90 men are employed'. J. S. Martin, Mines Inspector for the Southern district, reported on it in 1904, after the occurrence of a fatal accident: 'the supervision in the quarry rests in one of the workmen, who receives a small allowance for looking round occasionally during the day, the rest of his time he is employed on piece work the same as the other men'.[49] At the Dorothea Quarry, North Wales there were six officials in 1893 ('of whom three are slate examiners') to about 400 men.[50] At the nearby Cefn Dhu quarry, William Gladlys Williams had this to say about the management when he appeared before the Quarry Committee of Inquiry:[51]

> Are there no overlookers at your quarry as well as a local manager? Only the manager.
> There are no overlookers? No.

Even at a giant works the ratio of management to workers would be low, as some 1884 statistics from the Penmaenmawr granite quarries may suggest. Table 1.7 shows the comparative costs involved in forming a ton of paving setts when the quarries were in full work.[52]

Boy labour in the slate quarries for a long time escaped the authority of management entirely, being employed, on a more or less casual or family basis, by the quarrymen themselves. The boys came to the quarry as 'rubblers' or 'rubbishers', making slate out of debris, which the quarrymen sold to management as part of their monthly 'reckoning'. 'They come and go when and where they please' a Caernarvonshire

TABLE 1.7

Management cost in sett making

	Per ton of Setts	
	s.	*d.*
Quarrying, including removal of top rock	2	6
Sett making (average price)	9	0
Royalty	0	2
Powder and fuze	0	5
Management in and out of quarry	1	9
Trammers and labourers	1	3
Loading, smiths and contingencies	0	10
	15	11

owner complained in 1867, 'it will require a great change of organisation to exercise any control over their time and employment.'[53] At Mountsorrel, too, in the granite works, boy labour escaped control as Mr Hambly, a quarry manager, told the Factory and Workshops Commissioners in 1876:

> A large number of boys that are employed by the Mountsorrel Granite Company are not employed directly, but indirectly. They are employed by their fathers or their uncles or some relative, and they really do not appear on the books of the Mountsorrel Granite Company at all. We only know them by occasionally sending a clerk round and taking a kind of rough census. . . . What kind of work are the boys employed upon? – It is ordinary paving stone with which the streets are paved, which has to be roughly picked on the sides; that is one of the principal things that the boys from 10 to 13 are employed upon. It is by no means heavy work because they are only working part of the day there. Then another work they do is the ordinary breaking of stone with a hammer on a block. . . . That is all done by piece wages, and the boy is employed by his father, or by someone over him.[54]

In the smaller quarries the work was often farmed out to the quarrymen entirely, being taken on by individual workmen, or by a group. At Filkins in Oxfordshire quarrying jobs were put up for auction by the road surveyor.[55]

> In the old days the stone digging was done mostly piecework. Men used to contract for the job at so much a yard cube, which meant taking off the top, or ridding as it was called, down to the rock, then they dug the stone and wheeled it out and stacked it a yard high. . . . The price varied in different quarries. . . . The general rule was when stone was wanted for the roads, the road surveyor (Mr. Powell was then the surveyor) would put up printed notices asking for prices in different quarries, the men would send in their prices and the lowest price was accepted. There was great competition for this job, as it was a winter's job, and in those days there was not much work about, but it was very hard work to earn 10s a week.

On the Isle of Portland a company of four men would take on the running of the smaller quarries for themselves. An islander describes the system in a letter of 1896, preserved in the Dorset County Museum:[56]

> The Quarry is sometimes taken by a company of four men to clear away the rubble & cap for a certain fixed price for the stone underneath which would be about 10 or 12 shillings per ton on stone produced. These men would clear a *task* or a piece of ground, to get the stone under. This would occupy a time say of six months, in some cases they would not receive any money untill this *task* . . . had been worked out, and the account of the amount of stone they had produced sent in to the master, or owner of the land, one man keep the accounts but they all four would share alike. . . . The Quarrymen have nothing to do with the price of stone. . . . They . . . simply take a Quarry at so much per ton, and their accounts are squared quarterly.

There was a certain amount of co-operation between these different groups. Some of the rocks were very heavy – from 15 to 300 tons – and it needed seven or eight men to move them. Before the introduction of cranes it was usual to borrow men from the neighbouring quarries. 'When everything was ready one man would say "Stran all so-o ay-so-ay" when the rest would haul with all their strength as each syllable was uttered.'[57]

The weakness of management in the brickfields was reflected in the widespread use of indirect systems of employment. The moulder, who contracted with the proprietor for his 'stool', might keep the whole work to himself, or sub-let the preparation of the clay to a temperer, who in turn recruited a gang. Or both might be contracted on equal terms to the employer. The moulder was simultaneously worker and employer and recruited his helpers himself, or in the case of the Black Country yards, herself. The man or woman with a family to call on had a big advantage over those who had to pay their helpers,[58] but it was possible to cut costs by lodging helpers at one's house and charging them for board.

This system of indirect employment was maintained in larger works as well as small. When the young Will Thorne went to work at Burk's, clay wheeling at the rate of 3*s*. 6*d*. per day, he was employed by his cousin Jim Thorne, 'one of the fastest brickmakers in Birmingham and the champion cribbage player in the district'. It was a 'hard life' he found ('in addition to wheeling clay, I had to operate the press'), and when the manager attempted to reduce the price both he and his cousin walked off.[59] Ben Tillett also worked at a brickyard. At the age of six he took a job at Roache's brickyard, Easton (a proletarian suburb of Bristol). The man he worked for, a champion brick moulder 'making thousands of bricks a day', had two lads working for him, one of them his own son. 'The other lad . . . was stronger than myself. It was hard work for me to keep pace with him, but the handler of the moulding frames was a heartless person. If he was served with a bit too much clay, it hindered his work – and slap! with unerring aim, would come a lump of clay, possibly a pound in weight, on the back of one's stooping neck or body.' On Mondays there was no work. 'The bricks would be made only in conversation aided by liberal libations of cheap beer.' But this indulgence was paid for by working on Thursdays and Fridays from 5 am till dark 'if the weather was fine'. At the end of the week Tillett was paid eighteen pence for his labours; 'and a tired little fellow would creep away to find sleep against the exhaustion of pain'.[60]

The mineral worker was less likely than the factory operative to be dependent on a single employer or firm. The unsettled or seasonal nature of many of their occupations put a premium upon frequent movement from job to job. And the rural or half-rural settings in which they very often lived gave them access to alternative sources of livelihood or earnings. There was a very frequent interchange between quarrying and other classes of outdoor labour – navvying, walling and stonemasoning, for instance, on the more industrial side, field labour and farming on the agricultural. The quarryman was often a farm labourer looking for higher wages, and willing to undertake harder and more dangerous work for the sake of them. But at haymaking or harvest, when wage rates evened up, he was ready to desert the diggings for the farm – the case with coprolite diggers in the Cambridgeshire Fens, for instance,[61] and with the slate quarrymen whom Merfyn Jones writes about in North Wales. Others would take up farm labouring for longer periods, when work at the quarry was slack or if the earnings differential was scaled down. In the limestone district of Clopwell there was a general boycott of the quarries on this account in 1867:[62]

> A strike has been in force for some time for an advance of wages, as the lime merchants are giving the same rate for their labour as the agriculturists. They refuse to advance their rate, and the men, who would rather be employed as agricultural labourers, refuse the class of employment called 'stone getting'. It is a matter of serious consideration how to keep the supply of labour. Excepting the old hands, no men can be got to face the work.

Land was another possible resource, more particularly in hill or moorland districts, where smallholdings were comparatively easy to come by, and where it was possible to run sheep upon the waste. In the Aberdeen granite industry, according to an account from Rubislaw, numbers of the sett-makers were crofters; they worked comparatively short hours, and they took advantage of wet days to give a little extra time to their holdings, 'pleased to be able to devote . . . an occasional afternoon to their cultivation'.[63] On the Isle of Purbeck, when Walter White visited it in 1860, the quarrymen combined smallholdings with fishing.[64]

> They know how to be thrifty; and though earning but small wages with all their hard work, they contrive, by renting a plot of land, which feeds a cow and fowls and supplies them with vegetables, and occasionally catching fish, to maintain themselves in comfort and independence. You may frequently see them near the sea lifting up the thin loose slabs in search of sea-lice, which, congregated by thousands, afford a plentiful supply of bait; and when too many fish are caught, there is always a market for the surplus at Weymouth.

Brickmakers in West Middlesex (i.e. the regular brickmakers, who lived there all the year round) were well known as pig keepers (at West Drayton the pig was known as 'The Brickie's Bank', his protection against winter distress)[65] and some of them enjoyed a further 'secondary' income by taking 'lodgers' – giving board and lodging to the seasonal migrants who came to the work in the spring. Pigs were common, too, among the china clay diggers of Lee Moor, Dartmoor. 'Many own pigs, or a cow or two, and nearly all rear poultry', wrote William Crossing, 'while a number of them have ponies running on the moor.'[66] Similarly, in Cornwall, 'a big garden, with a pig sty if possible, was an important part of the clayworker's economy'.[67] Another pig-keeping group of mineral workers were the copper miners at Devon Great Consols, as Dr Bristowe notes in his report on Gunnislake in 1862: 'every house nearly has its pigsty, and rows of houses have often their rows of sties. . . . No one . . . can take even a cursory glance at the village without being struck by the unusual number of filthily kept pigs'.[68] Colliers could be great pig-keepers too, and indeed one of the wilder accusations launched against them before the Mines Commission of 1842 was that they traded their babies for pigs. G. A. W. Tomlinson has this to say about the collier pig-keepers of his native village in Charnwood Forest, Leicestershire.[69]

> Ours was a long row and at the top of every house-garden there was a ramshackle pigsty. It was the custom of the men, after they had taken their walk through the forest, to amble slowly down the 'backs' in groups of four or five, pausing for a time to examine with keen eyes the pigs in each sty. The proud owner of the pig would explain in detail what it had eaten during the last week, how it compared with the previous pig, and so on. 'That theer pig', Joe would say, ''as put over aife a stone on sin' last Sunday.'

The most privileged group of pig-keepers were the free miners of the Forest of Dean who, by ancient right, could feed them on the Forest. They also ran donkeys and sheep. 'The forest is essentially a mining district', a free miner told an inquiry of 1889: 'the work of the collier is very laborious, and sometimes, in consequence of foul air, very unhealthy; many of the colliers are prematurely old; it may be that at 55 years of age they are unable to follow their usual employment, but they are enabled to keep off the rates and from the workhouse, from the fact that they keep a few pigs or a few sheep; it may be, a horse or a donkey, and they find the right to "turn out" very valuable indeed.'[70]

The 'strange combination of miner and small farmer', as Tomlinson called it, was a familiar feature amongst the lead miners of the northern moors and dales. Numbers of them were cow-keepers, with little farms or smallholdings which they rented or owned.[71] The combination of mining and agricultural work led some to take a four-day week, rearranging their shifts and working longer hours in the early days of the week, so that they could get off early on Thursday. 'They frequently work on Monday, Tuesday, Wednesday and Thursday; and return home on Thursday, spending the remainder of the week in working on their small farms or in their gardens, or for the farmers in the neighbourhood.'[72] At Middleton in Teesdale 'comparatively few' of the miners were without land to cultivate. 'Some . . . are small "statesmen" or the sons of "statesmen" whose properties are inadequate to their maintenance, but the larger proportion of them are merely the tenants of the lords of the mines and other landowners in the neighbourhood, and a still larger proportion have not more than a cottage and a garden.'[73] In Allendale 'most of the farms' were occupied by miners, smelters and people connected with Beaumont and Blackett's works; 'They are partly occupied in the works, and partly in farming.'[74] In Derbyshire land was less easy to come by than in the northern Pennines, but there was more in the way of alternative employment on the farms. The independent miners, who took on the working of the 'poor men's mines' seem often to have done so only for the winter season ('in summer time they lend aid to agricultural purposes')[75] and there were numbers of men who went off to the fields for haymaking and harvest, even though they worked in the mines most of the year – 'half labourers and half miners' was how they were described to the Rating of Mines Committee in 1856: 'When a man can go off in harvest time for a month or six weeks, and get £1 or 25s a week, it is an object to him.'[76]

Lead miners were barely distinguishable from the moorland cottagers among whom they lived. In Derbyshire the 2,000 lead miners of 1851 were scattered among a population of some 36,000 people, and the family depended on more than a miner's earnings to survive; the women worked at lace making or in the upland mills; miners' sons (or it might be fathers) were agricultural labourers. In Montgomeryshire some of the lead miners lived at Llanidloes.[77] In 1851 it was a town of miners and flannel weavers all mixed up together – 'greasy weavers, in their blue shirt sleeves, taking rest at their doors . . . round-hatted miners . . . in their clay-bedaubed yellow jackets, with their dinner tins under their arms'. The men went out of town to the lead mines, their wives stayed at home weaving flannel.[78] In Swaledale many of the miners rented smallholdings, 'perhaps keeping a pig or a cow, and growing a hay crop'. At hay time, when the fields about Reeth were full of busy men and women, some of them left the mines to work as additional labour on the valley farms, 'making this their annual holiday'. Some miners undertook peat cutting for the smelt mills, 'in their spare time, as a little extra'. Some found occupation as wallers. 'All the miners', Arthur Raistrick tells us, 'were inveterate and very swift knitters.' This was a universal employment in the Dales 'and many of the miners knitted as they went to and from work'. Stockings were the main product. 'Wool could be bought from the agents and the finished stocking sold to the agent or in many cases paid in as part payment for groceries and goods.'[79]

Coal mining in mid-Victorian times was not yet an hereditary occupation, nor necessarily a complete and exclusive pursuit. A Flimby collier might go out with the fishing boats,[80] a Whitehaven one take ship with one of the brigs. 'It had become an axiom in Whitehaven,' Jack Lawson tells us, 'that no sailor was worth his salt who was not also a miner, and no man was a real miner unless he was also a sailor.'[81] In country districts, there was a considerable interchange between mining and outdoor labour (particularly navvying) and in summer when the pits were slack, there was some seasonal movement to the fields. 'The demand for domestic coal is quiet', runs a trade report from Staffordshire in July 1859,' . . . the colliers largely go out into the surrounding agricultural districts and work as harvestmen – a pleasant transition from the dark mine and its often noxious atmosphere . . .'[82] The miner was not yet wholly absorbed in a colliery world. Mining might well be his second occupation in life rather than his first, and his attitude both to job and to home was by later standards still to some extent provisional. In Lancashire, which in the 1850s had more miners than any other county, the pits were thoroughly mixed up with industry and the towns. At Rochdale 'Colliers and Weavers' followed the same rush-cart; at Wigan, they lived alongside each other in the Scholes. In the Black Country, too, miners were very far from being a race apart. 'The miners of Wolverhampton do not form so distinct a class, either as regards appearance or habits, as those of the Northern Coal Field, or of Cornwall', runs a report to the Medical Officer of the Privy Council in 1862. 'Many men, especially Irishmen, are met with among the miners of South Staffordshire, who have taken up the occupation somewhat late in life, after working for a time at some other kind of labour.'[83]

Job changing among miners was endemic, with new pits continually opening up (in 1893 as many as fifty-one were started up in the space of a single *month*),[84] and old ones being closed down as they were worked out, or became unprofitable for other reasons. In the 1840s a miner could change jobs most easily in coalfields like Lancashire, Yorkshire and South Staffordshire, where there was a multitude of pits in close proximity to each other (in 1851 there were 210 collieries in the Wigan district: almost as many as in the whole of Britain today).[85] The rapid development of the Aberdare valley in the 1850s, and later of the Rhonddas, produced a similar situation in South Wales, with an extra incentive to movement in the high wages offered to attract scarce labour to 'uncivilized' parts.[86] In the great Dowlais strike of 1853, for instance, so many of the colliers went off to work in Aberdare that the owners were warned that there would soon be no colliers left in Dowlais at all. A similar mass migration to the newer mines by striking miners happened in the Monmouthshire strike of 1873. In Northumberland and Durham, pits were fewer and larger than in other parts of the country, and free movement was for a long time restricted by the yearly 'Bond'. But the later nineteenth-century developments to the east of the county, about Tyneside and the coast, seem again to have produced a situation of immigration and flux. Boldon Colliery, at the time when Jack Lawson writes about it in the 1890s, was 'a sort of social melting pot', with a polyglot population drawn from Lancashire, Staffordshire, Cornwall, Ireland, Scotland, Wales, Northumberland and Durham. 'The older collieries were more settled in their personnel, but among the great coast collieries there was constant ebb and flow. . . . A new colliery or a new seam meant bigger money, and there was always an emigration, followed by the incoming of new people to take their place.'[87]

The mining population at mid-century was notoriously restless and unsettled. Indeed the 'roving disposition and migratory habits' of the miner were as much a commonplace of contemporary social comment as their heredity and fixity is today. 'Whatever is unsettled or lawless or roving or characterless among working-men . . . has felt an attraction to this district', wrote Jelinger Symons of the Monmouthshire colliers in 1847. 'It . . . contains a larger proportion of escaped criminals and dissolute people of both sexes than almost any other populace.'[88] At Coatbridge at this time, to judge by some remarkable figures produced by Alan Campbell, there was what amounted to a new mining population every ten years. His figures are based on the manuscript census returns. They show that of the coal and ironstone miners, drawers and redesmen recorded in the 1841 census, only 14·2 per cent were still in the same occupation at Coatbridge ten years later. The 'occupational persistence rate' was still very low in the 1860s, with only 23·3 per cent of Coatbridge miners appearing in both the 1861 and 1871 census.[89]

In the new districts opened up in the 1850s and 1860s, settlement at first was hardly less provisional, with camp-like living conditions, a population recruited from strangers, and a large plurality of single and unmarried young men. Here is what *The Times* Special Commissioner had to say about the Rhondda miners in 1873.[90]

> One cannot pass through the Rhondda Valley . . . without seeing that the colliers there are a more reckless . . . looking set of fellows than those in the Merthyr Valleys. The latter look like natives of the soil on which they are living; the Rhondda Valley men look like rough colonists. . . . . When those . . . valleys began to be worked the masters had to offer higher wages . . . in order to induce men to go there. Even with such an inducement only very young men . . . migrated there because there were but few houses near, and the colliers had to put up with many hardships. . . . Since then thousands of houses have been built . . . but life . . . is somewhat rougher than in the Merthyr district, and the colliers are still of the colonist stamp.

The Cornish miner was even more migratory than the collier, though the occasions for moving were not the same. Numbers of Cornish miners worked overseas even before the near collapse of their industry drove them out in thousands; in the 1820s there were some who had gone to work for a period in the Mexican and South American mines, and by the 1850s the to-and-fro movement of miners between Cornwall and the United States was well established,[91] while towards the end of the century Cornish miners played a big part in the opening up and exploitation of the Rand. (In a cohort of Cornish rock drill men recorded in 1904 as dying of phthisis, 27 had worked with the rock drill in Cornwall only, 25 in Cornwall and Transvaal and 46 in Transvaal only.)[92] Within the duchy itself movement was intense. Job changing was built into the very structure of Cornish mining. All the working places in the mine were put up to auction every month; the worker with a bad pitch could hope to better it by taking another, but a run of bad luck – and of low earnings – would prompt him to try his fortune elsewhere. The considerable variability in working conditions might play a part too, especially for the tutworkers, who did the driving of the levels, and who suffered most from miners' disease, congestion of the lung which made breathing and exertion difficult.

Here is the job career of a 58-year-old Cornish miner, who, though afflicted with miners' disease, was still working when the details were taken down in August 1862:[93]

> First worked at Bardon mine, near Blue Anchor, for 2 months, in the 40 level. Air good. Then went to Drakewells on tutwork, for 7 months, in 70 level. Then to Wheal Franco on tutwork, for 6 or 8 months, in the 30 level. Air not good there. Candle inclined. Then to Busy Pool for 9 months, in the 124 level. The air was 'dead'. He used to feel weakness in the legs in climbing, 'but nothing to hurt the body'. Then he went to adit's foot for 1½ years, sinking from grass, in good air. Then to Beckenwood, for 7 months, in good air. Then to Treburgat for 2 years, on tribute, in the 10 level. Air not bad there. Then to Singinnis for 3 years, in the 20 level, in middling air. Then back to Treburgat for 2½ years, in the 40 and 60 levels, in 'dead' air. Then to Wheal Mather for 12 months in 'dead' air, in the 50 level. Then to Holmbush for 3 years, in the 90 level. Air sometimes 'dead'. Then to Wheal Mather for 6 months in the 70 level, in 'dead' air after 'shooting'. Then to North Holmbush for 9 or 10 months on tutwork, sinking a grass shaft. Air was very well there. Then to Simonward for 5 years on tutwork, in 24 and 35 levels. The air was 'dead' there. Then came to Trewether for 4 years, on tutwork, in the 40 and 70 levels. Air very often 'dead' there. Then to Wheal Mary Anne for 7 months on tutwork, in the 110 level, in bad air. Then to Lidcut for 6 months, in good air, in the 40 and 60 levels. Then to Redmoor for 6 months, in the 110 level, in dead air. Then to Caradon Slades for 4 months, in a grass-shaft and in the 60 level. In the level the air was 'dead', but in the shaft it was good. Then came to South Caradon for 10 months, in the 70 level. Air good, by a machine. He is working there now.

Lead miners were more settled than colliers or tinners, but the uncertainties attending their individual place of employment were even more pronounced. Workings would very often come to a stop. In the famous plumbago mine at Borrowdale, for instance – the basis for the Keswick pencil industry – 20 tons were raised in 1875, but only one in the following year.[94] There seem to have been similarly sharp fluctuations earlier. When the *Morning Chronicle* Commissioner visited Borrowdale in 1849, the mine had been 'utterly barren' for six years; only six people were employed in it, and the Keswick workshops were supplying themselves with pencil lead by imports from abroad. 'The search for the mineral can be carried on only in very haphazard fashion' commented the Commissioner, 'the wad not lying in regular veins, but being found in "sops" or "bellies", formed by the intersections of strings of small rake veins, often lying at a considerable distance from each other.'[95] A smaller mine than this might stop entirely, if difficulties were encountered with water, or a vein was exhausted, or there was no capital to carry on. Again, the whole mining ground might shift if 'trials' indicated a better field to explore. 'The mines which used to be so prosperous here are now very poor', runs a note on Coalcleugh in 1871, 'so that many miners have had to remove with their families, and others go and work all the week in other mines, and only come home from Saturday till Monday.'[96] When a mine closed down the family might move, but often they stayed put, and the man would take up lodgings for the week if the journey to his new place of work was too far to make each day on foot. The larger companies met this need by building 'lodging-shops' – at Teesdale in 1864 two-thirds of the London Lead Company's labour force used them.[97] In Patterdale, some of the workers at the Greenside Lead Mine slept away for three nights rather than four: 'They generally come on the Monday morning and leave . . . Friday night, sometimes rather early; they will . . . perhaps go home say on Wednesday night, coming back on Thursday morning, and then finally leave on . . . Friday night', the manager told Lord Kinnaird's Commission in 1864.[98]

Lead miners, despite the remote parts of the country in which they worked, would travel far and wide, even if they retained a home village as an ultimate pivot and base. The chronicle of Fawcett of Swaledale, recently printed by Arthur Raistrick, enables us to follow an individual itinerary in some detail. Fawcett was a Muker man, whose father had originally been a miner-farmer, but who had been obliged to leave mining for work in a local woollen mill, where Fawcett worked with him as a boy. On his father's death he worked with his brother as a miner. He started first in a coal mine at Tan Hill, but later moved into the lead mines. Arthur Raistrick recounts his subsequent history as follows:[99]

> He found employment at the Oldfield Hush, Beldi Hill for many years. There his wages were 11/– a week and he became, like so many miners, a skilled poacher from which occupation at least his health greatly benefited. In 1864 he went to Sardinia with a Newcastle Company which was opening lead mines there. In less than two years he was back in Swaledale and married 'Mary o' Kisdon,' a member of an equally old and respected Swaledale family, the Keartons, and found work at Mukerside Mines.
>
> Soon he joined with another miner in an independent venture on

> the north-east side of Kisdon Hill, where they drove a level to a vein, only to find that the 'old man', as earlier miners are always called, had been in the veins before him, from a series of shafts on the hill top. In the old workings they found many tools, wooden shovels, and a hand barrow, but little ore. . . . In 1869 Fawcett removed to Askrigg, where after a short time of work in the Worton Mine, he found work with his brother on the construction of the new railway up the dale. . . . When the railway was opened the family moved back to Muker and father found work for a time in the Sir Francis mine, but this was not for long. Work became scarce and the family set out once more for a new place. This time work was found in Westmorland in the mines of the London Lead Company, but in 1881 a final move was made into Lancashire, where father and children found employment in the cotton mills.

Brickmakers and quarrymen had the reputation of being even more uncertain in their settlement than colliers. Dr Hunter, Medical Officer of the Privy Council, who visited a number of their villages in 1864, placed them firmly amongst those whom he called 'the migrating trades'. 'Brickmakers, builders, and quarrymen are', he wrote, 'though far richer, often more huddled together than rural hinds, because their occupation in a place is usually temporary, and it does not seem safe to erect cottages for them.'[100] At Woore in Shropshire they lived as lodgers. 'There is quarrying and brickmaking in the parish, but the workmen usually come for a week, returning to their families on Saturday'.[101] At Nunney, Somerset, Dr Hunter found 'some very bad cots. . . . They were intended for the workmen at a stone quarry, a temporary employment; they were built . . . about the quarry, and when the quarry moved on further off they were roughly patched up and let'.[102] Nash's Yard, Grimsby, which Dr Ranger notices in his inquiry into the sanitary state of Grimsby in 1850 ('a good deal of dysentery and fever' had prevailed there) seems to have been a rather similar affair, 'a temporary structure inhabited . . . by persons employed in making bricks for the docks'.[103] When coprolite fossils (which were ground for fertilizer) were discovered at Eversden in Cambridgeshire, 'about five hundred rough miners and labourers' poured into the neighbourhood to quarry them 'within a week'. Within a few years, however, the fossil strata were worked out 'so that only the labour of the regular population was required'.[104] Even more temporary were the quarrymen's settlements that appeared in the neighbourhood of a public works.

When the Keighley reservoir was building at Stanbury, for instance, there were what the village historian recalled as 'stirring times': 'Much stone was quarried from the Master Stones and the south side of Ponden. Navvies came to dig it from nearly all over the British Isles.' When the works were completed the village returned to the lonely setting described by Emily Brontë in *Wuthering Heights*.[105]

Even in the better established quarrying districts the working population was very far from settled. In the quarrying villages of Aberdeenshire a substantial proportion of the workers seem to have been weekly or seasonal lodgers[106] (in the 1871 Census there was a marked excess of males over females at Kemnay, Monymusk and Cruden – three of the quarrying villages – though in the county as a whole females outnumbered males by 129,000 to 116,000);[107] while some hundreds of stone cutters crossed to America each year, leaving Aberdeen in the spring and returning in the autumn.[108] An account from Mountsorrel in 1874 suggests that even in that large quarry the labour force was still provisional: 'many were tempted away a few years ago, by the offer of very high wages, to the Aberdeenshire quarries, but they have mostly returned'.[109] 'Barracks for workmen' – i.e. men who worked at a quarry but had not yet made their settlement there, or raised or brought a family – were still a feature of the slate quarrying districts of North Wales in the 1880s. 'One sees the men arrive on a Monday morning, carrying their provisions for the week on their backs; and they cook their food themselves by the common fire of the eating and sleeping apartments.'[110] At the time of the North Wales industry's most rapid expansion the proportion of lodgers and provisional residents was much higher. Here is a description by Dr Buchanan, after an investigation into an outbreak of fever at Festiniog in 1863:[111]

> The condition about the houses that calls for most serious attention is their overcrowding. The cottages were built for single families, and contain, for the most part, two or three rooms. Owing to the employment for the past two years of a much larger number of labourers in the quarries than formerly, there are not houses enough for all who want them, although many new cottages have been recently built. Every house, therefore, is sublet. As many lodgers as can be received are crowded into each room and the family of the house share the same rooms. . . . Labourers are attracted from some distance by the profitable employment that is to be had. Many of them lodge in the district all the week, and return to their families between

Saturday and Monday. This high rate of wages in the quarries makes building labour scarce and dear. . . .

Now in one cottage of Treddol ten people sleep at night; five of the family and five lodgers. Three lodgers sleep in the lower room; the family and two lodgers in the loft. There has been one case of fever here. In another of the cottages, twelve people have slept by night in the two rooms. A man, his wife, two grown sons, a daughter of eleven, and three lodgers occupied the upper room. A married daughter, her husband, and two children, slept downstairs. In this house, there were six cases of fever, and a child died of it; and so on for other houses in this row. In every house of the row there has been fever.

But the whole tale of overcrowding in the slate villages is not even yet told. The quarry labourers work by relays; some by day, others by night. As a day labourer leaves his bed in the morning he makes room for a night worker who has just returned home. . . . All the quarry labourers are not lodged in these cottages. Others of them sleep in a sort of barracks at the quarries. These are not so seriously crowded as the private lodging-houses; but they too are tenanted by day sleepers as well as night sleepers, and it does not appear that there is adequate provision for a proper renewal of air.

V

In the later nineteenth century mineral works developed in a much more unambiguously capitalist direction, with ownership becoming increasingly associated with control. The scope of sub-contract was progressively narrowed and 'bargain' systems of payment steadily gave way to more modern systems of piece-work. The decline of agriculture as an alternative source of employment left the mineral worker more dependent on a single line of work; so did the settling down of communities – pit villages, instead of being camps, became the habitat of clans, though the process was long-drawn-out, and hereditary recruitment to the industry was offset by a flow of immigration to the coalfields which did not stop until 1914. The biggest change was in brickmaking, where the removal of child labour undermined the whole system of labour recruitment, whilst continuous kilns, machine moulding and covered premises, though slow to spread, increased the amount of all the year round work. The local character of the industry was also undermined, first in the 1890s by the rise of the nationally orientated Fletton-Peterborough industry; then by the building slump of the 1900s, which put many local yards out of business.

In coal mining the scale of enterprise was transformed by the coming of 'deep' mining, which in the third quarter of the century spread from Northumberland and Durham to the other coalfields. In South Wales it led to the dense peopling of the Rhonddas, and the rise of steam coal collieries serving an export trade and a wide home market. In South Yorkshire with the discovery of the 10 foot Barnsley seam, it produced what was virtually a new industry. 'Most of the colliers are laid out on a scale capable of producing 1,000 tons of coal per working day, and giving employment to from 500 to 800 men.' In Nottinghamshire, too, a new type of coalfield appeared with an active class of owners taking over the running of the pits from the butties. No such dramatic change took place in the scale of quarrying, but from the 1880s there was a pronounced tendency to amalgamation in ownership – Bath Stone Firms was formed in 1887 to take over the innumerable small workings on Combe Down, and at Buxton thirteen of the seventeen different limestone firms formed themselves into a combine, the Buxton Lime Firm Company, in 1891. In 1910 there was a similar amalgamation amongst the quarrying enterprises in the Forest of Dean.

The later years of the nineteenth century saw much more aggressive forms of capitalist intervention and management, and mineral workers were affected by it no less than, say, engineers. The formation in 1888 of the Salt Proprietors' Union, which Brian Didsbury discusses, was a deliberate attempt to capture the trade: the small proprietor was eliminated, and the independent boatmen, the lumpmen and wallers at their pans, all found their fundamental autonomies threatened. Merfyn Jones shows a comparable change at work in the slate quarries of North Wales, with an active management transforming the nature of the bargain system, and eventually imposing its will, though only after a three-year lock out. In South Wales the face of the coalfield was changed with the rise of a powerful class of coalowners, and of big companies like Cambrian Combine and Powell Duffryn with whom three generations of miners were to do battle.

Trade unionism was not necessarily hostile to the tendency of these changes, despite some famous strikes; indeed, so far as mineral workers were concerned, they were the very soil in which it grew. Trade unionism was as hostile to 'bargain' systems of payment and sub-contract as the most imperialist management; both preferred to settle for standard rates. The quarrymen's union in the 1890s – like their counterparts in boot and shoe making – were engaged in active agitation to free the worker from responsibility for providing his own tools. In the movement towards more regular hours, too, trade union agitation coincided with the managerial drive towards greater accountability and control. In fact, the whole tendency of trade union intervention in the late nineteenth century, in mining and quarrying as in many other fields, was to confirm the status of the worker as a wage earner and to limit those ambiguities which, if they had given him certain latitudes, also laid him open to exploitation by the master.

Notes

I

1 W. G. Armstrong (ed.), *The Industrial Resources of . . . the Tyne, Wear, and Tees*, Newcastle upon Tyne, 1864, p. 188. The annual wholesale value of Whitby jet ornaments at this time amounted to £125,000.

2 J. Fairfax Blakeborough, *Life in a Yorkshire Village*, Middlesbrough, 1912, pp. 28–9.

3 *Geological Survey*, pp. 136–7 gives a general account; Arthur Cossons, 'The villagers remember', *Trans. Thoroton Society*, LXVI, 1962, p. 70 for the making of 'presents from Niagara' at East Bridgford; Arthur Aikin, *Illustrations of Arts and Manufactures*, London, 1841, p. 97, for the 'considerable manufacture' of ornamental vases and statuettes at Derby; 85,888 tons of gypsum were raised in Derby and Notts in 1886: P.P. 1890–1 (C6455) LXXVIII. Return of Wages . . . in Mines, p. xv.

4 For 'Purbeck squares' in Clerkenwell: Finsbury Archives, St James and St John Vestry Letter Books, In-Letters, 1856–7, contractors' tenders, March 1856.

5 P.P. 1867 (321) XIII, S.C. Mines Assessment Bill, Qq. 776–81.

6 J. B. Hill and D. A. MacAllister, *The Geology of Falmouth and Truro*, London, 1906, p. 261.

7 P.P. 1912–13 (Cd 6390) XLI, R.C. Metalliferous Mines and Quarries, App. A, p. 306. There was a slight recovery after 1898 and numbers increased to 8,533 in 1907.

8 Ibid., p. 299.

9 P.P. 1893–4 (C 7237) LXXIII, Quarry Committee of Inquiry, p. iii; 'clayworkers including brickmakers' amounted to a further 54,536 in the 1891 census, and some of these would have come under the Quarry Act.

10 P.P. 1912–13 (Cd 6390) XLI, R.C. Metalliferous Mines and Quarries, App. B, p. 309.

11 Beattie Scott, Mines Inspector for the Stafford District, reported that 1,900

quarries and pits had been returned to him under the Quarry Act, but that he had excluded 1,200 of them as being under 20 feet in depth. P.P. 1896 (C 8074) XXII, Annual Report Mines and Quarries, Scott, p. 20.

12 P.P. 1912–13 (Cd 6390) XLI, R.C. Metalliferous Mines and Quarries, Q. 10,081.

13 This point was made in the report on Bridgwater brickmaking to the Children's Employment Commission in 1866.

14 P.P. 1866 (3678) XXIV, Children's Employment Commission, 5th report, p. 146. Some idea of the seasonal influx of strangers may be seen from the 1851 Census enumerator's MS. note on Yiewsley, Middlesex. The population was recorded as 200, 'but during the Brickmaking season the temporary residents would be nearly 300 in a busy time'. PRO, HO 107/1697.

15 E. J. Connell, 'Hertford breweries', *Industrial Archaeology*, IV, February 1967, p. 42.

16 E. J. Hobsbawm, 'British gas workers', in *Labouring Men*, London, 1964.

17 *Industrial Resources of . . . the Tyne, Wear, and Tees*, p. 197.

18 P. G. H. Boswell, 'British glass-sands', *Jnl of the Society of Glass Technology*, I, 1917, p. 8; to give the glass additional toughness, broken fragments of Bridgwater brick were added to the batch.

19 P. G. H. Boswell, *A Memoir on . . . Refractory Sands*, London, 1918, pp. 158–66; for the canal traffic in sand, Charles Hadfield, *The Canals of the West Midlands*, Newton Abbot, 1969, p. 92.

20 J. Beete Jukes, *The South Staffordshire Coalfield*, London, 1859, pp. 145–60; C. Lapworth, *A Sketch of the Geology of the Birmingham District*, Birmingham, 1907, p. 48; Walter White, *All Round the Wrekin*, London, 1860, pp. 245–6; T. J. Raybould, *The Economic Emergence of the Black Country*, Newton Abbot, 1973, pp. 23, 172–82. There is a dramatic photograph of the limestone caverns at Wren's Nest in V. L. Davies and H. Hyde, *Dudley and the Black Country, 1760–1860*, Dudley, 1970, p. 54.

21 White, *Wrekin*, p. 270. The *Morning Chronicle* Commissioner put it more dramatically: 'The earth hereabouts seems, as far as the eye can reach, to be literally turned inside out', *Morning Chronicle*, 'Labour and the Poor: Manufacturing Districts, XXIII', 3 January 1850, p. 5, col. 2.

22 Oliver Jones, *The Early Days of Sirhowy and Tredegar*, Tredegar, 1969, pp. 26–9.

23 Ibid., pp. 40, 55; Arthur Gray-Jones, *A History of Ebbw Vale*, Risca, 1971, pp. 44, 51; Mathew Owen, *The Story of Breconshire*, Cardiff, 1911, p. 120 (horses were still pulling the trams down the mountain at Ebbw Vale in 1899: Gray-Jones, op. cit., pp. 184–5).

24 Frank Machin, *The Yorkshire Miners*, Barnsley, 1958, for a Sheffield miners' strike of the 1840s.

25 John Holland, *The Tour of the Don*, London, 1837, vol. II, p. 327; *British Ass. Guide to Sheffield and the District*, Sheffield, 1879, p. 104; ibid., p. 107 for

some other local quarries; Joseph Hunter, *Hallamshire*, London, 1869, p. 396.

26 *The Quarry*, February 1896, p. 35 for complaints of the lime-burning nuisance at Deepcar.

27 J. Radley, 'Peak millstones and Hallamshire grindstones', *Trans. Newcomen Soc.*, 1963–4, XXXVI, p. 172.

28 A. H. Green, C. Le Neve Foster and J. R. Dakyns, *The Geology . . . of North Derbyshire*, London, 1869, p. 10; Holland, op. cit., vol. I, pp. 95–6.

29 *British Ass. . . . Sheffield*, 1879, pp. 107–8.

30 R. A. Hadfield 'On the early history of crucible steel', *Jnl Iron & Steel Inst.* LXVI, 1894, p. 234; cf. also J. S. Jeans, *Steel*, London, 1880, pp. 369, 382; C. B. Holland, 'The manufacture of bessemer steel', *Jnl Iron & Steel Inst.*, 1878, p. 105; *British Ass. . . . Sheffield*, 1879, pp. 107–8.

31 White, *Wrekin*, p. 272; according to Robert Hunt it was generally computed in South Staffordshire that, for the production of one ton of pig iron, about 2½ tons of iron ore, 1 ton 5 cwt of limestone and 3 tons of coal were needed: *Geological Survey, Mineral Statistics for 1856*, p. 107.

32 *Morning Chronicle*, 21 March 1850, Labour and the Poor, Mining and Manufacturing Districts of South Wales, Merthyr Tydfil, III, p. 5, col. 6. For a pencilled drawing of the Dowlais work-girls in 1865, see *Munby, Man of Two Worlds*, ed. Derek Hudson, London, 1972, p. 210.

33 *Geological Survey, Mineral Statistics for 1856*, p. 134.

34 Ibid., p. 73; P.P. 1857, Sess. 2 (241) XI, S.C. Rating for Mines, Qq. 2962–5; Hadfield, *W. Midlands Canals*, p. 181.

35 R. H. Campbell, *Carron Company*, Edinburgh, 1961, pp. 46, 203.

36 Barrie Trinder, *The Industrial Revolution in Shropshire*, Chichester, 1973, pp. 71, 240, 248–9.

37 David C. Davies, *A Treatise on Earthy and other Minerals*, London, 1884, p. 33. For the use of Rowley Rag in glassmaking, *Geological Survey*, p. 134n; for the use of Derbyshire spar, White, *Wrekin*, p. 235.

38 Samuel Oldknow among others, who set up lime kilns at Marple. George Unwin, *Samuel Oldknow and the Arkwrights*, pp. 206–10; 215–21. Cf. also Ian Donachie, 'The lime industry in south-west Scotland', *Trans. Dumfries and Galloway Nat. Hist. Soc.*, XLVIII, 1971.

39 George R. Burnell, *Rudimentary Treatise on Limes, Cements . . . Plastering*, London, 1850.

40 In the 'Great Stink' of 1858, the windows of the Houses of Parliament were draped with curtains soaked in chloride of lime so that members could breathe. N. J. Barton, *The Lost Rivers of London*, London, 1962, p. 111.

41 Louis J. Jennings, *Rambles Among the Hills*, London, 1880, p. 44.

42 Philip K. Boden, 'The Limestone Quarrying Industry of North Derbyshire', *Geographical Journal*, CXXIX, March 1963; for a detailed account of the canal, Charles Hadfield and Gordon Biddle, *The Canals of North West England*, Newton Abbot, 1970, vol. II, pp. 300, 306–13.

43 For Devon, 'The Rolle canal', *Industrial Archaeology*, VII/1, February 1970; 'Coastal limekilns', Ibid., V/2, May 1968; for Sussex, W. V. Cooper, *A History of the Parish of Cuckfield,* Haywards Heath, 1912, pp. 158–9; for Montgomeryshire, Hadfield, *W. Midlands Canals*, pp. 189–90, 192–3.

44 David Williams, *The Rebecca Riots*, Cardiff, 1955, pp. 75–6.

45 J. Fenwick Allen, *Some Founders of the Chemical Industry*, Manchester, 1906, p. 24.

46 Dona Torr, *Tom Mann and His Times*, London, 1956, p. 274.

47 P.P. 1912–13 (Cd 6390) XLI, R.C. Metalliferous Mines and Quarries, App. B. p. 310; according to these figures gravel production increased from 1m to 1·8m in the 1890s, while limestone was comparatively stable; the differences may have to do more with the way the statistics were collected than with changes in quarrying.

48 Fenwick Allen, op. cit., p. 26.

49 W. Whittaker, *The Geology of London*, London, 1889, vol. II, p. 501.

50 Boswell, *Refractory Sands*, p. 5.

51 P.P. 1873–4 (C 1071) XXV, Local Government Board Rep. for 1873–4; I am grateful to Anna Davin for this reference. Sand, gravel and shingle were also used extensively by contractors in street cleansing. See, for example, Holborn Archives, Holborn Board of Works M.B., Vol. IV, 1860–2.

52 Boswell, *Refractory Sands*, p. 34.

53 Holland, op. cit., vol. II, p. 325; Ivor J. Brown 'Notes on the sand workings in the Doncaster area', *Bull. Peak Dist. Mines Hist. Soc.*, IV/3, May 1970.

54 Boswell, *Refractory Sands*, p. 169.

55 D. T. Ansted, *The Applications of Geology to the Arts and Manufactures*, London, 1865, p. 96.

56 Boswell, *Refractory Sands*, p. 65.

57 Ibid., pp. 155–7.

58 Whittaker, op. cit., vol. I, p. 503.

59 Aikin, op. cit., p. 52; Clement Reid and Aubrey Strahan, *The Geology of the Isle of Wight*, London, 1889, p. 122. Alum was of course, also extensively used by the master bakers to adulterate their bread.

60 Andrew Pearson, *Wilmslow Past and Present*, Stockport, 1901, pp. 49–50. Sand was also used as a blotting paper, though by the 1880s this was apparently considered old-fashioned. George Gissing, *The Nether World*, Everyman edn, p. 73.

61 Ansted, op. cit., p. 97; Boswell, *Refractory Sands*, p. 3; Gray-Jones, op. cit., p. 218. Bath bricks were manufactured out of river sand at Bridgwater and were extensively exported abroad: *Meliora*, XII, 1869, p. 147.

62 'The mysteries of the courts', *Porcupine*, 21 March 1863.

63 Frank G. G. Carr, *Sailing Barges*, London, 1951, pp. 62–3.

64 Henry Mayhew, *London Labour and the London Poor*, London, 1861, vol. II, pp. 90–1. For sand diggings at Hampstead Heath, J. J. Sexby, *Municipal Parks*, London, 1898, p. 383.

65 'Autobiography of Samuel Fielden', *Knights of Labour*, Chicago, 19, 26 February, 5 March 1887. For a similar domestic industry at Mountain Ash, Glamorgan, Joseph Keating, *My Struggle for Life*, London, 1916, p. 34. For grit-dealers in Liverpool, *Porcupine*, XV, 5 July 1873, p. 214.

66 On the Birmingham canals in 1845 'general merchandise' made up 1m of the tonnage of 4m; all the rest was minerals (Hadfield, *W. Midlands Canals*, p. 92); on the Rochdale canal in 1819 corn, merchandise and wool amounted to £66,000; coal, stone, lime and salt to £157,000 (Hadfield and Biddle, op. cit., vol. II, p. 277).

67 Henry T. De la Beche, *The Geology of Cornwall*, London, 1839, p. 479.

68 Elihu Burritt, *A Walk from London to Land's End*, London, 1865, pp. 354–5. Still larger amounts were taken from the estuary sands at Padstow, and it was estimated in the 1850s that some 4m cubic feet of sea sand were annually employed to dress the lime-deficient soils of Cornwall. A. Voelcher, 'The use of . . . shell-sand in agriculture', *Jnl Bath and W. of Eng. Ag. Soc.*, VI, 1858.

69 Hadfield, *W. Midlands Canals*, p. 110.

70 M. C. Ewans, *The Haytor Granite Tramway*, Newton Abbot, 1966; a special quay was built at Teignmouth in 1821 to facilitate the shipping. Carr, op. cit., p. 172.

71 The Chapel-en-le-Frith quarries were started by the Peak Forest Canal Co.; the Caldon Low quarries were owned by the Trent and Mersey. Quarries also played a big part in the growth of the railways (Duke's quarries, Derbyshire, provided 100,000 charr stones for the London and Birmingham) while the spread of railways in their turn opened up new districts for quarrying, hitherto inaccessible because of difficulties of supply.

72 By 1871 the salt traffic on the Weaver reached 1m tons. Hadfield and Biddle, op. cit., vol. II, p. 381.

73 Charles Hadfield, *The Canals of the E. Midlands*, Newton Abbot, 1970, pp. 113, 223 for the brickyards which sprang up alongside the Grand Union Canal; cf. Sybil Marshall, *Fenland Chronicle*, Cambridge, 1967, pp. 39–40: 'one reason for there being so many brickyards in Ramsey Heights was undoubtedly the fact that the brick and tile could be took by water. . . . Each brickyard had its own gang of barges, or lighters . . . and some men were employed entirely as watermen, to take the brick and tile up to the towns by water, and to bring back coal on the return journey.'

74 *Capital and Labour*, III, 19 June 1876, p. 467.

75 George E. Diggle, *A History of Widnes*, Widnes, 1961, pp. 17–20; Carr, op. cit., p. 171. Vitriol was one of the chief cargoes on the Worcester Canal in the 1870s, according to the lock-keeper at Diglis. P.P. 1876 (C 1443–1) XXIX, Rep . . . Factories and Workshops Act, App. C, p. 122. For the importance of vitriol in the bleaching industry, A. and N. Clow, 'Vitriol in the industrial revolution', *Economic History Review*, XV, 1945.

76 Durham coals continued to be largely shipped at the staithes well after the coming of the railways. In 1857 coals seaborne to London from Northumberland and Durham amounted to 2,905,854 tons; coals sent by rail were 109,175. *Geological Survey*, p. 94. By 1867 the proportions had evened up for the London coal trade as a whole, 3,033,193 tons coming by sea, 2,980,072 by railway and canal. *City Press*, 5 January 1867, p. 3, col. 6.
77 F. C. Mather, *After the Canal Duke*, Oxford, 1970, pp. 258–9.
78 Hadfield and Biddle, op. cit., vol. II, p. 367. Some 350 boats were engaged in the trade from the Potteries northwards. P.P. 1876 (C 1443–I) xxx, Factories and Workshops, Q. 10,503.
79 Basil Greenhill, *The Merchant Schooners*, Newton Abbot, 1968, 1/178.
80 Carr, op. cit.; Mayhew, op. cit., vol. II, p. 171; R. C. Dixon, *Paddington*, p. 110 for the dust end of the trade; Herbert W. Tompkins, *Marsh Country Rambles*, London, 1904, p. 25 for the Essex end. For an excellent recent account of this trade in Essex, Frank G. Willmott, *Bricks and 'Brickies'*, Deal (privately printed), 1972.
81 Cyril Noall, *The Story of Cornwall's Ports and Harbours*, Truro, 1970, p. 29, quoting a description of the 1870s.
82 R. M. Barton, *A History of the China Clay Industry*, Truro, 1966, p. 133n for the statistics; pp. 33, 74, 76, 131 for the ports.
83 John Sydenham, *The History of . . . Poole*, Poole, 1839, pp. 400–3. In 1856 the total quantity of clay shipped at Poole was 57,613 tons. Clay sent from Poole to London by the London and South Western railway amounted to 582 tons. *Geological Survey*, p. 127. For a description of the seaborne trade, *The Clay Mines of Dorset, 1760–1960*, London, 1960, p. 12.
84 J. A. Bulley, 'The beginnings of the Devonshire ball-clay trade', *Trans. Devonshire Assoc.*, 1955.
85 Hadfield, *W. Midlands Canals*, p. 205.
86 John Tomlinson, *From Doncaster into Hallamshire*, Doncaster, 1879, p. 165. For the shore-side picking of 'blue flints' for the Potteries, see *Sussex County Magazine*, 1940, XIV, p. 71; 1955, XXIX, p. 356.
87 Carr, op. cit., p. 173 for the stone barges of Devon and Cornwall; C. E. Robinson, *Rambles in the Isle of Purbeck*, London, 1882, pp. 155, 157–9 for those of Purbeck.
88 Peter Perry, 'The Dorset ports and the coming of the railways', *Mariner's Mirror*, LII, 1967.
89 D. T. Ansted, *The Channel Islands*, London, 1862, pp. 502–4.
90 Greenhill, op. cit., *passim*; D. C. Davies, *A Treatise on Slate and Slate Quarrying*, London, 1878, pp. 164, 169.
91 *Illustrated London News*, 13 December 1879. The reference is to the Latimer road brickfields. Cf. Patricia Malcolmson, 'Getting a living in the slums of Victorian Kensington', *London Journal*, 1/1, May 1975, pp. 36–7. For an account of them in the 1840s, *Kensington, Notting Hill and Paddington . . . by an Old Inhabitant*, London, 1882, p. 27. The district between Dalston and

Canonbury contained at this time what is remembered as 'a nest of brickfields', Louis Bamberger, *Bow Bell Memories*, London, n.d., p. 33.

92 Robert P. Smith, *A History of Sutton, A.D. 675–1960*, prtd Thornton Heath, 1960, p. 84.

93 Thomas Hughes, *The Practice of Making and Repairing Roads*, London, 1838, pp. 34–6. The brick pits at Finchley and Muswell Hill made them a Mecca for the Saturday afternoon geologist. Henry Walker, *The Glacial Drifts of Muswell Hill*, London, 1874, pp. 8, 14, 15.

94 Westminister City Archives, 2495, Westminster Board of Works, Street Cleansing Comm. M.B., 18 September 1867.

95 Sexby, op. cit., p. 237.

96 For Plumstead Common, ibid., p. 70, and Greater London Record Office, Met. Bd. of Works, Parks, Commons Comm. M.B. 16 March 1871; for Charlton Sands, Aikin, op. cit., p. 31, Boswell, *Refractory Sands*, p. 141. Hammersmith was another good source of builder's sand, cf. *The Builder*, 7 December 1878, XXXVI, p. 1287.

97 Edward Dobson, *A Rudimentary Treatise on . . . Bricks*, London, 1850, pp. 122–3. Surrey Record Office, Acc. 219, Bundle B, 'Little Heath Brickfield' for Thames sand as an ingredient of bricks; Carr, op. cit. for ship's ballast from the Thames; Hughes, op. cit., p. 48 for Thames gravel for the roadmakers. In Mayhew's time fifty-two lighters and fourteen barges were employed to dredge ship's ballast from the Thames, Mayhew, op. cit., vol. III, p. 271. For a description of sand-getting at Charlton about 1770, W. Gilby, George Morland, *His Life and Works*, London, 1907, p. 25.

98 Thomas Allen, *A History of Surrey*, London, 1831, vol. I, p. 35; cf. K. W. E. Gravett and E. S. Wood, 'Merstham limeworks', *Surrey Arch. Colls*, LXIV, 1967. Papers for both the Merstham and Dorking limeworks are in the Surrey Record Office.

For the use of Dorking lime in Clerkenwell: Finsbury borough archives, St James and St John Letter Books, 1856–7, Tenders, March 1856.

99 Burnell, *Treatise on Limes*, pp. 78–9.

100 *The Builder*, 24 August 1878, XXXVI, p. 877. For lime exports from Rosherville and Northfleet, *City Press*, 13 May 1871, p. 3, col. 1.

101 Henry Walker, *Saturday Afternoon Rambles round London*, London, 1871, p. 67.

102 *The British Clayworker*, October 1898, VII, p. 195.

103 Dobson, op. cit., p. 252.

104 Donald Young, 'Brickmakirg at Weymouth', *Industrial Archaeology*, IX/2, May 1972, p. 188.

105 P.P. 1873 (C. 743) XIX, Factory Inspectors' half yearly report, p. 157.

106 Arthur B. Searle, *Refractory Materials*, London, 1917, for a general account; Armstrong, *Industrial Resources of . . . the Tyne, Wear, and Tees*, for a contemporary account.

107 Davies, *Earthy and Other Minerals*, p. 53.

108 E. D. Lewis, *The Rhondda Valleys*, London, 1959, pp. 143–4.
109 *Morning Chronicle*, 21 March 1850, Labour and the Poor, Mining and Manufacturing Districts of South Wales, III, p. 5, col. 5.
110 George Head, *A Home Tour through the Manufacturing Districts*, London, 1836, pp. 171–3.
111 J. E. Wilson, 'Geology', in *British Association Handbook to Bradford*, Bradford, 1900, p. 130.
112 A. R. Byles, 'Stone trade', in ibid., p. 83; Hadfield and Biddle, op. cit., vol. I, p. 181, and vol. II, pp. 412–15 for the stone trade and the Bradford canal.
113 Wilson, loc. cit.
114 Byles, loc. cit.
115 'The workshops of the West Riding', *Leeds Mercury*, 20 March, 1875, p. 12.
116 J. A. Howe, *The Geology of Building Stones*, London, 1910, p. 151; Lewis, *Rhondda*, p. 144.
117 Gray-Jones, *Ebbw Vale*, p. 45.
118 'Town halls and exchanges' became a special heading in *The Builder* in 1865.
119 For some of the quarries which supplied stone: A. Smith, *A New History of Aberdeenshire*, 1875, vol. II, p. 781; I. Donachie, *The Industrial Archaeology of Galloway*, Newton Abbot, 1971, p. 113; Howe, op. cit., p. 77.
120 William Knight, 'Granite quarries of Aberdeenshire', *Trans. Highland and Ag. Soc.* 2nd ser. IV, 1835, pp. 66–7; for a later account, William Diack, 'The Scottish granite industry', *The World's Work*, 1903, p. 637. I am grateful to Bob Duncan for this reference.
121 For the export trade in granite statuary, ibid., p. 638.
122 *The Builder*, 20 July 1878, XXXVI, 752. Sir Gilbert Scott liked using stone in this fashion. In the steps leading to the Albert Memorial, blue Groby granite was largely used, 'forming an admirable contrast with the white mountain limestone from Derbyshire and red Permian sandstone from Mansfield' (W. J. Harrison, *A Sketch of the Geology of Leicestershire*, Sheffield, 1874, p. 12). At St Pancras station he used Shap Fell granite for the pillars, Ketton stone for the stairs, plinths and mullions, and, among other materials for the station, Mansfield limestone and Groby grey slate (Howe, op. cit., p. 198; Harrison, op. cit., p. 12; K. Hudson, *Building Materials*, London, 1972, p. 76; G. A. T. Middleton, *Building Materials*, London, 1905, p. 46).
123 Davies, *A Treatise on Slate and Slate Quarrying*, p. 170.
124 Hughes, op. cit., pp. 28–9; H. Law and D. Kinnear Clark, *The Construction of Roads*, London, 1887, p. 153. Thornhill Local Board used rubblestones from Scarr Quarry (Dewsbury Archives, Thornhill Local Board M.B. 10 December 1863; 6 July 1868; 6 May 1872).
125 Law and Clark, op. cit., p. 109.
126 For a detailed analysis, *Rep. Highway and Sanitary . . . Joint Comm. of the*

Paddington Vestry on Wood and other Pavements, London, 1878, pp. 13–15, 19–20, 51, 53–5.

127 Ibid., p. 9.

128 Ibid. p. 14. In St Luke's, Finsbury, 12,783 yards of roadway in granite pitching were laid in 1868 compared with 2,793 yards of granite macadam (St Luke's Vestry, Surveyor's Ann. Rep. for Lady Day 1868, p. 11). Pebble stones accounted for a further 7,987 yards and York Paving for 7,122 yards. The leading thoroughfares of the East End of London – City Road, Kingsland Road, Hackney Road, Bethnal Green Road, Whitechapel Road – were being converted from macadam to granite in 1867–8 (*Shoreditch Advertiser*, 14 December 1867, p. 2, col. 1). Asphalt, first laid in Threadneedle Street in 1869 (Sir J. J. Baddeley, *Cripplegate*, London, 1921, pp. 319–20) was a more potent threat to granite than macadam. For public controversy about their comparative merits, *City Press*, 27 November 1869, p. 5, col. 6; 28 May 1870, p. 4, col. 5; 26 August 1871, p. 3, col. 1; 14 October 1871, p. 3, col. 3; 4 November 1871, p. 2, cols. 4–6; *Camberwell and Peckham Times*, 21 October 1871, p. 4, cols 4–5; *Islington Gazette*, 6 June 1871, p. 3, col. 3. For the loyalty of Clerkenwell and St Luke's to granite, Finsbury borough archives, St Luke's, Surveyor's *Ann. Rep.*, 1868, London, 1869, p. 11; St James and St John, Clerkenwell, *Ann. Rep.*, 1873–4, p. 8; *Clerkenwell Diall*, 29 October 1864, p. 2, col. 5.

129 Maude E. Davies, *Life in an English Village*, London, 1909, pp. 49–50 for the 'new occupation . . . quarrying of stones' which resulted at Corsley, Wilts; Oxfordshire Record Office J. IX/J/1–2 for an example from Bicester.

130 Thomas Codrington, *The Maintenance of Macadamised Roads*, London, 1879, pp. 33–4; *The Quarry*, 10 October 1896, I, p. 209.

131 William Wood, *A Sussex Farmer*, London, 1938, pp. 49–51.

132 W. C. Williamson, 'The natural history of paving stones' in *Science Lectures for the People*, 2nd series, Manchester, 1871, p. 8.

133 Helen Harris, *The Industrial Archaeology of Dartmoor*, Newton Abbot, 1968, p. 76.

134 Helen Harris, *The Industrial Archaeology of the Peak District*, Newton Abbot, 1971, p. 60.

135 Knight, op. cit., p. 74.

136 Ibid., pp. 72–3.

137 Ibid., pp. 69–71.

138 Law and Clark, op. cit., p. 179.

139 Ibid., pp. 185–6.

140 Ibid., pp. 174–81.

141 For some East London examples, S. Hughes, 'A Survey of Metropolitan Roads', in Henry Law, ed., *Rudimentary papers on the Art of Constructing . . . Roads*, London, 1862, pp. 4–8.

II

1 P.P. 1896 (C8074–X) xxii, Annual Report Mines and Quarries, p. 53.
2 W. J. Harrison, *A Sketch of the Geology of Leicestershire*, Sheffield, 1874, p. 10.
3 P.P. 1896 (C 8074–XLI) xxii, Annual Report Mines and Quarries, Foster, pp. 40–7.
4 P.P. 1839 (574) xxx, Rep. . . . Commissioners on Building Stone Quarries, p. 27.
5 J. Arthur Phillips, *A Treatise on Ore Deposits*, London, 1884, pp. 166–7; T. Baines, *Yorkshire Past and Present*, London, n.d. (1874?), vol. I, p. 64.
6 Cyril Hart, *The Free Miners of the Forest of Dean*, Gloucester, 1953, contains a large number of documents about the legal position of the free miners.
7 *Victoria County History of Gloucestershire*, vol. II, p. 233.
8 A. J. Taylor, 'The Wigan coalfield in 1851', *Trans. Hist. Soc. of Lancashire and Cheshire*, 1955.
9 P.P. 1845 (670) xxvii, Rep. . . . State of Population in the Mining Districts, p. 36.
10 Baines, op. cit., vol. I, p. 42, quoting Charles Morton's report for 1852.
11 J. W. F. Rowe, *Wages in the Coal Industry*, London, 1923, App. III, p. 147.
12 *Geological Survey*, pp. 88, 106–7. There were twice as many pits in Staffordshire in 1856 as there are today in the whole of the British Isles.
13 P.P. 1854 (277 and 325) ix, S.C. Accidents in Coal Mines, 3rd and 4th reports, Q.2995.
14 *Morning Chronicle*, 3 January 1850, Labour and the Poor, Manufacturing Districts, xxiii. Cf. also V. L. Davies and H. Hyde, *Dudley and the Black Country, 1760–1860*, Dudley, 1970, p. 57. John Burnett, *Useful Toil*, London, 1974, for the autobiography of a butty man of the 1840s.
15 *The Busy Hives Around Us*, London, n.d. (1858?), pp. 70–2; Taylor, loc. cit., for a helpful discussion of the size of collieries at this time.
16 Frank Peel, *Spen Valley Past and Present*, Heckmondwike, 1893, pp. 302–3.
17 P.P. 1866 (231) xiv, S.C. Mines, Qq. 5354–5. For earlier years T. J. Raybould, *The Economic Emergence of the Black Country*, Newton Abbot, 1973, *passim*.
18 *Morning Chronicle*, 6 May 1850, Labour and the Poor, Mining and Manufacturing Districts of South Wales, X, p. 5, col. 6.
19 D. F. Schloss, *Methods of Industrial Remuneration*, London, 1898, pp. 199–200 for a later nineteenth-century example. It is curious that writers on sub-contract in mining have not explored more closely the links with ironworking.
20 H. Scott, 'Colliers' wages in Shropshire, 1830–50', *Trans. Shrops. Arch. Soc.* LIII, 1949–50; Alan Griffin, *Mining in the East Midlands*, London, 1970, pp. 28–30.
21 In the early days of the Butterley Iron Company, limestone-getting, too, had been put in the hands of 'bargainmen', Jean Lindsey, 'The Butterley

Coal and Iron Works, 1792–1816', *Derbyshire Archaeol. Jnl*, LXXXV, 1965, p. 40; the Company's coal-getting was still in the hands of sub-contractors at the end of the century. P.P. 1892 (C 6795–VII) XXXVI, Pt III. R.C. on Labour, Group 'A', Answers to correspondents, p. 207.

22 P.P. 1866 (231) XIV, S.C. Mines, Qq. 7022–4.

23 Arthur Gray-Jones, *A History of Ebbw Vale*, Risca, 1971, pp. 43, 111; Oliver Jones, *The Early Days of Sirhowy and Tredegar*, Tredegar, 1969, pp. 34–5.

24 Gray-Jones, op. cit., p. 85. For a sub-contractor at Crawshay Bailey's coal and iron works, charged with employing two women underground, *Bethnal Green Times*, 8 August 1866, p. 4, col. 2.

25 J. R. Leifchild, *Cornwall, its Mines and Miners*, London, 1855, pp. 150–1, 173.

26 Joseph Yelloly Watson, *Cornish Notes*, 1st ser, London, 1861, p. 21.

27 *Mining Journal*, 10 November 1866; Geach and Webb, *A Brief Review of the British and Foreign Mining Markets for the Year 1861–2*, London, 1862, pp. 24–5.

28 Walter Graham, 'Tin, tin plate and tin alloy', in G. Phillips Bevan (ed.), *British Manufacturing Industries*, London, 1876, p. 161.

29 A. C. Todd, *The Cornish Miner in America*, Truro, 1967, pp. 18–19.

30 For an example, Cyril Noall, *Levant, The Mine Beneath The Sea*, Truro, 1970, pp. 28–32.

31 For a good example, P.P. 1864 (3389) XXIV, Pt II, Commissioners on Non-Inspected Mines (hereafter Kinnaird Comm.), Qq. 733–4.

32 Warrington Smyth, 'On the mining district of Cardiganshire and Montgomeryshire', in *Memoirs of the Geological Survey*, London, 1848, II, pt 2, pp. 680–4.

33 *Geological Survey*, pp. xi, 28.

34 C. Le Neve Foster, *Notes on the Van Mine*, London, 1879; T. A. Morrison, 'Some notes on the Van Mine', *Industrial Archaeology*, VIII/1, February 1971.

35 P.P. 1864 (3389) XXIV Pt II, Kinnaird Comm., App. B., p. 329. For an account of Talargoch mine, and a strike of the 500 workers there, P.P. 1857 (241) XI Sess. 2, S.C. Rating of Mines, Qq. 791–6.

36 Ibid., Pt I, p. xxxiv.

37 *Morning Chronicle*, 25 April 1851, Labour and the Poor, Mining and Manufacturing Districts of South Wales, XVIII, p. 6.

38 *Morning Chronicle*, 19 January 1850, Labour and the Poor, Rural Districts, XXVII, Cumberland and Westmorland, p. 5, col. 2.

39 P.P. 1857, Sess. 2 (241), XI, S.C. Rating of Mines, Q. 126.

40 Ibid., Q. 20.

41 A. Raistrick and B. Jennings, *A History of Lead Mining in the Pennines*, London, 1965, p. 249.

42 P.P. 1856 (346) XVI, S.C. Rating of Mines, Qq. 3004–6.

43 P.P. 1864 (3389) XXIV Pt 1, Kinnaird Comm., p. xxxix.

44 P.P. 1856 (346) XVI, S.C Rating of Mines, Qq. 2878–86, 2901–3.

45 Warrington W. Smyth, 'Metallic mining', in G. Phillips Bevan (ed.), op. cit., p. 12.
46 Walter White, *A Month in Yorkshire*, London, 1858, p. 152.
47 M. R. G. Cozen, 'The growth and character of Whitby', in G. H. J. Daysh (ed.), *A Survey of Whitby*, Eton, 1958, p. 68.
48 W. G. Armstrong (ed.), *The Industrial Resources of . . . the Tyne, Wear, and Tees*, Newcastle upon Tyne, 1864, p. 188.
49 George Head, *A Home Tour through the Manufacturing Districts*, London, 1836, p. 284.
50 Ibid., p. 283.
51 N. J. G. Pounds, 'The discovery of China clay', *Economic History Review*, 2nd ser. 1, 1948.
52 R. M. Barton, *A History of the China Clay Industry*, Truro, 1966, pp. 31, 33.
53 Ibid., p. 53n.
54 Kenneth Hudson, *The History of English China Clays*, Newton Abbot, 1969, p. 23.
55 P.P. 1866 (3678) xxiv, Children's Employment Commission, 5th report, p. 160.
56 Ibid., p. 153.
57 *Geological Survey*, p. 99.
58 Ibid., pp. 124–5.
59 P.P. 1873 (C 745) xix, Factory Inspectors' half yearly report, p. 101.
60 Donald Young, 'Brickmaking at Broadmayne', *Proc. Dorset Nat. Hist. Soc.*, vol. 89, 1967.
61 Frank Garnett, *Westmorland Agriculture, 1800–1900*, Kendal, 1912, p. 60.
62 For examples of these last two in East Lancashire, John Travis, *Notes (Historical and Biographical) mainly of Todmorden and District*, Rochdale, 1896, pp. 35, 360–3.
63 P.P. 1876 (C 1443–1) xxx, Rep. . . . Fact. and Workshops, II, Q. 8720.
64 P.P. 1866 (3678) xxiv, Children's Employment Commission, 5th report, *passim*.
65 P.P. 1912–13 (Cd 6390) xii, R.C. Metalliferous Mines and Quarries, vol. 1, Q. 236.
66 For a very small-scale example, Oxfordshire Record Office, J vl/n/1, Somerton quarry accounts, February–October 1837.
67 P.P. 1914 (Cd 7478) xlii, R.C. Metalliferous Mines and Quarries, vol. III, Q. 24,435.
68 Anon., *Reminiscences of a Stonemason*, London, 1908, pp. 195–6, 207.
69 P.P. 1912–13 (Cd 6390) xii, R.C. Metalliferous Mines and Quarries, vol. I, Q. 617.
70 Alfred Williams, *Villages of the White Horse*, London, 1913, pp. 165–6. At Liddington, Williams tells us (p. 12), it was an ancient privilege of the parishioners to go on the Downs flint-digging. 'This they did in spare time, and when other work was slack. The flints, when unearthed, were sold for

roadmaking and repairing, and the money was very often used to buy a pig for the cottager's sty.'

71 C. E. Robinson, *Rambles in the Isle of Purbeck*, London, 1882, p. 88.

72 Ibid., p. 86. The quarrymen styled themselves Freemen of the Ancient Guild of Purbeck Marblers. For an autobiographical reference to them in the 1870s, Fred Bower, *Rolling Stonemason*, London, 1936, pp. 17–18.

73 *Morning Chronicle*, 23 January 1850, Labour and the Poor, Rural Districts, XXVIII, 'The stone quarries of Swanage', p. 5, col. 3.

74 Robinson, *Purbeck*, p. 86; Eric Benfield, *Purbeck Shop, a Stoneworker's Story of Stone*, Cambridge, 1948, for a valuable autobiographical account of more recent times.

75 *Morning Chronicle*, loc. cit.

76 C. G. Harper, *The Dorset Coast*, London, 1905, pp. 100, 102.

77 James Blaikie, 'The slate quarries of Aberdeenshire', *Trans. Highland and Ag. Soc.*, 1835, 2nd ser. IV, pp. 104–5.

78 *Oxford Chronicle*, 21 August 1847, p. 1.

79 P.P. 1914 (Cd 7476), R.C. Metalliferous Mines and Quarries, 2nd Report, p. 31.

80 My account of quarrying in Clydach vale is based on interviews with elderly villagers there, collected in August 1974. My thanks are due to Mr David Powell, Mr Harry Powell, Mr Ivor Probert, Mrs Reynalt Parry, and Mr Albert Thomas.

81 Mathew Owen, *The Story of Breconshire*, Cardiff, 1911, p. 289.

82 *Morning Chronicle*, 3 January 1850, Labour and the Poor, Manufacturing Districts, XXIII.

83 It was introduced at Tredegar in 1829: Oliver Jones, *Tredegar*, p. 50 and cf. pp. 49, 57.

84 *Morning Chronicle*, 21 March 1850, Labour and the Poor, Mining and Manufacturing Districts of South Wales, Merthyr, III, p. 5, col. 6.

85 P.P. 1866 (3678) XXIV, Children's Employment Commission, 5th report, p. 160.

86 J. T. Arlidge, *The Diseases of Occupations*, London, 1892, p. 489. In the fireclay districts more of the work was under cover. P.P. 1876 (C 1443–1) XXX, Rep. . . . Fact. & Workshops, vol. II, Qq. 5527, 5548–9, 5561, 5620, 5932.

87 *Morning Chronicle*, 23 January 1850, Labour and the Poor, Rural Districts, XXVIII, 'The stone quarries of Swanage', p. 5, col. 5.

88 G. Joan Fuller, 'Lead mining in Derbyshire', *East Midlands Geographer*, III, 1965.

89 H. Osborne O'Hagan, *Leaves from my Life*, London, 1929, I, p. 89.

90 Rowe, op. cit., p. 147.

91 Edward Dobson, *A Rudimentary Treatise on . . . Bricks*, London, 1850, p. 87. The most temporary brickworks of all were those of the contractors: they would be set up on land laid waste for new building and temporarily

dug up for clay; house builders often supplied themselves with bricks in this way, and so did the railway contractors – several millions of bricks would be needed in the course of constructing a new line, for the viaducts, tunnels and cuttings.

92 I. Donachie, *The Industrial Archaeology of Galloway*, Newton Abbot, 1971, pp. 111–12.

93 P.P. 1912–13 (Cd 6390) XLI, R.C. Metalliferous Mines and Quarries, Qq. 457, 9479.

94 For a late example, cf. Schloss, op. cit., p. 201.

95 The Greystone Lime Co., Dorking, seems to have relied entirely upon outside hirings of horses and horsemen, to judge by the frequent entries in its Strap Book. The following are a few of the recorded payments to Edward Bennett (Surrey Record Office Acc. 1053).

22 June 1865	Horse at siding 5 days £1.5.0.
30 June 1865	Horse at siding ½ day 2/6
30 June 1865	Carting 16 yds lime to station 8
1 July 1865	Horse and lad ¾ day carting
1 July 1865	3 Horses and 2 men ½ day carting bricks
4 July 1865	Horse at siding half day 5
4 July 1865	Carting 8 yds lime to station ½ price 2

96 P.P. 1857 Sess 2 (241) XI, S.C. on Rating of Mines, Q. 317. 'Most small farmers in our neighbourhood keep horses only for the use of the colliery', a Swansea ironmaster told the same commission, ibid., Q. 1229.

97 P.P. 1864 (3389) XXIV, Pt I, Kinnaird Comm., pp. xxxix, 293–4.

98 Ibid., Pt II, Q. 14,164; Raistrick and Jennings, op. cit., p. 286.

99 James Dunn, *From Coal Mine Upwards*, London, 1910, pp. 10–12 for a vivid account of the butty system in the Leicestershire coalfield in the 1840s, where the work was sub-divided among a number of butties.

III

1 Joseph Yelloly Watson, *A Compendium of British Mining*, London, 1843, p. 15.

2 John Rowe, *Cornwall in the Age of the Industrial Revolution*, Liverpool, 1953, p. 70. J. R. Liefchild, *Cornwall, its Mines and Miners*, London, 1885, p. 173.

3 For the increased employment of women and girls, Rowe, op. cit., p. 8, n. 1. Ore dressers at Dolcoath in the 1880s, according to Captain Thomas, the mine manager, were the daughters of the miners 'but none of them wives. There is not a single married woman in our mines.' P.P. 1887 (257) XII, S.C. on Stannaries Act Amendment Bill, Q. 1822.

4 See *The Western Chronicle of Science*, Falmouth, 1872, pp. 1–4, 115–20, for some of these 'improvements'.

5 C. Le Neve Foster, *A Text-book of Ore and Stone Mining*, London, 1894, pp. 544–5.
6 P.P. 1861 (161) XVI, 3rd Rep. Med. Officer Privy Council, pp. 130–1.
7 P.P. 1864 (3389) XXIV, Pt 1, Kinnaird Comm., Qq. 226, 353.
8 3rd Rep. Med. Officer Privy Council, p. 131.
9 Ibid. and Kinnaird Comm., pp. xiv–xv.
10 Ibid., p. viii.
11 P.P. 1890–1 (7) LXXVIII, Return of Metalliferous Mines . . . Devon and Cornwall.
12 F. C. Mather, *After the Canal Duke*, Oxford, 1970, pp. 324–5. The labour force in coal mining was tailored to the times and seasons of maximum rather than minimum demand, a very unusual feature in nineteenth-century industry. The *Morning Chronicle* Commissioner noted this when he visited South Wales in 1850: 'In periods when the trade is depressed, as it has been of late, the men only work half or three-quarters of their proper time. It is the policy of the masters, in the continual hope and expectation of an improved demand, to keep on the same number of hands as when trade is good, so that when the better times come there may be sufficient workmen available.' *Morning Chronicle*, 27 March 1850, Labour and the Poor, Mining and Manufacturing Districts of South Wales, p. 5, col. 2.
13 The Registrar General's figures differ from the totals of those of the Mines' Inspectors and the Geological Survey, though their upward-moving tendency is the same.
14 Not in the Midlands coalfields, according to Rowe, where even in the 1920s there were no subsidiary classes of specialists. Rowe, op. cit., p. 64.
15 *The Times*, 5 July 1873, p. 5; *Iron and Coal Trades Review*, 13 March 1872, p. 212.
16 Warrington W. Smyth, 'Mines and Mining' in G. Phillips Bevan (ed.), *British Manufacturing Industries*, London, 1876, II, pp. 122–7.
17 Unnamed writer quoted in *Stourbridge Times*, 15 March 1862.
18 Haulier lads (in Durham, putters) appear as a strong and independent force in trade unionism in the years 1888–1914. For a good description of their combativity, Jack Lawson, *A Man's Life*, London, 1932, pp. 98–102; for the South Wales Hauliers' strike of 1893, Ness Edwards, *History of the South Wales Miners*, London, 1926, pp. 108–17; C. J. Down and A. J. Warrington, *The History of the Somerset Coalfield*, Newton Abbot, 1971, p. 102 for the carting boys' strike of 1908.
19 *Manchester Examiner*, 26 July 1865, p. 5, cols 2–3; cf. also *Barnsley Chronicle*, 1 September 1866; *Capital and Labour*, 15 April 1874; *Iron and Coal Trades Review*, 15 May 1872, p. 386.
20 J. E. Williams, *The Derbyshire Miners*, London, 1962, p. 174.
21 P.P. 1873 (313) X, S.C. on Coal, Qq. 2198–9.
22 J. H. Morris and L. J. Williams, *The South Wales Coal Industry, 1841–75*, Cardiff, 1958, pp. 71–2.

23 Paul de Rousiers, *The Labour Question in Britain*, London, 1896, p. 122.
24 Rowe, op. cit., p. 16. For a general discussion, A. J. Taylor, 'Labour productivity and technological innovation in the British coal industry', *Economic History Review*, XIV, no. 2, August 1961.
25 'The strike in South Wales', *The Times*, 22 January 1873, p. 12; Williams, op. cit., pp. 284, 374–5; Rowe, op. cit., p. 219.
26 *Morning Chronicle*, 'Labour and the Poor: Mining and Manufacturing Districts of S. Wales IV', 27 March 1850, p. 6, col. 3.
27 David C. Davies, *A Treatise on Metalliferous Minerals and Mining*, London, 1880, p. 315.
28 Joseph H. Collins, *Principles of Metal Mining*, London, 1875, p. 64
29 Davies, op. cit., p. 315.
30 Ibid., p. 321.
31 P.P. 1881 (C 2876) XXVI, Two Special Reports on the Use of Gunpowder in . . . Mines, p. 6.
32 P.P. 1892 (C 6709–IV) XXXIV, R.C. on Labour, Group 'A', vol. I, Qq. 1573–4, 1586.
33 P.P. 1864 (3389) XXIV Pt II, Kinnaird Comm., App. B, I, p. 23; P.P. 1912–13 (Cd 6390) XLI, R.C. Metalliferous Mines and Quarries, Qq. 6513–6.
34 J. D. Marshall and M. Davies-Shiel, *The Industrial Archaeology of the Lake Counties*, Newton Abbot, 1969, p. 133.
35 Collins, *Principles of Metal Mining*, p. 58.
36 'The Mountsorrel granite quarries', *The Quarry*, October 1896, p. 201.
37 P.P. 1912–13 (Cd 6390) XLI, R.C. Metalliferous Mines and Quarries, Qq. 9283, 9167–70; for an Aberdeenshire exception, ibid., Q. 9524.
38 P.P. 1897 (C 8450–XII) XX, Ann. Rep. Mines and Quarries, Foster, p. 52.
39 R.C. Metalliferous Mines and Quarries, loc. cit., Q. 8231.
40 Interviews with David Powell, Ivor Probert and Harry Powell, Clydach, August 1974. For the 'jumper' in other places, Joseph H. Collins, *A First Book on Mining and Quarrying*, London, 1872, pp. 33–4; C. Le Neve Foster, *Quarrying*, London, 1894, p. 4; *The Quarry*, May 1896, pp. 86–7.
41 P.P. 1892 (C 6795–VII) XXVI, Pt III, R.C. on Labour, Group 'A', answers to questions, p. 82.
42 Foster, *Ore and Stone Mining*, p. 546.
43 Eye accidents where splinters of rock damaged the eyes were a major occupational health hazard for both slate dressers and sett-makers. P.P. 1912–13 (Cd 6390) XLI, R.C. Metalliferous Mines and Quarries, Qq. 1148, 9283, 8805.
44 Sydney B. J. Skertchley, *On the Manufacture of Gun Flints*, London, 1879, pp. 27–34.
45 'Tools – II', *The Quarry*, May 1896, p. 86. Two other hammers much in use at Mountsorrel at this time were the 'chopper' for starting a line or channel in breaking up the larger blocks; and the 'spalling' hammer used for

reducing the stone to manageable shape before it was passed through a crushing mill for macadam.

46 William Knight, 'Granite quarries of Aberdeenshire', *Trans. Highland and Ag. Soc.*, 2nd Ser. IV, 1835, p. 67.

47 M. Powis Bale, *Stone-working Machinery*, London, 1884, p. 75. The granite used in the Chelsea Embankment was hammer-dressed, 'which will, it is believed be effective, and give it a character of massiveness', *Chelsea News*, 12 August 1871, p. 5, col. 2.

48 'The workshops of the West Riding', *Leeds Mercury*, 20 March 1875, p. 12.

49 Sandstone was generally detached by crowbar and wedge (rather than being shaken down by blasting), and worked into shape ('sculptured') by mallet and chisel: D. T. Ansted, *The Applications of Geology to the Arts and Manufactures*, London, 1865, p. 147; 'Sandstone Tools', *The Quarry*, May 1896, p. 87; Middleton, op. cit., p. 83; George Smith, 'Account of the quarries of sandstone in the Edinburgh and Glasgow Districts', *Trans. Highland and Ag. Soc.*, 2nd ser. IV, 1835, pp. 83, 94.

50 P.P. 1914 (Cd 7477) XLII, R.C. Metalliferous Mines and Quarries, vol. II, Qq. 11,405, 11,407, 11,417, 11,445, 21,310; 'Hand quarrying', *The Quarry*, August 1897, p. 173.

51 Knight, op. cit.

52 The 'Blondin' was invented by Fyfe of Kenmay, one of the chief granite centres of Aberdeenshire: William Diack, 'The Scottish granite industry', *The World's Work*, 1903, pp. 636–7.

53 P.P. 1912–13 (Cd 6390) XLI, R.C. Metalliferous Mines and Quarries, Qq. 7578, 7559.

54 Fred Bower, *Rolling Stonemason*, London, 1936, pp. 40–1.

55 David Cook, *A Treatise on China-clay*, London, 1880, pp. 37–41.

56 Collins, *Metalliferous Mining*, p. 60.

57 Kenneth Hudson, *The History of English China Clays*, Newton Abbot, 1969, p. 35.

58 Walter White, *A Londoner's Walk to Land's End*, 2nd ed., London, 1861, p. 128.

59 Hudson, op. cit., p. 36.

60 *Morning Chronicle*, 28 November 1849, Labour and the Poor, Rural Districts, XII, Dorset, p. 5, col. 3.

61 Charles Tomlinson, *Illustrations of Trades*, London, 1860, p. 25.

62 Edward Dobson, *A Rudimentary Treatise on . . . Bricks*, London, 1850, pp. 22–3.

63 Ibid., p. 184 and A. B. Searle, *Modern Brickmaking*, London, 1911, p. 43.

64 P.P. 1865 (3473) XX, Factory inspectors' half yearly reports, p. 121.

65 Ibid., pp. 121–2. For a personal reminiscence, W. F. Wescombe, 'Brickmaking', *The British Clay-Worker*, 1893, p. 4.

66 P.P. 1873 (C 745) XIX, Factory inspectors' half yearly reports, p. 19.

67 Ibid.

68 P.P. 1876 (C 1443–I) xxx, Rep. . . . Fact. and Workshops, Q. 9317.
69 A. B. Searle, *The Clayworker's Handbook*, London, 1911, p. 106; Kenneth Hudson, *Building Materials*, London, 1972, p. 35.
70 'A brickmaker on brickmaking', *Brick, Tile and Builders' Gazette*, 12 April 1887, II, 295; Searle, *Modern Brickmaking*, pp. 424–5.
71 *The British Clay Worker*, January 1895, II, 214; December 1898, VII, 250; Searle, *Modern Brickmaking*, p. 39.
72 P.P. 1876 (C 1443) XXIX, Rep. Fact. and Workshops, vol. 1, App. D, p. 148; P.P. 1876 (C 1443–I) xxx, ibid., vol. II, Q. 4918.
73 Ibid., Q. 8368.
74 'The brickmaking industry in 1892', *The British Clay-Worker*, January 1893, I, 200.
75 Searle, *Modern Brickmaking*, pp. 20–3; and *The Clayworker's Handbook*, p. 57.
76 A. B. Searle, *Cement, Concrete and Bricks*, London, 1913, p. 323.
77 A. B. Searle, *An Introduction to British Clays, Shales and Sands*, London, 1912, p. 409.
78 A. B. Searle, *Refractory Materials*, London, 1917, pp. 308–9.
79 *The Industrial Resources of . . . the Tyne, Wear and Tees*, Newcastle upon Tyne, 1864, pp. 204–5; Anon., *Bricks and Brickmaking*, Birmingham, 1878, p. 16; London School of Economics, Webb T.U. MSS., Sect. A. Vol. X, fol. 386.
80 P.P. 1876 (C 1443–I) xxx, Rep. Fact. and Workshops, Qq. 5529, 5644.
81 Searle, *Refractory Materials*, p. 144.
82 For the clay-diggers' spade, G. G. Andre, *A Descriptive Treatise on Mining Machinery*, London, 1877, 1/61; Searle, *Modern Brickmaking*, p. 38.
83 'Each knapper has a certain peculiar style which enables him to distinguish his own flints. He tells them at once by feeling them with his left hand, and though the differences are very slight I have always found the knapper correct, though he cannot say more than that a gun-flint is or is not his own workmanship. This judgement is irrespective of sight and can be equally well used in the dark', Skertchley, op. cit., p. 33.
84 Warrington W. Smyth, 'Metallic Mining' in G. Phillips Bevan (ed.), op. cit., II, p. 39.
85 'Tools', *The Quarry*, April 1896, pp. 63–4.
86 W. G. Renwick, *Marble and Marble Working*, London, 1909, p. 28; *Crossing's Dartmoor Worker*, Newton Abbot, 1966, pp. 70–1; J. A. Howe, *The Geology of Building Stones*, London, 1910, p. 99.
87 Warrington W. Smyth, 'Mines and Mining' in G. Phillips Bevan (ed.), op. cit., vol. II, p. 113.
88 All but one of the machinery accidents in the North Wales slate industry in 1897 were attributed to cranes, hand winches and derricks. P.P. 1898 (C 8819–XII) XVII, Ann. Rep. Mines and Quarries, Foster, p. 14.

IV

1 Carter L. Goodrich, *The Miner's Freedom*, New York, 1926.
2 Cf. annual reports of the Mines Inspectors from 1894.
3 L. L. Price, 'West Barbary . . . Notes on . . . work and wages in the Cornish mines', *Journ. Roy. Stat. Soc.*, LI, 1888, p. 535n quoting a lecture by Warrington Smyth in 1867.
4 C. J. Hunt, *The Lead Miners of the Northern Pennines*, Manchester, 1970, pp. 35, 46–7.
5 Surrey Record Office, Acc. 566, Merstham Lime Works Piece Rate book, 1892–3; Merstham Lime Works Wages Book, week ending 22 January 1886.
6 Richard Heath, *The English Peasant*, London, 1893, p. 94. For an excellent illustration of the speculative element in lead miner's earnings, and the sharp variations which took place in them, A. Raistrick and B. Jennings, *A History of Lead Mining in the Pennines*, London, 1965, p. 288.
7 *Morning Chronicle*, 9 January 1850, Labour and the Poor, Rural Districts XXIV, pp. 5–6; J. R. Featherston, *Weardale Men and Manners*, Durham, 1840, pp. 64–5; Hunt, op. cit., pp. 44–6; John Lee, *Weardale Memories*, Consett, 1950, p. 54.
8 Joseph H. Collins, *A First Book on Mining and Quarrying*, London, 1872, p. 22.
9 A. Harris, *Cumberland Iron*, Truro, 1968, p. 38.
10 P.P. 1912–13 (Cd 6390) XLI, R.C. Metalliferous Mines and Quarries, vol. 1, Q. 8065.
11 *Morning Chronicle*, 28 November 1849, Labour and the Poor, Rural Districts XII, Dorset, p. 5, col. 3.
12 Interview with Mr David Powell, Clydach, August 1974.
13 P.P. 1864 (3389) XXIV Pt II, Kinnaird Comm., Qq. 2713–18. For Monday absenteeism among the North Wales quarrymen, and its alleged diminution as a result of Sunday closing, P.P. 1890 XL, Rep. . . . Welsh Sunday Closing, Qq, 12,192–3; 12,558; 12,716–7; 12,736; 12,788; 12,818.
14 *Thomas Burt . . . an Autobiography*, London, 1924, p. 91.
15 *Morning Chronicle*, 3 January 1850, Labour and the Poor, Manufacturing Districts, XXIII. The Staffs Collieries, p. 5, col. 4; ibid., 27 March 1850, Mining and Manufacturing Districts of South Wales IV, p. 6, col. 1. For an example of this among the clay miners of Wareham, ibid., 28 November 1849, Labour and the Poor, Rural Districts, XII, Dorset, p. 5, cols 3–4.
16 *Morning Chronicle*, 24 December 1849, Labour and the Poor, Manufacturing Districts, XX, Northumb. & Durham, p. 5, col. 1.
17 P.P. 1866 (231) XIV, S.C. on Mines for many examples.
18 P.P. 1844 (592) XVI, Rep . . . Popn in Mining Districts, p. 13.
19 Frank Machin, *The Yorkshire Miners*, Barnsley, 1958, pp. 280, 290.
20 L.S.E., Webb T.U. MSS., Coll. E. Sect. A, XXVIII, fol. 26.

21 P.P. 1866 (431) xiv, S.C. on Mines, Qq. 1418–19.
22 R. F. Wheeler, *The Northumbrian Pitman, his Work and Ways*, Newcastle upon Tyne, 1885, p. 19.
23 P.P. 1892 (C 6795–VII) xxxvi Pt III, R.C. on Labour, Group 'A', answers to questions, p. 41. There is a mass of evidence for a similar state of affairs in metalliferous mining in mid-Victorian times; it has been omitted for reasons of space.
24 *Colliery Guardian*, 23 June 1866, p. 463 for four-day week in answer to a threatened wage cut; P.P. 1867–8 (3980–II) xxxix, R.C. on Trade Unions, 6th Report, Qq. 11,794–8 for 'darg' as prelude to strike; *Capital and Labour*, 28 November 1877, IV, 638 for 'darg' as answer to owners' refusal of concessions.
25 In some pits the 'darg' was regularly imposed whenever coal stacks began to accumulate at the pit-head.
26 'The strike in South Wales', *The Times*, 22 January 1873, p. 12.
27 P.P. 1847 (844) xvi, Rep. . . . Popn in Mining Districts, pp. 12–14 for Staffs; P.P. 1867–8 (3980–III) xxxix, R.C. on Trade Unions, 7th report, Q. 15,225 and *Capital and Labour*, 8 April 1874, p. 126 for 'darg' in the Wigan coalfield. For 'darg' in Yorkshire, Machin, op. cit., pp. 55, 83, 280, 290–1, 342–3; *Trade Societies and Strikes*, London, 1860, p. 21; P.P. 1867–8 (3980–IV) xxxix, R.C. Trade Unions, 8th report, Qq. 16,169, 16,181.
28 *Iron and Coal Trades Review*, 28 February 1872, p. 171; 3 April 1872, p. 273; 17 April 1872, p. 309; *Ironmonger*, 1 March 1874, XVI, p. 280; *Capital and Labour*, 22 August 1877, IV, p. 442; *The Miner*, February 1887, p. 31; October 1887, p. 160.
29 P.P. 1866 (231) xiv, S.C. Mines, Q. 7873.
30 P.P. 1844 (597) xvi, Rep. . . . Popn in Mining Districts, p. 33.
31 *Good Words*, 1 January 1869, p. 44; *Morning Chronicle*, 8 April 1850, Labour and the Poor, Mining and Manufacturing Districts of South Wales, VI, Merthyr, p. 5, col. 1; *Merthyr Express*, 15 September 1860, p. 3, col. 6. for 'Pay Monday' at Rhymney.
32 P.P. 1892 (C 6795–VII) xxxvi Pt III, R.C. Labour, Group 'A', Answers to questions, p. 256.
33 *Morning Chronicle*, 24 December 1849, Labour and the Poor, Manufacturing Districts, XX, Northumberland and Durham, p. 5, col. 2.
34 P.P. 1881 (C 2839) xxiv, Rep. . . . Penygraig colliery explosion, pp. 1–3.
35 C. Le Neve Foster, *A Text-book of Ore and Stone Mining*, London, 1894, p. 637.
36 'The strike in South Wales', *The Times*, 28 January 1873, p. 7.
37 'Bowden's Heading' at Wyndham Western Colliery, Ogmore, Glam., was so called because four brothers worked the stall together in pairs on alternate shifts (conversation of the writer with one of their sons on the Newport-London train, 25 August 1974).

38 *Morning Chronicle*, 24 December 1849, Labour and the Poor, Manufacturing Districts, XX, Northumberland and Durham, p. 5, cols 1–2.
39 Manchester Reference Library, M. Thomason, 'Leigh of Yesterday', part VII.
40 *Morning Chronicle*, loc. cit.
41 J. R. Leifchild, *Cornwall, its Mines and Miners*, London, 1885, p. 146; P.P. 1864 (3389) XXIV Pt II, Kinnaird Comm., Pt B, p. 446.
42 Ibid., App. B, XII for examples of this in 'Bal' or pay bills.
43 Joseph Yelloly Watson, *A Compendium of British Mining*, London, 1843, p. 14; Kinnaird Comm., App. B, pp. 452–3 for examples.
44 Leifchild, op. cit., pp. 173, 179.
45 Kinnaird Comm., part I, Q. 8731.
46 C. Le Neve Foster, op. cit., pp. 644–5. For an account of the miners' conspiracies to keep down the price of their bargains, A. K. Hamilton Jenkin, *The Cornish Miner*, Newton Abbot, 1972 ed., pp. 226–9. I have not read John Rule's important doctorate on the Cornish tinners but, like others, expectantly await its appearance in print. For a brief statement of one aspect of his work see 'The Tribute System and the weakness of trade unions in the Cornish Mines', *Bulletin of the Society for Labour History*, autumn.
47 P.P. 1914 (Cd 7478) XIII, R.C. Metalliferous Mines and Quarries, Q. 24,435.
48 P.P. 1893–4 (C 7237) LXXIII, Quarry Comm. of Inquiry, Q. 3294.
49 P.P. 1904 (Cd. 2119–XI) XII, Ann. Rep. Mines and Quarries, Martin, pp. 33, 44.
50 Quarry Comm. of Inquiry, Qq. 1426–40.
51 Ibid., Qq. 1125–6.
52 David C. Davies, *A Treatise on Earthy and other Minerals*, London, 1884, p. 30.
53 P.P. 1868–9 (4093–I) Factory Inspectors' half yearly reports, p. 170.
54 P.P. 1876 (C 1443–I) XXX, Rep. Fact. and Workshops, Qq. 7350, 7353.
55 Bodleian Lib., MS. Top. Oxon. d. 475, History of Filkins, fol. 3.
56 Dorset County Museum, Moule Notebooks, IX, pp. 146–8.
57 A. M. Wallis, 'The Portland stone quarries' *Proc. Dorset Archaeol. and Nat. Hist. Soc.*, XII, 1891, p. 190.
58 W. F. Wescombe, 'Brickmaking, 1850–90', *The British Clay-Worker*, April 1893, p. 4. 'I soon began to keep company with another young woman, whose name was T. . . . She used to work in her father's brickfield, for he was a brickmaker; but the end of it was, she killed herself with it. She was the only woman I ever knew killed herself by hard work. She used to do just the same as a man, and she was nearly as strong.' *Autobiography of a Working Man*, ed. Emily Eden, London, 1861, pp. 10–11.
59 Will Thorne, *My Life's Battles*, London, n.d.
60 Ben Tillett, *Memories and Reflections*, London, 1931, pp. 32–3.
61 Frederic Clifford, *The Agricultural Lock-out of 1874*, London, 1875, pp. 45–7.

For coprolite digging, R. Samuel, 'Village Labour' in *Village Life and Labour*, ed. R. Samuel, London, 1975, pp. 21–2.
62 *Building News*, 29 March 1867, XIV, p. 232.
63 P.P. 1914 (Cd 7476) XIII, R.C. Metalliferous Mines and Quarries, II, Q. 10, 2211.
64 Walter White, *A Londoner's Walk to Land's End*, 2nd ed., London, 1861, p. 34.
65 G. F. L. Packwood and A. H. Cox (eds), *West Drayton and District during the nineteenth century*, West Drayton, 1967, p. 67.
66 Brian Le Mesurier (ed.), *Crossing's Dartmoor Worker*, Newton Abbot, 1966, p. 80.
67 Kenneth Hudson, *The History of English China Clays*, Newton Abbot, 1969, p. 39.
68 P.P. 1862 (179) XXII, 4th Rep. Med. Officer, Privy Council, App. 1, p. 38.
69 G. A. W. Tomlinson, *Coal Miner*, London, 1937, pp. 144–5.
70 Cyril E. Hart, *The Commoners of Dean Forest*, Gloucester, 1951, p. 129. I am grateful to Sean Hutton for this reference.
71 For a Swaledale example, see Heath, op. cit., pp. 94–5.
72 P.P. 1864 (3389) XXIV, Pt II, Kinnaird Comm., App. B, p. 15.
73 Ibid, p. 18.
74 Hunt, op. cit., pp. 146–7.
75 P.P. 1856 (346) XVI, S.C. Rating of Mines, Q. 2721.
76 Ibid., Q. 2908.
77 G. Joan Fuller, 'Lead mining in Derbyshire in the mid-nineteenth century', *East Midlands Geographer*, III, Pt 7, 18 June 1965.
78 *Morning Chronicle*, 22 April 1851, Labour and the Poor, Mining and Manufacturing Districts of Wales, XX, The Welsh Lead Mines, p. 6, col. 1.
79 Arthur Raistrick, *Mines and Miners of Swaledale*, Clapham, Yorks, 1955, pp. 84–6.
80 Jack Lawson, *A Man's Life*, London, 1932, p. 27.
81 Ibid., p. 15.
82 *Mining Journal*, 16 July 1859, p. 507.
83 P.P. 1862 (179) XII, 4th Rep. Med. Officer Privy Council, App. IV, p. 158.
84 *Labour Gazette*, July 1893, I, p. 6.
85 A. J. Taylor, 'The Wigan Coalfield in 1851', *Trans. Lancs. and Cheshire Hist. Soc.*, CVI, 1955.
86 J. H. Morris and L. J. Williams, *The South Wales Coal Industry, 1841–75*, Cardiff, 1958, p. 221.
87 Lawson, op. cit., pp. 56–7.
88 P.P. 1847 (871) XXVII Pt II, Rep. . . . State of Education in Wales, p. 393.
89 Alan Campbell, 'Honourable men and degraded slaves', paper at miners' conference of Society for the Study of Labour History, 10 November 1973.
90 'The strike in South Wales', *The Times*, 3 February 1873, p. 9.
91 See A. C. Todd, *The Cornish Miner in America*, Truro, 1967, *passim*; P.P.

1887 (252) XII, S.C. Stannaries Act, Qq. 510, 2540–2, for transatlantic migrations in the 1880s.

92 P.P. 1904 (Cd 2091) XIII, Rep. . . . Health of Cornish Miners, Table 16.

93 P.P. 1864 (3389) XXIV, Pt II, Kinnaird Comm., App, p. 150.

94 Davies, op. cit., p. 194.

95 *Morning Chronicle*, 19 January 1850, Labour and the Poor, Rural Districts, XXVII, p. 5, col. 3.

96 Hunt, op. cit., p. 162.

97 Ibid., p. 163.

98 P.P. 1864 (3389) XXXIV Pt II, Kinnaird Comm., Q. 14,580.

99 Raistrick, op. cit., pp. 12–13.

100 P.P. 1865 (3483) XXVI, 7th Rep. Med. Officer Privy Council, Dr Hunter's Rep., App. 6, p. 225.

101 Ibid., p. 261.

102 Ibid., p. 262.

103 W. C. E. Ranger, *Rep. . . . General Board of Health . . . Grimsby*, London, 1850, pp. 17–18.

104 Lilian M. Birt, *The Children's Home Finder*, London, 1913, p. 9.

105 Joseph Craven, *A Bronte Moorland Village and its People: a History of Stanbury*, Keighley, 1907.

106 For an example, P.P. 1912–13 (Cd 6390) XLI, R.C. Metalliferous Mines and Quarries, vol. I. Qq. 9432–53.

107 James Valentine, *The People of Aberdeenshire in Account with the Census*, Aberdeen, 1871, p. 43.

108 Raphael Samuel, 'Comers and goers' in H. J. Dyos and M. Wolff (eds), *The Victorian City*, London, 1973, vol. I. For seasonal migration across the Atlantic by Scottish coal miners, as well as for their generally migratory character, see A. J. Youngson Brown, 'Trade Union policy in the Scots Coalfields, 1855–1885', *Economic History Review*, 2nd ser. VI/1, pp. 37–40.

109 W. J. Harrison, *A Sketch of the Geology of Leicestershire*, Sheffield, 1874, p. 10.

110 C. Le Neve Foster, op. cit., p. 676.

111 P.P. 1864 (3416) XXVIII, 6th Rep. Med. Officer Privy Council, App. 18(e).

Part 2

Y chwarelwyr: the slate quarrymen of North Wales

Merfyn Jones

Slate-quarrying in North Wales is of uncertain origin. Slate scratched from the surface was probably used locally for roofing purposes as early as the Middle Ages but it is doubtful whether anything more than a very part-time quarryman appeared until much later. By the middle of the eighteenth century quarrying on a steadier basis had become an important item in the commerce of the area and of several small ports. Inheriting considerable wealth from interests in Jamaica, Richard Pennant in 1765 married his way into possession of the extensive Penrhyn estates. He found the quarrying works on his land let out to some eighty quarrymen exporting, via small ports on the Menai Straits, some thousand tons of slate annually. In 1782, a year before he was created Baron Penrhyn of Penrhyn, Co. Louth (an interesting use of the prevalence of identical Celtic names in Ireland and Wales), Pennant bought the lessees out, kept them on as hired workers and managed the quarry himself. In the next ten years the new Lord Penrhyn created roads out of the narrow paths that had previously existed and built the quay known as Port Penrhyn from which to export the slate. Before the depression of the late 1790s set in, his quarry was producing 15,000 tons of slate annually and employed some 400 workmen. Following depression, recovery and recession, the industry entered a half-century boom after the repeal of the slate duty in 1831; by mid-century Penrhyn was employing nearly 2,000 men.[1] Apart from the Penrhyn Quarry near Bethesda the other main slate-producing areas were Dyffryn Nantlle and Llanberis in Caernarvonshire and Blaenau Ffestiniog in Merionethshire.

Until the end of the nineteenth century Penrhyn remained the largest slate quarry in the world; hacked into galleries, gigantic steps in the mountainside, it employed some 2,800 men who with gunpowder and crowbar, hammer and chisel and skill, blasted and coaxed the slate from the mountain. Its galleries varied in height from 36 to 66 feet, with an average of 54 feet; in breadth they varied from wide platforms of 45 feet to narrow ledges no wider than 6 feet.[2] When levering blocks from the face, the rockmen hung over the precipitous and shattered sides with a hemp rope looped around their thighs. Every hour a bell rang out to signal them to shelter. One minute later on the signal of another bell fuses would be lit all over the quarry; then after the roar of rock shattering in the explosions, four minutes later another bell would ring summoning them back to work. The massive slabs loosened by the blast had to be split and pushed in trucks running along the rails to the head of the gallery where they were lowered down an incline to the great sheds below. There they were sawed and split into slates and then taken on Penrhyn's own railway down to Port Penrhyn whence they went to roof the industrial centres of England and Germany.

There were two other major quarries in North Wales, the Dinorwic Quarry at Llanberis, and the Oakeley at Blaenau Ffestiniog, the largest slate mine in the world. Between them these three concerns employed half of the slate workers of North Wales. The remainder laboured in over fifty smaller mines and quarries scattered over the hillsides of the counties of Caernarvonshire and Merioneth. In the last quarter of the century the total workforce fluctuated between some 13,000 and 15,000 employees. The importance of the industry, however, must not be judged by the relatively small number of men involved. It was officially judged in 1882 that 'after coal and iron slate is the most valuable mineral raised in the UK'.[3] It certainly made enormous fortunes for a few families and dominated economically and culturally as well as physically the many slate-quarrying communities of Welsh-speaking north-west Wales. The Industrial Revolution may have been founded on textiles and powered by steam; it was roofed with slates skilfully wrenched from the Welsh hills.

The men who worked in the slate quarries lived mainly on bread and butter, and, typical of many industrial groups, sustained themselves by drinking great quantities of stewed tea. A doctor complained that:[4]

> They could get plenty of sweet milk or butter milk to drink, but they have got into the habit of drinking tea, at every meal. They have no relish for any liquid but tea. It is tea all day long.

Their working hours were determined largely by the weather and the seasons, their own working day lengthening with the coming of summer.[5] Wages depended on which category of work was executed. The quarrymen proper, making up just over 50 per cent of the workforce, organized themselves into crews consisting normally of three or four men, each crew coming to a separate working agreement or 'bargain' with the management. A crew usually consisted of two men '*yn y twll*', rockmen working on the rock face, and two men working in the sheds or slate mills dressing and splitting the slates.

The other main occupational groups in the quarry were bad-rockmen and rubbish men. The bad-rockman worked usually in a crew of three taking a bargain which, unlike that of the quarrymen proper, would be of bad-rock, that is, rock from which no slates could be worked. The agreement worked out with the management would give him so much per ton for removing the bad-rock.

The rubbish men were divided into two groups, those who cleared the rubbish (waste rock) from the galleries where the quarrymen's bargains lay, and those who were responsible for building the giant tips of waste rock around the quarry. They were paid by the ton or by the yard of materials removed.

Also at work in the quarry were the 'rybelwrs', boys who were in the first stage of learning the craft; their job was to wander along the galleries offering assistance whenever there was need for an extra hand. Sometimes they would be given an extra slab of rock to split for which they would be paid by the crew. From this stage the rybelwr could hope to become a journeyman and then a quarryman proper. The quarry also employed a number of time-workers such as weighers, hauliers, brakesmen, stationary enginemen, locomotive engine drivers, engineers, blacksmiths, saw-sharpeners, carpenters, platelayers, storekeepers, timekeepers and general labourers – but their total number was small.[6]

The quarrymen proper were the élite group of the quarry and though they did not earn much more than the bad-rockmen their status was always treated with some awe. There was no real apprenticeship in slate-quarrying, and a man became a quarryman proper through his skill and his connections in a fairly informal way. But a quarryman was conscious of his superior skill and status, and relations with other workers could be strained, sometimes exacerbated by the fact that the quarrymen usually lived together in the quarry village while lower grades often came from outside the immediate area. The North Wales Quarrymen's Union, established in 1874, was, however, an industrial union and was open to all those employed in slate quarries, and wage claims, while respecting differentials, were almost always made on behalf of all the workers. In Penrhyn in 1874, in fact, the men went so far as to put in a claim, along with their own, on behalf of the journeyman whom they themselves, and not the quarry owner, employed. This fairly amicable relationship between quarry workers was reflected in the relatively small difference in wages between the higher grades. Only the labourers, very often men from the surrounding agricultural areas, were left out of the consensus; their lives were not staked in the quarries and they knew of other ways to live.

Industrial relations were not unnaturally therefore dominated by the claims and the concerns of the quarrymen proper, and their main preoccupation was the wages system under which they worked, the bargain system, the effects of which were profound. For the bargain system was central to the method of working and to the consciousness of the quarryman.

It was a system which in its different forms was known in other extractive industries[7] though perhaps in no other did it survive on such a scale into the twentieth century. The problem which the system was supposed to deal with was the one posed by the tremendous unevenness in the nature of the rock worked, from one part of a quarry to another. Thus one crew's stretch of rock might be buckled or the slate imperfect while another's would be finely grained and easily worked; as it was quarried, moreover, the nature of the rock was constantly changing. Any wages system which merely remunerated men for the number of slates produced was therefore clearly inoperable. The bargaining system meant that each crew of four or five quarrymen would negotiate a monthly contract with the management, the terms of which contract depended on the assessed ease or difficulty of extracting slates from the face in question; on the basis of the monthly bargain the men were paid a sum of 'poundage' per pound's worth of slates produced. Men working on inhospitable rock would be paid a high poundage, compensating for the low yield or poor quality of the slates produced, those quarrying good rock received a low poundage.

Theoretically, therefore, the bargaining system recognized each crew of quarrymen as independent contractors who could argue about the terms of the contract before coming to an agreement. The traditional independence of the quarrymen was thus formalized into a wages system. In practice it need not – and usually did not – work but it maintained and indeed encouraged the feeling among quarrymen that they were equals in some sense with the quarry owners. In practice, of course, the bargain was rarely equal; it could not be. The men usually had to accept the terms offered by the setting steward, for if they refused they were simply out of a job. The haggling on setting days continued, however, if only formally, and occasionally a crew could convince a steward that his assessment of the rock was mistaken. In good times, indeed, and in the smaller quarries, the system could become a genuinely bargaining one again with the crews holding out for a higher poundage and the management, eager to meet the demands of the market, yielding readily. But in the larger enterprises any element of free bargaining had long since passed. What remained – a wages system which reminded the men constantly of their equal and independent position but at the same time treated them like wage slaves – was a persistent source of friction. In the Penrhyn strike of August 1865,[8]

> the common complaint was that the manager always determined the price, and that therefore the men had no choice but to either accept his offer or turn their backs on the works and go to seek work elsewhere, this is considered by them to be the greatest oppression.

And in Dinorwic the situation had degenerated so much by 1885 that the men were not even allowed to state their case during the bargain-setting, but merely to accept the steward's ruling; they appealed in that year that a bargain-taker 'should be allowed to advance his reasons when he considered the offered terms of the manager unreasonable and insufficient'.[9]

This breakdown of the ideal bargaining system was a crucial factor in collective action. For although, as some historians have argued, the bargain system encouraged an individualistic, anti-union sentiment in the work force,[10] and although such attitudes, as we shall see later, were certainly prevalent among a large section of slate quarrymen, nevertheless it was also a powerful force in the creation of a collective consciousness. As a wages system which had lost its genuine bargaining element it invited a collective response while at the same time stressing and encouraging not necessarily an individualistic, but certainly an independent, spirit among the men: a combination which could produce a willingness to indulge in collective struggle.

As a reporter from the *Pall Mall Gazette* noted, therefore, in 1885, 'there exists in the system upon which the quarry is worked a permanent source of difference – i.e. the bargain system'.[11]

In some of the more inefficiently managed quarries the system could be manipulated by the men to their own advantage; they could deliberately hold back output, pleading hidden complications in the rock, thus pushing the poundage up. They could then either enjoy good wages for less work or, after winning a high poundage at the beginning of the month, the rock could miraculously improve, the slates would come off *fel ymenyn* (like butter) and the month's wages could be considerably higher than the setting steward had anticipated. An inexperienced agent could therefore be 'in their hands and they know how to handle him'.[12] The opportunity for this kind of anti-managerial action which the bargaining system afforded, coupled with its essential imprecision (bargain-setting time was known as 'The Guess' to the Penrhyn quarrymen) meant that the setting stewards were hostile and suspicious and that most quarrymen at one time or another suffered from a sense of grievance. Every month the system invited distrust and hostility and the result was continuous friction between men and management.

The system was jealously guarded by the quarrymen and in 1876 the defence of the bargain led to a riot at the Hafod y Wern Quarry in Betws Garmon. When the management tried to introduce a payment by the hour scheme the men walked out, were sacked and then replaced by Cornish blacklegs. This roused the quarrymen of a wide area into fury and a large body of between 500 and 1,000 men converged on Hafod y Wern, many of them, it seems, from the Nantlle Valley quarries some miles away. They marched into the quarry, scattering the Cornishmen and assaulting the agent and engineer. (The latter, having been 'knocked down with a stone and brutally kicked, was, it is stated, dragged some distance on a wagon, and pitched out upon a heap of iron pipes'.[13]) The men remained in control of the quarry for an hour, leaving before the police could arrive, panting, from Caernarvon.

Quarrymen were not often moved to take such action and the incident shows how much they treasured the bargain system and how determined they were to protect it, not just in their own quarry but wherever slate was worked in North Wales.

The union had nothing officially to do with the incident (following which sixteen men, only one of whom was a union member, were fined £5 each[14]) and it roundly condemned the behaviour of the men. In the defence of the bargain system, however, the union too was determined and in December 1877 the union council, for the first and only time during the nineteenth century, called a strike. The scene was the small quarry of Rhos in Capel Curig where the management tried to introduce new regulations which the union felt to be 'destructive of the principle of contract which has worked so well in slate quarries';[15] the strike involved very few men but lasted for two and a half years and cost the young and small union £1,100 to sustain. Such was the readiness of slate quarrymen to prevent any breach in the bargain system.

The real threat to the bargain system was not, however, to come from payments by the hour schemes but from another variant of 'contract' work. In May 1879 the Penrhyn quarrymen's committee decided for the first time to look at an issue which was to preoccupy them for twenty years, to 'investigate the case of those who had taken "contracts"'.[16] For though a bargain was, in a sense, a contract, and that term was frequently used to describe it, the form of contract which had now come to the committee's notice was completely different and posed a very real threat to the bargain system. The new contract system was one under which parts of the quarry were sub-contracted to men who then themselves employed a gang to do the work; under such a system a quarryman was not the contractor himself but was merely the employee of one; and a part of what would have previously been the quarryman's wages now went into the middleman's pocket. Far more important than this financial consideration was the assault on the independence of the quarryman implicit in the contract system, for under contract a man worked not for himself in his own style but for a wage and under the direct instruction of a contractor.

By March 1881 there were twenty-six contracts in the Penrhyn Quarry and the main victims were boys. The choice of boys as the labour force for this system was itself an interesting one, suggesting as it does that quarrymen of any experience were loth to lose their former status but also, perhaps, pointing to a management strategy of trying to breed a new generation of workmen unused to the freedoms of the bargain system and trained to accept the indignities of contract work. The boys working under contract in November 1884 complained bitterly that they were 'not allowed the same amount as boys in general', and felt that 'they are bound to carry out that which is ordered that they do by their masters whether or not it is right or wrong'.[17]

The contract system was consistently opposed by the men and the threat of its further extension was a central underlying cause of the 1896–7 lock out in Penrhyn, for it was 'believed to be the cause of great injustice to a large number of quarrymen'.[18] It was also the immediate cause of the 1900–3 Penrhyn struggle during which one of the men's most important demands was 'the introduction, experimentally, of a system of co-operative piece work in place of work hitherto done under contract'.[19]

In 1900, in fact, hatred of the contractors once again led quarrymen to violence and it was the beating up of two contractors which led directly to the lock out. Fourteen men who had earlier been suspended for three days for refusing to work on a Saturday were informed in October 1900 that their bargains were to be let in one contract to a big contractor. The contractors in question, Richard Hughes and Edward Williams, were warned to keep clear of the bargains. Williams ignored these warnings and the threats thrown at him when he arrived at the quarry; he was then attacked, chased round the galleries, beaten up and thrown physically out of the works.

The following week events escalated. Richard Hughes, the other contractor, made the unfortunate slip, in a newspaper interview, of calling the men 'loafers'. That night a poster went up on walls in Bethesda which read:[20]

> To the Loafers of Chwarel y Penrhyn – take notice. Monday night October 29th, there will be a procession starting at 10 o'clock from Adwy y Pant, Bethedsa, to visit a certain place in the district when we shall pay our debts to the arch loafer. It is hoped that all will be over at midnight. Everyone with an interest in the present disturbances is invited to attend with the appropriate weapons.

Hughes fled and escaped the planned attack; three days later, when he returned to the quarry accompanied by his three sons, he was beaten up and escorted, bleeding, back to Bethesda by a singing crowd of several hundred. Twenty-six men were subsequently arrested and 300 dragoons entered the area to prevent further disturbance. On 5 November the whole body of quarrymen marched to Bangor to the trial of the twenty-six (of whom only six were found guilty); and a furious Lord Penrhyn suspended them all from the quarry for a fortnight. The Penrhyn lock out had started.

The bargaining system was defended not merely for the good wages it could bring with effective organization, though that was not to be overlooked, but also, and more important, because of the style of working for which it allowed; for the very organization of their work allowed to quarrymen a considerable degree of independence and control over their labour. Once a bargain had been settled the crew could work it as it saw best. As Robert Parry, one of the leaders of the Penrhyn men in 1874, explained, 'contracts should be let according to the nature of the work, and after that is done no further meddling with the industry of the contractors should be tolerated in any way'.[21] And twenty years later a Ffestiniog rockman was adamant that 'when they let me a bargain I do not want them to interfere with me in my work until I have finished my contract'.[22] So strong was this feeling that it was generally considered that a bargain, the actual place in the quarry not the settlement, was in a sense the property of those who worked it, not just for the month of any agreement's life, but for good. Morgan Richards, a small quarry owner sympathetic to quarrymen, advised managers that, 'the customary or prescriptive right of a crew to their bargain is so sacred and well established that no wise manager wishing to be at peace with his men, will venture to interfere with it'.[23] Disputes over the manning and location of bargains were therefore not uncommon, managers attempting to arrange the working of the quarry as they saw fit and quarrymen protecting their rights in the bargain. For such a system was not one that suited the employers: a degree of mobility, moving crews from bargain to bargain depending on their abilities and the demands of the rock, would have been more to their advantage. Joseph Kellow, a quarry engineer with twenty-six years of practical experience, complained bitterly in 1868 of this practice of one crew staying permanently on the one bargain; the best bargains, he argued, should be given to the best crews and the poorer crews should be moved if their bargain improved, whereas instead, 'Unfortunately it is the rule for each party to retake the same bargain (regardless of their general fitness) at the monthly letting, however much it may militate against the employer'.[24] Such a system did not encourage close supervision. Each crew worked its bargain in its own way, advice was not sought and interference coldly received. This tradition also militated against adequate safety supervision as well as work discipline and though the union was in the 1890s calling for government inspection there still existed a residue of resentment against any such interference amongst many of the men. A bargain, in all its aspects, was seen to be the responsibility of the crew concerned and it was up to them to ensure that it was safe.

Supervision was also discouraged by the fact that as the men were working a kind of piece work it was in their interest as well as that of the employers to work as hard as possible. Indeed there were complaints that it drove some men to work too hard. On the other hand it also meant that the pace of work was to some extent determined by the men themselves and in the boom of the 1870s it was apparently a 'great and constant complaint . . . that quarrymen do not work as they ought to do the first and second week in every month'.[25]

The work itself was therefore largely out of the management's control though the situation varied considerably from quarry to quarry, some claiming to inspect each bargain daily others not really inspecting them at all. Supervision was not made any easier by the huge scale of some of the workings and by the remoteness of others. William Jones, M.P., claimed in the House of Commons that the Penrhyn Quarry was 'four miles long, three wide and nearly a mile deep'[26] which is a bit of an exaggeration but it must have seemed that big. Even in the more efficient quarries supervision was not at all tight compared to factory production; in 1886 there were nineteen foremen and fifty slate inspectors in the open quarries keeping an eye on the work of over 6,000 men.[27] In the smaller and more remote mines supervision and management of any kind seems, in the 1860s and 1870s, to have been virtually absent. When Her Majesty's Inspector of Mines came to investigate conditions in the Merionethshire slate mines for the first time in 1875 he encountered considerable difficulty as he could rarely find, on his visits to the mines, managers or agents in charge of them. It was in some consternation that he complained in his report about this startling absence of management for 'it is manifest that an idea prevails that a mine is able to manage itself'.[28] In a great many mines this indeed seems to have been the case; a correspondent to the *Mining Journal* in 1865 listed twenty-two Welsh slate quarries and mines, some of them quite substantial, which showed a profit and 'are all worked under the quarryman with a clerk, without even a Secretary, no engineers, no directors, not even an office'.[29]

The nature of employment at the quarries sustained this independence of the men. Not only, as we have seen, were the bargain-takers in a sense sub-contractors themselves employing journeymen, but the 'rybelwrs' were also relatively free of managerial control (though not of managerial vindictiveness), wandering as they did around the quarry in search of someone to assist, a kindly soul to help them out, or an unfulfilled debt. These men, and the journeymen, worked in reality for the quarrymen themselves rather than for the quarry owners, though they were also attached to and policed by the quarry officials to whom they sold their slates. It was a system, as one can imagine, which could rapidly get out of the control of the management, especially a management not familiar with the men, their language or their ways.

Not only the actual work of slate-quarrying but also the hours and days worked were only tenuously controlled by the management and until the 1880s the men seemed to have established a relatively free and easy system of working hours, treating them more as guide lines than as imperatives. This is not to say that working hours were ever short, but attempts to regularize them, especially to extend them into the hours of darkness before dawn and after dusk do not seem to have been particularly successful. Thus even in the mines, where underground work made considerations of daylight on the whole irrelevant, the managers encountered considerable difficulty and resentment in their attempts to enforce a pre-dawn start to the working day in winter; and a Ffestiniog rockman explained in 1893 how he might ignore the whistle and 'follow the day'.[30] Draconian disciplinary measures intended to regularize hours of working were an essential part of the employers' offensive in the 1880s and 1890s.

Equally disturbing for them were the disruptive effects of quarrymen taking unofficial days off. In 1878 this had become such a problem in Penrhyn that a new code of discipline, causing much opposition, had to be introduced. The manager intended to impose a system of fines for late-comers and to have, as the *Carnarvon and Denbigh Herald* explained:[31]

> Some arrangement by which what are called 'extraordinary holidays' shall not exceed a certain number of days during the year. The latter is a matter respecting which the managers wish more particularly to have an understanding with the men for as many as 150, we are told, have frequently announced their intention of going away for an excursion for a day, thus causing inevitably a considerable stoppage of work. We have heard it stated that as many as 2,300 days have been lost to the proprietor in one month, the workmen taking holidays in this way.

Despite rules and regulations the problem was not easily solved and in the June of 1889 the 2,323 men employed in Penrhyn lost between them 5,789 working days.[32]

There seem also to have been some 'official' unofficial holidays accepted by the community but not by the quarry management, such as Ascension Thursday which was taken as a holiday by Penrhyn quarrymen. It was said that a serious accident was bound to take place if the men worked on that day and 'in order to strengthen this belief it has been ensured that accidents have happened every time there has been an attempt to break the holiday'.[33]

Not only the seasonal demands of small holdings, but also social demands such as funerals (always in the quarrying areas attended by hundreds of men) and fairs, could and did claim priority for time which would otherwise have been the quarry's. The right to take the day off if he considered it necessary was held by the quarryman to be a fundamental one, so it is not surprising that the question of regularity of hours and of days worked loomed as a central issue in most of the major disputes in the industry. The demand that they obtain permission from the manager before absenting themselves from their work was considered by the men to be a monstrous imposition. 'The permission paper', in the opinion of the locked-out men of Dinorwic in 1885, 'places us on the same ground as the black slaves in the South of America used to be on.'[34] This defence of the freedom to decide, within limits, when one worked, to some extent reflected the rural patterns which still exercised some influence over the quarrymen. Many came from farming backgrounds and a significant number themselves rented some land. There were in Bethesda, for example, 168 cottages owned by the Penrhyn estate 'with land sufficient to keep a cow' and many quarrymen also had 'certain rights on the mountains for grazing according to the amount of rent, they are allowed to graze so many sheep or ponies on the mountain'.[35] Hay-harvest time was notorious as a period when quarrymen left their work. Nathaniel F. Robarts, the manager of the New Welsh Slate Company of Ffestiniog, complained in 1892 that, 'in the summer, for instance in the hay harvest, a good many go away so that our men are rather more irregular in the hay harvest'.[36] And one of the main points in dispute in the Dinorwic lockout was the management's attempt to curtail this practice of leaving the dust and noise of the quarry for the fields, replacing the hammer, for a while, with the pitchfork. The farmer's calendar was therefore also the quarryman's.

This influence should not, however, be exaggerated; most quarrymen held little or no connection with the land and lived their lives in the terraced streets of the quarrying villages. The skilled men, especially the slate-splitters, were normally drawn from these villages and their attachment to freedom seems to have been as great as the rockmen's, and greater than the labourers', who were, very often, recent recruits from the land (and who were, in any case, often attracted to the quarries by the shorter hours worked there as compared to hours on the land).

The bargain system imposed its own pattern of wage negotiation in which power and personalities mattered more than skill or hard work, for the only way that the quarrymen could hope to improve their month to month bargains and thus their wages was to in some way influence the setting agent or the higher management which appointed him. The setting steward by his assessment of the rock could give excellent wages or pitiful ones, and variations in wages could be immense. In Penrhyn in February 1889, for example, a well-paid quarryman earned almost £10 for a month's work, while one less fortunate bargain-taker earned only just over £2.[37] This kind of difference could be partly explained by distinctions of skill, labour and the nature of the rock, but in a functioning bargain and poundage system the partiality of the setting agent could also be central.

Influencing the agents could be done in two ways, individually or collectively. The individual method created a widespread system of bribery and flattery by which the crews hoped to grease or buy their way into the favour of the management and thus to easy bargains, good rock or high poundage. The other way, attempted less frequently, but with more dramatic consequences, was to collectively pressurize the management or even to attempt at times to remove the existing management altogether and replace it by one more favourable to the men. Both methods could be effective and both pursued a similar goal – higher wages. As Henry Jones of the Alexandria Lodge told the NWQU annual conference in May 1882,[38]

> there is a particular likeness between the brave and manly Union man and the cowardly and flattering 'cynffonwr' [see below]. Both want good recognition for their labour. Both struggle, not in the same way perhaps, but equally energetic, for more of the gold than the silver. They are one in their aim but they separate on the methods used to achieve it.

He went on to uphold the union method of collective struggle and to condemn the man for whom 'the general voice is quite meaningless . . . the centre point of all his impulses is himself'.

Such men were not uncommon in the slate quarries of the nineteenth century and the bribery and flattery system which they worked remained as long as the bargaining system itself. Robert Parry, the first quarryman president of the NWQU, explained that,[39]

> when once a workman felt that an interest was being taken in him greater than his position as workman claimed, there soon would set in a system of peace offerings, and once this began it would, like that of Israel's sins, demand sacrifices every morning and every evening and never be satisfied.

Sacrifices of bacon, beer and self-respect. The bribery system could only operate satisfactorily for a minority, a minority which invited the severest moral censure; and the hatred and bitterness which existed between them and the union party is a central part of the story of the quarries, for the difference was hardened by differing allegiances in religion and politics.

This hatred is, perhaps, best summed up by the very term with which the union men lashed their opponents – 'cynffonwyr'. Literally it translates as 'flatterers'; but *cynffon* is also the Welsh for tail. The 'cynffonwyr' had grown tails and thus betrayed their animal natures. As a Penrhyn striker explained in 1901 they were,[40]

> creatures of a man's shape with tails, yet they were not men. If Prof. Darwin were living in Bethesda now he would not have to go far to find something to prove his point that man is descended from the ape.

The derision expressed an intensity of hatred which even affected the local Liberal press. One paper wrote that they knew of, 'no class crueller, more loathsome, sicker, dirtier or more dangerous to live or work with than the cynffonwyr of our quarries'. After much more abuse underlining the inhuman nature of the 'cynffonwyr' the paper gave the quarrymen of Penrhyn and Dinorwic and all other quarries afflicted by such creatures this advice:[41]

> 1. do not accept them into your houses . . .
> 2. put your hand on your lips in their presence . . .
> 3. let them and their property be accursed things to you . . .
> 4. let not one of them, nor any of their sons, marry a Welsh woman. Let their lineage be forever seen as foreigners to our land and our language.

For those others who chose collective struggle and the self-sacrifice and threat of repression it invited, the greatest success was the Penrhyn strike of 1874 and the agreement which came of it. The mechanics of that dispute are revealing.[42] The men initially made a demand for a 'standard' wage of 30*s*. a week for all skilled quarrymen. Penrhyn replied that he could not agree to this but that he would, as a concession to the men, raise the 'making price' of slates, i.e., the price paid to the quarrymen for the slates they produced. In a properly functioning system this could have been acceptable to the men but they rejected the offer because, though it was generous, it really did not amount to much if the management persisted in their low-wage policies. For wages could still be kept down despite the rise in prices by the agent offsetting the rise by cutting down on the poundage. The men therefore replied that 'owing to our distrust of the management we are compelled to decline the offer as it now stands'.[43] Management therefore became the central issue of the dispute. 'Grant us our demands as they were laid before you at first,' the men asked, 'or a change in the chief manager.'[44]

As the dispute proceeded the men came to see a formula which might effectively grant them both demands, an agreement on a 'standard' of wages and the appointment of a 'supreme manager' who would oversee the quarry managers and to whom appeal could be made in cases of unfair treatment or unduly low wages. The men even recommended the man for the job, Mr Pennant Lloyd, the Penrhyn estate agent; they also recommended as a second acting chief manager a Mr T. H. Owen who would keep a further watch on the existing management.

Given such changes in the structure of management they were willing then to modify their earlier demands for 30*s*. a week 'minimum standard' for quarrymen. Their new plan differentiated crucially between bargains set on the men's terms and those taken by men 'compelled by the manager to work on his price'. In cases of the first type the bargain was to stand whatever the resulting wages, in cases of the second, however, when the crew 'fail to realise proper wages on that taking, and they have worked honestly the whole month, that the wages of quarrymen [bargain-takers working slate] in such case, and in such a case only be made up to 28/- per week'. This was to ensure that managerial control of wages was kept to an absolute minimum. The men's demands were eventually granted in substance; the test was to be their implementation, and the men's suspicions of the managers were well-founded for when they returned to work they found that the agents were simply ignoring the agreement. The men therefore came out on strike a second time and an arbitration board studied their complaints against the management. Out of nineteen cases brought forward seventeen were sustained, and a mass meeting of the men decided that they would 'return to our work under the agreement come to with Mr Pennant Lloyd and to take lettings from others than the present managers'.[45] The three managers (J. Francis, R. Morris and O. P. Jones) were forced to resign and a new chief manager was appointed.

A struggle for improved wages thus became a struggle to limit the power of management, to check their control on the bargaining system. And the battle in 1874 ended not only with the new structure of management brought about by the appointment of a supreme manager acceptable to the men but also with a built-in counter to the setting agent's previous ability to force bargain agreements on to unwilling crews. A managerial clique who had been in power for forty years were thus swept out of the quarry, and a quarry committee was elected by the men to investigate 'all complaints about letting'.[46]

The committee was able to maintain an effective pressure on the managers, and it was in the wages field that they were most active. The committee worked very much like a shop stewards' committee in that it consisted of the elected representatives of the galleries in the quarry, though there was also an executive committee which met regularly. In later years it was virulently denied by the quarrymen that the committee had been a union body and its defenders insisted that it had been open to receive complaints from all workers employed in the quarry, irrespective of union membership. They did indeed receive complaints from all quarrymen, but they certainly did not act on any except those coming from union members, and crew no. 158 from Sebastopol Gallery were not alone in having their complaint on wages rejected in November 1881 with the note that 'the case was thrown out because they were not Union members'.[47]

The committee dealt with a wide range of questions and took up with the managers matters relating to the location and manning of bargains, disciplinary offences against quarry rules, craft questions, disputes between crews, costs payable to the quarry and, above all, questions of wages and poundage. It is difficult to assess how much power the committee actually exerted but, acting as it did as a watch dog on conditions in the quarry, ready to press all issues it considered of importance, it certainly prevented the management, between 1874 and 1885, from having a free rein in the works and from introducing any major new regulations or reorganization. The number of cases actually taken up was not very great but this may well have been because the managers hesitated to carry out unpopular measures which would invite the committee's attentions. Of forty-one cases taken up between 1881 and 1884 only four were definitely settled in the management's favour, and while a good many were settled by verbal compromises, sixteen were definitely settled in the men's favour; of the forty-one cases, twenty-six concerned wages in some way.

The quarry committee, while certainly a union committee, being in effect the Penrhyn Lodge of the union, was democratic and sensitive to all the happenings in the quarry, concerning itself at times even with such mundane matters, normally left to management, as controlling petty thefts. When a complaint about the thieving of slates was received from the Red Lion Gallery in August 1881 the committee did not inform any authority either in the quarry or outside but decided rather that, 'the representatives of the place where there is complaint about the thieves meet together in order to arrange something to meet it'.[48] The committee had authority among the men, and could enforce discipline in the quarry as well as face up to management every time they slipped up on the Pennant-Lloyd agreement or tried to push any underhand change.

The management and Lord Penrhyn himself came to make many charges of interference and even intimidation against the committee, but what angered them most was the effect the committee was having on wages, pressing every case in which they thought the men were not being given their due; 'the worst part of it is,' the manager of the Penrhyn Quarry tetchily complained in 1885, 'the Committee has endeavoured to interfere with the bargain setting, which virtually means managing the quarry . . . interference such as I mean is practically intolerable – it means pressure on the management in their favour'.[49]

Herein lay the strength of the committee and the danger of union organization to the owners in general; the combining of an individualized wages system with a collective consciousness created a situation hard for management to control; for the whole power of the quarrymen, organized in the committee, could be brought to bear not on a general wages front but on any individual bargain where there was complaint. In such a situation the bargaining strength of the men was much enhanced for a crew no longer confronted the agents alone but rather with the knowledge, and the threat, that the whole weight of the workforce could be swung on to their side if the agents failed to set to the required standard. In such a situation the committee was indeed a severe limitation on the power of management.

And when men and management held different interpretations as to fundamental rights and obligations then this limitation could be exercised with subtlety as well as determination. In defence of the right to work, for example, quarrymen exhibited an array of tactics.

By the end of the nineteenth century slate-quarrymen were not often given to wandering; once they settled in one area they tended to stay. This parochialism was, of course, cemented by homes and gardens and pensions but an attitude to the place of work was also important. Men grew to belong to certain quarries and would rarely move. Morgan Richards commented that a 'Bethesda quarryman . . . would almost rather live on bread and water at Bethesda, were that necessary, than go to Nantlle or Ffestiniog where perhaps he would earn more and live better'.[50] Once established a family would expect employment for its sons at the quarry; for kinship ties were important determinants in recruitment: of forty-five applicants for work in the Penrhyn Quarry in 1896, thirty-nine had fathers already at work there, four had brothers, one an uncle and one a grandfather.[51] Apart from rejecting unfit or quite unsuitable boys, management's ability to freely choose recruits was therefore limited. In some quarries, the situation appears to have been so much in the men's control that it was claimed that rockmen were 'taken on without any reference at all to management'.[52]

The quarry, therefore, was expected to give continuing employment to the sons of the community, and though the men accepted that depression obviously limited the scope for such employment they would not tolerate any attempt to recruit skilled labour from outside their number. One of the underlying causes of the Dinorwic lock out of 1885 was the resentment felt by the men towards the tendency of the management to recruit 'children of others who were not tenants, nor children of old workmen';[53] a resentment inflamed by the suspicion that recruits were being chosen for their political and religious allegiances.

The quarrymen also rejected the idea that an immediate reduction in the labour force was the right reaction to depression. They accepted, as they had to, that some unemployment was inevitable, but maintained firmly that 'as it is the worker who brings in the true profit to all works he should be the last to suffer',[54] and they fought determinedly to uphold this principle. In the Penrhyn quarries the initial reaction of the men to the management's proposals for redundancies in November 1878 was total opposition to any sackings and they expressed their willingness to bargain away other rights in order to maintain the right of all to work. They passed a resolution which pointedly expressed their concern 'that we as workmen make an appeal to be allowed to suffer together',[55] and asked on 'what grounds can they meet us without turning anyone away from the works?'[56]

The management appears to have been unusually sensitive to the feelings of the men on this question and proposed a scrupulous plan for 'turning away from the end of the book', and dismissing only those who had come recently to the quarry, and of those to choose the ones who lived furthest away. Such a scheme, coupled with protection for skilled men, was, in fact, accepted by the quarry committee, but was rejected by a vote of the men as a whole who still opposed all redundancies though they must by then have appeared inevitable.

Other tactics were also employed and enforced to beat the effects of redundancy. In February 1879 the committee called for 'compelling crews in which there are only two partners to take an additional partner'.[57] It is not clear whether the initiative for short-time working, four days a week, came from the men or the management; both sides could benefit from such a scheme, the men because it was a means of keeping more men at work and the management by being saved from having to turn away a number of skilled men who could not be easily replaced come a revival in the trade. Whoever was responsible for introducing it in May 1879, the quarry committee was certainly involved in enforcing it. In January 1880 they held meetings throughout the quarry 'to set forth the voice of the works namely disapproving of those who worked on Fridays and Saturdays' and in December 1879 they had talked severely of 'the harm to the general body of workmen that some persons come to the works six days instead of four'. It seems, however, that some men persisted in working full time, and were allowed to do so by the management, which would suggest that the committee was more interested than the management in the protection to employment offered by short-time working, and in March 1881 it had to call again for all the men to work 'but four days a week unless unavoidable circumstances arise in the work'.[58]

Work-sharing by short-time working was therefore one method used by the organized quarrymen in Penrhyn to cushion the effects of unemployment. Another tactic which one might have expected to have been used would have been output control: for the workforce to discipline itself to produce less, thus preventing over-production and redundancies. Short-time working itself was, of course, one way of doing this, the deliberate slowing down of work could have been another. There is no mention in the committee minutes of any such practice but the quarry management seems to have been in little doubt that something like this was going on and blamed an otherwise inexplicable drop of 600 tons of manufactured slate in September 1879 on to the men's 'unconcern'.[59]

Employment itself, therefore, was something which was expected from the quarry and during a severe depression extreme measures would be taken to try and ensure the highest possible level of employment. To a certain extent managements and quarry owners respected this basic function, but the men's defensive actions drove the irritated quarry-owner Morgan Richards to lecture them on the laws of the economy:[60]

> the working class should pay more attention to the governing laws of trade than they do at the moment . . . the worker is naturally slow to

> believe in and realise the disadvantages and difficulties of quarry owners when the market is low and unsettled, but if he had to sell his produce himself he would soon come to understand the effects of the law of 'supply and demand' on the value of labour.

Quite so; but the quarryman was not a quarry owner and however much his union leaders might appear to acquiesce in the theories of supply and demand he himself found no difficulty in interfering in the 'natural law' and defending his right to a job.

The quarry was expected to be sensitive to needs other than the quarry's own and to respect the men's independence. The men expected respect too for their status and their skill for they had a high sense of the dignity of that skill and of the respect which it deserved; and they considered that it deserved better than the bullying of ignorant managers and agents. As the president of the NWQU explained in 1901, 'If Mr. Young [the manager of the Penrhyn quarries], thinks that he can work the quarry on the same principles of government as dockwork or brickwork, he is greatly mistaken.'[61] Quarrymen expected and demanded something better; expected, in fact, 'to be treated like men'.[62] For on the foundation of the bargain system had been built a whole structure of customs and attitudes and beliefs which added up to the only definition of a quarryman which most of the men could find acceptable. This was a definition resting on a consciousness, almost a mystique, of skill; for slate-quarrying was an extraordinarily skilled craft. As the correspondent of the *Pall Mall Gazette* noted in 1885,[63]

> slate quarrying is not a matter of mere manual labour but an art which years of patient practice will hardly acquire . . . a slate splitter is like a poet . . . and contends with the poet on an equal footing at the National Eisteddfod where slate splitting, music and poetry are stock subjects of rivalry.

In the normal crew of quarrymen there would be two rockmen, a splitter and a dresser. The crew might also employ one or two journeymen, young men or boys learning the craft. The rockmen worked on the face itself and were primarily responsible for the first stage of placing the explosive charges in the best place and then, with a crowbar, levering off the giant slabs loosened by the blast. The huge slabs thus prised from the rock were then split into manageable sizes before being sent up to the sheds where the splitters would split the rock with hammer and chisels into fine slates which were then dressed to various sizes.

In the slate mines of Merionethshire a man was usually expected to master either the underground, rock-face skills or the splitting and dressing processes of the sheds; in the open quarries of Caernarvonshire, however, while each member of the crew had his special responsibilities, each was also expected to be able to carry out all the processes, to follow the rock from the face to the finished slate. What was vital in the work of the quarryman was not so much his skill at using various tools and explosives but rather his understanding of the nature of the particular bit of rock he was dealing with. Thus a rockman, before laying his charge, had to know exactly how big a slab he wanted bringing down and which way it would fall; too big or too small a charge, placed in the wrong place, could shatter rather than loosen the rock or could make the getting of further slabs more difficult. As a Ffestiniog rockman pointed out in 1893, being a technically good workman was not the point: 'To bore a hole is one thing, but to know where to put it is quite a different matter.'[64]

The splitter, the real aristocrat of the quarry craft, had to be able to tell at a glance what size and quality slate he could coax out of a particular block. It was this understanding of the rock which gave the quarryman's craft such a mystique; being able to glance at a slab and recognize 'posts, crychs, bends, sparry veins, faults, joints and hardened rock'[65] which would affect the work was perhaps the most important part of the quarryman's skill.

This acquaintance with the rock was, moreover, bred of a lifetime of familiarity. There was no defined apprenticeship in the industry and different men took very different periods to learn; some could never learn, 'some learn in two what others would not in twenty-two years'.[66] What was important was an early start. 'They must begin young', declared the manager of the New Welsh Slate Company, 'to understand the rock, to thoroughly work it. A man does not take to slate rock unless he is brought up to it.'[67] This was particularly true of the above-ground slate-splitter. One quarryman had 'never known of anyone who learnt slate making after 17 or 18'.[68] The rockmen on the other hand often learnt their skills later though it was generally agreed that 'it requires a man to be working for four or five years before he can be considered a good practical rock man', and one rockman of thirty-three years' experience considered that he had 'been at it long enough and I am learning even now'.[69] Ideally a rockman should have spent some years as a slate-maker before going on to the face, not because the particular techniques of the sheds would be very useful to him but rather because he could thus learn the 'nature and proclivities of the slate'.[70]

With such an emphasis on growing up with the rock it was generally agreed by many outside commentators, as well as by all quarrymen, that no one really understood slate-quarrying except slate quarrymen. Morgan Richards warned, 'Let a man be in a quarry ever so long and let him pay all the attention possible to his duties, yet, if not a brought up quarryman, he can never properly and thoroughly understand quarrying.[71] And a quarry manager confessed in the *Mining Journal* in 1865 that,

> I have read the best German, American, English authors on geology, and I have not seen one single passage in any one of their works that can help, assist or enlighten a quarryman in any one of his operations. It is all very well to talk of things, and compile large volumes but bring these great authorities face to face with Nature or to a slate quarry and I will be bold enough to affirm that I can point to more than one hard working Welshman that will shame the best of them.

Another correspondent to the *Mining Journal*, writing three years later, dismissed the claims of engineers and surveyors to any knowledge of quarrying, and went on to remind 'Solicitors', 'Geologists' and 'Oxonians' that they had yet to learn that 'a simple quarryman has more real knowledge of slate quarries than they will acquire in a lifetime'. In the same correspondence a Dr Bower went so far as to declare that 'an honest quarryman knows more of the appearance of the genuine laminating features, for every working purpose, than all the members of the Geological Society put together'. With the introduction of more complex machinery and electricity into some of the quarries towards the end of the century, engineering and technical expertise was not so easily frowned at.[72]

Higher management were not often skilled quarrymen and a considerable number, even of those of local origin, had risen to their positions through the quarry office rather than the rock face or the splitter's stool. Their competence to manage at all was therefore viewed with some contempt by most quarrymen. 'The agents with whom I have been working', a Ffestiniog quarryman told a committee of inquiry in 1893, 'are as incompetent as a child three years old.'[73]

The annual conferences of the NWQU regularly accused the various managers of 'gross mismanagement' and incompetence, a charge which was also forcefully made in Dinorwic in 1885 when the men passed a vote of no confidence in the management. John Davies, the resident manager, had worked at the quarry for forty years and had been manager for eleven, and was heartily detested. His claims to proficiency were dismissed by the men with the comment that his first twenty-nine years had only given him 'experience in figures and nothing more' while during the following eleven years 'Mr Davies' opportunities to gain experience in slate quarrying consisted mainly of a daily walk from his house to the various offices in the works'. As to the qualifications of the haughty principal manager of the quarry, the Hon. W. W. Vivian, a man of some business experience in Lancashire, the men pointed out that 'an extensive experience in a Manchester Mercantile House would qualify a man for a managership of a slate quarry just as much as a knowledge of farming would qualify a young man to be the captain of a ship'.[74] Managers should, it was held by the men, be either skilled men themselves or have passed an examination in 'practical quarrying' (which presumably only a skilled quarryman could do); business expertise did not impress – what the men considered essential was skill at the job.

Their distrust of management was deepened by the inability of many managers to speak Welsh. Slate-quarrying was a Welsh-speaking industry, very few quarrymen having any real grasp of the English language; consequently the whole terminology of the craft was Welsh[75] and it seemed impossible to the men that it could be practised in English. John Williams, a quarryman, recalled in 1942 how an English manager visiting his quarry saw a man smoking, and asked, 'Do you allow this idleness?' The accompanying agent explained that the man was, in fact, studying the rock as well as smoking. This episode, concluded Williams, proved that 'a quarry cannot be worked in English'.[76] A hundred years earlier, in the 1840s, a David Jones had sung,[77]

Os bydd eisiau cael swyddogion,
Danfon ffwrdd a wneir yn union,
Un ai Gwyddel, Sais neu Scotsman,
Sydd mewn swyddau braidd ymhobman.

Mewn gweithfeydd sydd yma'n Nghymru
Gwelir Saeson yn busnesu;
Rhaid cael Cymry i dorri'r garreg,
Nid yw'r graig yn deall Saesneg.

If there should be a need for officials,
Then they are immediately sent away for,
Either an Irishman, an Englishman or a Scotsman,
Is in the job almost everywhere.

In workplaces here in Wales
See Englishmen interfering;
But you must get Welshmen to break the stone,
For the rock does not understand English.

'The English element', explained Robert Parry to the annual conference of the union in May 1882, 'was . . . very damaging to the success of the quarrymen.'[78] Few present would not have nodded in agreement.[79]

The quarrymen were the masters of an immensely complicated and delicate craft, a craft which they well knew only they could exercise, and they had no doubt but that they were in all respects 'the best class of workmen in the United Kingdom',[80] a description they often adopted for themselves. And the very 'attitudes' of managements and agents caused them much pain and were as much in dispute as wages and conditions. Most of the arguments of the men during the Penrhyn lock out complained of these attitudes and one of their ten demands was the 'punishment of unjustifiable conduct on the part of foremen and officials towards the men'.[81] The agents they considered to be insufferably arrogant and rude, refusing to acknowledge the quarryman's superior morality and skill: 'We have to suffer', one of the quarrymen wrote, 'the vanity, harshness, arrogance and injustice of the "stiwardiau bach"'.[82] Tales about the harshness of the stewards and of their insolence were legion. The kind of taunt they had to suffer was retold in April 1882: a quarryman explained to a manager that he could not face his creditors because of his low wages, a plea to which the manager replied by advising him that if he could not face them he should 'walk towards them backwards'.[83] Such a comment, and many worse, cut their pride like a knife. An elegy for such a manager expressed the resentment; there was nothing to rejoice in except his unpopularity,[84]

Oes faith ddi-broffit dreuliodd hwn
Gan chwyddo rhif gormeswyr
Ac yma mae ei arch a'i fedd,
Ond ple y mae'r galarwyr?

A long and profitless life he spent
Swelling the number of oppressors
And here are his coffin and grave,
But where are all the mourners?

As we have seen, the degree of supervision varied considerably from quarry to quarry, for slate-quarrying itself entailed a great many supervisable processes, and a degree of what could be called 'policing' activity had been a feature of the industry since the early decades of the century – checking, for example, such important details as whether the men were actually producing as many slates as they said they were. Large-scale quarrying, however, called for more extensive organization and its effective management needed a considerable array of skills, pre-eminently in engineering, and not only a practical knowledge of actually working the rock. Quarrying had to be carefully organized and planned in advance, and one authority lamented that plans could not be worked out beforehand to last for a hundred, or even two hundred, years.[85] The reason is fairly obvious. A rapid concentration on one section of the quarry would soon bring higher parts crashing down; areas which could be long unprofitable to work would still have to be cleared in the long-term interest; rubbish tips had to be placed in parts which would not in the future need to be quarried, and so on. The continuing difficulty for both management and men was that a long-term plan was not always apparent and the short-term attractions of profitability were always tempting; when a period of reorganization was finally forced on the working of the quarry both sides accused the other of having sacrificed the long-term interests of the quarry for easy money.

The Penrhyn management claimed in 1885 that the men had hitherto controlled the organization of the quarry by insisting on working only good rock, leaving the unproductive rock to be cleared at some future date when it would be absolutely necessary, and expensive, to remove. Without 'proper supervision, many a bargain may be spoilt through over-anxiety of the men to make good wages',[86] complained the manager of the Aberllefenny Quarry in 1893; and a quarry was not much more than a collection of bargains.

The men, of course, hotly denied that there was ever such a tendency in their work, complaining bitterly that all the difficulties were due to incompetent management; managers were 'too keen to make big profits', complained Dewi Peris in 1875, and consequently quarried too deep into the mountainside without clearing the tops of the galleries, thus causing both danger and trouble for the future.[87] The quarrymen's claim that it was more profitable on some low poundages to throw away good slate as rubbish than to make slates out of it was probably justified. But the men's defence briefs were not always so convincing and there was probably some truth in the managements' charge of 'short sighted' working by the men. It is, however, difficult to assess the truth of the accusations and counter-accusations as most of the arguments vied with each other in drama rather than in detail: the Dinorwic Quarry, the locked-out men of 1885 claimed, would 'completely fall in' if the existing manager continued in control, a prophecy which subsequent years did not substantiate.

To the new men in control of the larger quarries in the 1880s and 1890s – at Dinorwic, for example, the manager W. W. Vivian, at Penrhyn the new Lord Penrhyn himself and his manager E. A. Young – it was obvious that to exploit the industry profitably it would be necessary to break the independence and the control of the men, and if they overstated the degree of that control there is no reason why we should underestimate it.

Gazing out in the early 1880s from the turrets of his father's monstrous Norman revivalist folly, Penrhyn Castle, the future Lord Penrhyn must often have wondered about how much control he really had over his quarries. They produced the profits regularly and generously enough,[88] but virtually all the 2,500 men at work there had spent their whole working lives among the galleries and sheds; they knew infinitely more about the place than he or any of his managers, and their skill, the source of his wealth, was a mystery to him. They spoke a foreign language, worshipped in Nonconformist chapels, and had voted *en bloc* against him when he stood for Parliament. In 1874 they had unseated the whole management and set up their North Wales Quarrymen's Union (with the assistance of prominent Radicals), and a quarry committee which leapt on his present management whenever it acted contrary to the quarrymen's wishes. His charities no longer bought allegiance. He was unsure what the men were doing much of the time, and they seemed to be coming and going as they pleased. Their independence made them an uncontrollable workforce, unwilling to obey the capricious demands of a by-now disturbed market.

For W. W. Vivian, trained in Manchester business, the rules which he was trying to impose on the Dinorwic men seemed exceedingly reasonable and sensible; real factory rules, he claimed, were much harsher. 'If the men', he charged with some exasperation, 'will take the trouble to get a copy of factory rules, or of those in use in any real business company, they will at once see the great leniency of the new rules.'[89] The Dinorwic quarrymen, however, would accept nothing that smelt of factory discipline without a fight and they resisted for five months the attempt to bring[90]

> the works under the same rules as the works in England, that is the factories, a thing which is quite impossible and the attempt to do so betrays a lack of experience and common sense. Because to bring such a big, wide and open works such as this under the strict rules of the factories is lunacy,

for that would mean 'the best class of Welsh workmen being trodden upon as mere slaves'.[91] For them slavery and factory discipline were synonymous.

We can only guess at the masters' strategy but it seems fairly certain that they intended to pursue a two-pronged policy: the bargain system, on which so much else depended, could not be dropped at will, its hold was too extensive, but an attempt was made on the one hand to minimize the effects of the system while on the other hand gradually introducing an altogether new wages system. The rules and regulations which pinched the quarrymen so hard were the means to achieve the former aim while the latter was to be satisfied by the extension of the contract system. We have already noted the attempt to introduce boys into contract work in the early 1880s and there was a strong suspicion by the end of the century that Penrhyn's long-term aim was to completely re-create his skilled workforce, to bring up a whole generation of quarrymen in a disciplined fashion and, with the already existing cynffonwyr, to 'nurture a new generation of obsequious quarrymen'.[92] Penrhyn in particular saw the independence of the men as the main threat not only to his profits but also to his pride. It was this feeling which fuelled his obsessive hatred of trade unions and explained his willingness to let his quarries lie idle for so long and his steely determination never to take back into his employment anyone with a spark of independence in him.

Commenting on the Dinorwic lock out of 1885–6, a reporter noted that the dispute was not a matter of wages; it was 'simply a matter of sentiment; whether the quarry should be worked under one set of rules or another set of rules and whether the men or the managers shall be permitted to regulate the business'.[93] The question was nicely stated; the quarrymen did not claim that the quarries were theirs and they were always extremely respectful towards the owners, but what they did assume was that they had the right to regulate the way they were worked, the right to have some control over their own working lives. This was an assumption, rarely a claim; the men were usually adamant in their denials that they wished in any way to interfere in the 'rightful' concerns of their employers. Rule 2 of the union caught the contradiction, the union's aim was to [94]

> resist any infringements that may be attempted by employers upon the established rights and privileges of the quarrymen of North Wales . . . . it is not intended that this Union is in any way to interfere in the management in any works.

Protecting their 'established rights and privileges' usually entailed a degree of 'interference' in management which the quarry owners found unpalatable.

And at times such interference was admitted and its rationale eloquently articulated; R. Jones, a Dinorwic quarryman, explained to a mass meeting in 1885,[95]

> there were some who denied the rights of the workmen to have a voice in the management of the works, and maintained that that was the privilege of the master, who had invested his money, his capital, in the quarries . . . this was partly true, but when the master was not careful in his selection of proper agents the workmen under such circumstances had a right to raise their voices and object because the appointment of incompetent managers endangered their lives. If the argument that the master, because he invested his capital, had the right to appoint the managers, was logic, then for the very same reason, the workmen ought to have a voice in their appointment. What were the workmen's labour and their lives, but their capital? How many workmen had lost their capital in the Dinorwic Quarries, and how many orphans and widows were there in the neighbourhood of Llanberis who had seen their capital brought home in pieces upon a bier. These facts . . . justified their present action in raising their voices against the incompetence of the head manager.

Such feelings were not often expressed but the quarrymen did feel that they had invested their labour and their lives in the quarries and that they therefore had a stake in the way they were run. This consciousness did, of course, find political articulation, the nationalization of the quarries seems to have been discussed by Nantlle quarrymen in the early 1890s. In 1886 a delegate from Penmachno had declared, to the approval of the NWQU annual conference, that if the masters could not work the quarries without losing money then, 'should they not be forced to sell their works and mines for reasonable royalties and let the working class try their hand'?[96] But such schemes made little headway. More popular was the ideal of co-operative quarries and two attempts were made to launch such ventures: in 1880 the NWQU invested £2,000 in a co-op quarry and one of the results of the 1900–3 lock out was the setting up of another trade union backed co-operative quarry. The nationalization of the land, on which, of course, the quarries stood, was another popular notion, for if the ownership of the quarries, of the productive units, was not in dispute, the ownership of the mountains in nineteenth-century Wales certainly was. 'Do not go back to Egypt my people,' a speaker urged a mass meeting of Dinorwic quarrymen in 1885. 'Demand the Elidir again. . . .'[97] Egypt was the quarry, the Elidir was the mountain itself.

Politics and religion sharpened and soured industrial relations in the quarries; every squabble was defined as a clash of cultures and traditions, of allegiances and values. The battle line was clear; on the one hand the quarrymen consciously upheld their brand of Radical Liberalism, their Nonconformity and their Welshness; on the other side the masters not only jealously guarded their profits, but also defended the ideology and institutions of an English squirearchy's Toryism and Anglicanism. Politics coloured every wage negotiation, every struggle; and religion provided that massive sense of self-righteousness which characterized both quarryman and master. When they went into battle slate-quarrymen fought not only as workers but as Radicals and as Nonconformists, they carried with them the whole cultural apparatus of the communities they had created.

And in a profound sense community and social pressures were important in the industrial struggle in the quarries. For though the quarrymen created a flourishing quarry culture which made a unique contribution to Wales, and in such a way extended the literary, political and religious preoccupations of the community into their work, nevertheless the structures of community and quarry were largely incompatible. Though it is dangerous to oversimplify it is largely true that the quarry villages were dominated by the Nonconformist chapels, by the norms of hard work, thrift and respectability, while the quarries on the other hand were where the shifty, the unprincipled, the flatterer ready to sell his religion and his character, often rose to acquire dictatorial powers. The setting agents, the stiwardiau bach, the men who got the best bargains, were often men of this type, men who could work the patronage system and who owed their position least of all to their skill at the job. It was this inverted structure which made conditions in the quarry so intolerable for the hierarchy of the village: God-fearing and respectable men, finding themselves so often spurned and insulted by the hierarchy of the quarry. As the Dinorwic quarrymen complained in 1885,[98]

> no high price is put at the Dinorwic Quarries for manly independence, honesty, faithfulness to work and good workmanship . . . [but] that which degrades [the workman] is covered with favour, and receives as reward that which should only be given to the best and most honourable of the class.

The village was dominated by the mainstream of Welsh culture, Nonconformity, Liberalism and the principles, if not of trade unionism, then of collective, democratic organization. The values which ran the major quarries of North Wales, however, were fashioned by the Tory Party, paternalism, Anglicanism and the English language. It was this perpetual conflict between two systems and two cultures which ensured the bitterness and persistence of the battle for the quarries. The aim of the union, the Tory minority in the Penrhyn quarries claimed, was to change the name of Caebraichycafn, the Welsh for the quarries, to Caebraichcalvin: to bring the quarry, as well as the village, under the narrow rule of the deacons.

The slate quarrymen, therefore, fought consciously as members of a well-defined community with its own values and aims and this perspective should not be lost. But it needed only a translation of terms to turn the decades-old struggle against Anglicanism into an industrial battle, for the enemy, an alien squirearchy, remained the same. For the men involved the disputes were essentially about work, about what it meant, about how and when it should be done and about how much should be paid for it. For the activist Welsh middle-class Radicals, usually the publicists of the quarrymen's struggles, the perspective was different and they consistently drafted the strikes and lock outs of the quarries to their own continuing campaign against landlordism and the Anglican Church. For them, caught between a suspicion of the quarrymen and an enduring hatred of landlordism, the battles in the quarries were seen, invariably, as blows for religious and political emancipation rather than as battles for freedom at work.

The quarrymen themselves were prone to extravagant language and to an Old Testament turn of phrase; and people who speak in the language of the pulpit, in moral absolutes, in terms of justice and basic human rights and principles, as the quarrymen consistently did when discussing their industrial relations, do invite misunderstanding. But whatever such words might mean to us, they seemed to them appropriate terms for the concrete realities of their own working lives. The nature of the work men and women must do, a central experience in the lives of the vast majority, might well deserve such a vocabulary. The principles the quarrymen fought for were intimately concerned with their working lives, with working hours and holidays, with wages and wages systems, with work discipline, rates for the job and managerial prerogatives. They were no less noble principles for that. For the quarrymen sensed accurately enough that what was at stake was fundamental: 'We have been told', wrote the Dinorwic quarrymen's committee in 1885, 'that we are too independent.'[99] In the face of the growing and intolerable threat to their independence and self-respect the North Wales quarrymen fought bravely and at times fiercely, albeit inconsistently, to defend their rights in their quarries. 'There is a danger that we lose our independence,' cried a locked-out quarryman from Clwt-y-Bont at a mass meeting in Llanberis in December 1885.[100] 'We must fight like men or fall lower than men.'

Notes

1 For the economic background see A. H. Dodd, *The Industrial Revolution in North Wales*, Cardiff, 1951; D. Dylan Pritchard, 'The early days of the slate industry', *Quarry Managers' Journal*, July 1942, p. 30; 'New light on the history of the Penrhyn slate quarries in the eighteenth century', ibid., September 1942, p. 117; 'Financial structure of the slate industry 1780–1830', *Quarry Managers' Journal*, December 1942, p. 211; and 'Aspects of the slate industry', and other articles *passim* in ibid., May 1943 to October 1946.

2 P.P. 1893–4 (C 7237) LXXII, Report by the Quarry Committee of Inquiry, August 1893, App. III, p. 47.
3 *Mineral Statistics*, 1882, quoted in P.P. 1884 (C 4078) XIX, 1, Report of H.M. Inspector of Mines for North Wales, 1883, p. 207.
4 Report by the Quarry Committee of Inquiry, p. 76.
5 P.P. 1890–1 (C 6455) LXXVIII, Return of Wages in Mines and Quarries, pp. 667–8. 'In a full week, exclusive of meal times, about 46% of the total number of Men and Boys returned worked 52 hours, 36% From 54¼ to 57 hours, 17% about 50 hours; and of the remainder, some worked 46 to 49 and others about 58½ hours. The full time worked at open quarries would be reduced by bad weather and short days in winter.' These figures, which applied at 1 October 1886, are averages covering all of North Wales' quarries.
6 P.P. 1890–1 (C 6455) LXXVIII, Return of Wages in Mines and Quarries in the United Kingdom, with Report, p. 58; these thirteen groups of workers added up to only 9·7 per cent of the total quarrying force of North Wales.
7 Caernarvonshire copper mines knew the system in the eighteenth century and the Cornish 'tribute' system was very similar.
8 *Yr Herald Cymraeg*, 12 August 1865.
9 *The Lock Out at the Dinorwic Quarries*, Caernarvon, 1885, signed by the Lock Out Committee.
10 See, for example, D. Dylan Pritchard, 'Trade unionism', *Quarry Managers' Journal*, January 1945, and for a discussion of this problem in another industry, J. G. Rule, 'The tribute system and the weakness of trade unionism in the Cornish mines', *Bulletin*, Society for the Study of Labour History, autumn 1970.
11 University College of North Wales, Bangor, Coetmor MSS., 46, p. 30.
12 Morgan Richards, *Slate Quarrying and How to Make it Profitable*, Bangor, n.d. (1881?), p. 19.
13 *Bulletin*, no. 3, Caernarvonshire Record Office (CRO), 1970, p. 16.
14 Ibid.
15 National Library of Wales (NLW), W. J. Parry MSS. (4), 8736C, 29 December 1877.
16 CRO M/622/11, 19 May 1879.
17 NLW, North Wales Quarrymen's Union Collection, *Cofnodydd* Penderfyniadau Pwyllgor y Gwaith, Caebraichycafn, 14 November 1884.
18 W. J. Parry, *The Penrhyn Lock Out*, London, 1901, p. 74.
19 Ibid., pp. 172–3.
20 *Yr Herald Cymraeg*, 6 November 1900.
21 *Carnarvon and Denbigh Herald*, 15 August 1874.
22 P.P. 1895 (C 7692), XXXV, Report of the Departmental Committee upon Merioneth Slate Mines, Q. 1186.
23 Richards, op. cit., p. 20.

24 Joseph Kellow, 'The slate trade in North Wales', *Mining Journal* Reprint, 1868, p. 9.
25 Richards, op. cit., p. 71.
26 William Jones, House of Commons, 5 March 1903, Hansard, *Parl. Deb.*, vol. 118 (4th ser.), 1651.
27 P.P. 1890–1 (C 6455) LXXVIII, Return of Wages in Mines and Quarries, pp. 569, 667.
28 P.P. 1876 (C 1499), XVII, Annual Rep. H.M. Inspector of Mines, 1875, p. 385.
29 *Mining Journal*, 15 April 1865.
30 P.P. 1895 (C 7692) XXXV, Report on Merioneth Slate Mines, Q. 3140.
31 *Carnarvon and Denbigh Herald*, 12 January 1878.
32 Penrhyn Quarry wages book, June 1889. CRO 55/27.
33 *Y Genedl Gymreig*, 24 May 1882.
34 Ibid., 5 August 1885.
35 P.P. 1892 XXXVI, R.C. on Labour, Minutes of Evidence, Group A11, Q. 16,846.
36 Ibid., Q. 9449.
37 CRO, Penrhyn Quarry wages book, 1889.
38 *Y Genedl Gymreig*, 24 May 1882.
39 *Carnarvon and Denbigh Herald*, 28 May 1881.
40 *Yr Herald Cymraeg*, 6 August 1901.
41 Coetmor MSS., no. 45, no date.
42 See Parry, op. cit., pp. 7–24.
43 Ibid., p. 12.
44 Ibid., p. 16.
45 Ibid.
46 Ibid.
47 *Cofnodydd . . .*, 25 November 1881.
48 CRO, M/622/11, 18 August 1881.
49 Parry, op. cit., p. 109.
50 Richards, op. cit., p. 49.
51 CRO, Penrhyn Quarry letter book, 11 June 1896.
52 Report on Merioneth Slate Mines, 1895, p. 316.
53 *The Lock Out at the Dinorwic Quarries*, Lock Out Committee, Caernarvon, 1885.
54 *Y Genedl Gymreig*, 2 February 1882.
55 Copy, Minutes of the Quarry Committee, CRO M/622/11, 19 November 1878.
56 Ibid., 21 November 1878.
57 Ibid., 17 February 1879.
58 Ibid., 29 December 1879; 14 January 1880; 25 March 1881.
59 Ibid., 8 September 1879.
60 *Y Genedl Gymreig*, 12 January 1882.

61 *Yr Herald Cymraeg*, 24 December 1901.
62 *Daily News*, 8 January 1901.
63 Coetmor MSS., no. 46, p. 30.
64 Report on Merioneth Slate Mines, 1895, Q. 2477.
65 D. C. Davies, *A Treatise on Slate and Slate Quarrying*, London, 1880, 2nd ed., p. 118.
66 Report on Merioneth Slate Mines, 1895, Q. 2473.
67 Minutes of Evidence, R.C. on Labour, Group A II P.P. 1892, Q. 9387.
68 Report on Merioneth Slate Mines, 1895, Q. 2677.
69 Ibid., Qq. 1596, 3117.
70 Ibid., Q. 1977.
71 Richards, op. cit., p. 29.
72 *Mining Journal*, 22 April 1865; 11 January 1868.
73 Report on Merioneth Slate Mines, 1895, Q. 2677.
74 *Carnarvon and Denbigh Herald*, 23 January 1886.
75 Some words were originally English, for example, 'rybelwrs' derived from 'rubbelers', and some English terms were used – the classification of slate sizes according to the noble hierarchy, for example (duchesses, etc.) – but most quarry words were Welsh. For a glossary see Emyr Jones, *Canrif y Chwarelwr* (Gwasg Gee, Denbigh, 1963).
76 John Williams, 'Atgofion chwarelwr', *Y Llenor*, winter 1942, pp. 129–35.
77 British Museum, *Welsh Songs 1767–1870*, p. 69.
78 *Y Genedl Gymreig*, 24 May 1882.
79 The feeling was mutual: 'these Welshmen', wrote E. A. Young in 1900, 'are so childish and ignorant.'
80 See, for example, *The Lock Out at the Dinorwic Quarries*, p. 1.
81 Parry, op. cit., pp. 172–3.
82 *Yr Herald Cymraeg*, 5 February 1901.
83 *Y Genedl Gymreig*, 5 April 1882.
84 Ibid.
85 Dewi Peris, 'Chwarelyddiaeth', *Y Geninen*, XIV, January 1896, pp. 52–4.
86 Report on Merioneth Slate Mines, 1895, Q. 4500.
87 Dewi Peris, op. cit., July 1896.
88 It was estimated in October 1894 that Lord Penrhyn, 'who is one of the richest men in the Welsh Principality' had been making about £100,000 a year from his quarries. (*Our Gazette*, National Association of Slate Merchants & Slaters, Hull, October 1894, p. 2). In 1899 he made £133,000 net profit. (CRO Penrhyn Quarry letter book, 7 January 1899.)
89 *Carnarvon and Denbigh Herald*, 30 January 1886.
90 *Y Genedl Gymreig*, 29 July 1885.
91 *The Lock Out at the Dinorwic Quarries*, p. 7.
92 *Yr Herald Cymraeg*, 19 November 1901.
93 *Pall Mall Gazette*, Coetmor MSS., no. 46, p. 30, no date.
94 Richards, op. cit., p. 140.

95 *Carnarvon and Denbigh Herald*, 2 January 1886.
96 *Y Genedl Gymreig*, 26 May 1886, p. 7.
97 Ibid., 16 December 1885.
98 *The Lock Out at the Dinorwic Quarries*, p. 7.
99 Ibid.
100 *Y Gendl Gymreig*, 16 December 1885.

Part 3

Cheshire saltworkers

Brian Didsbury

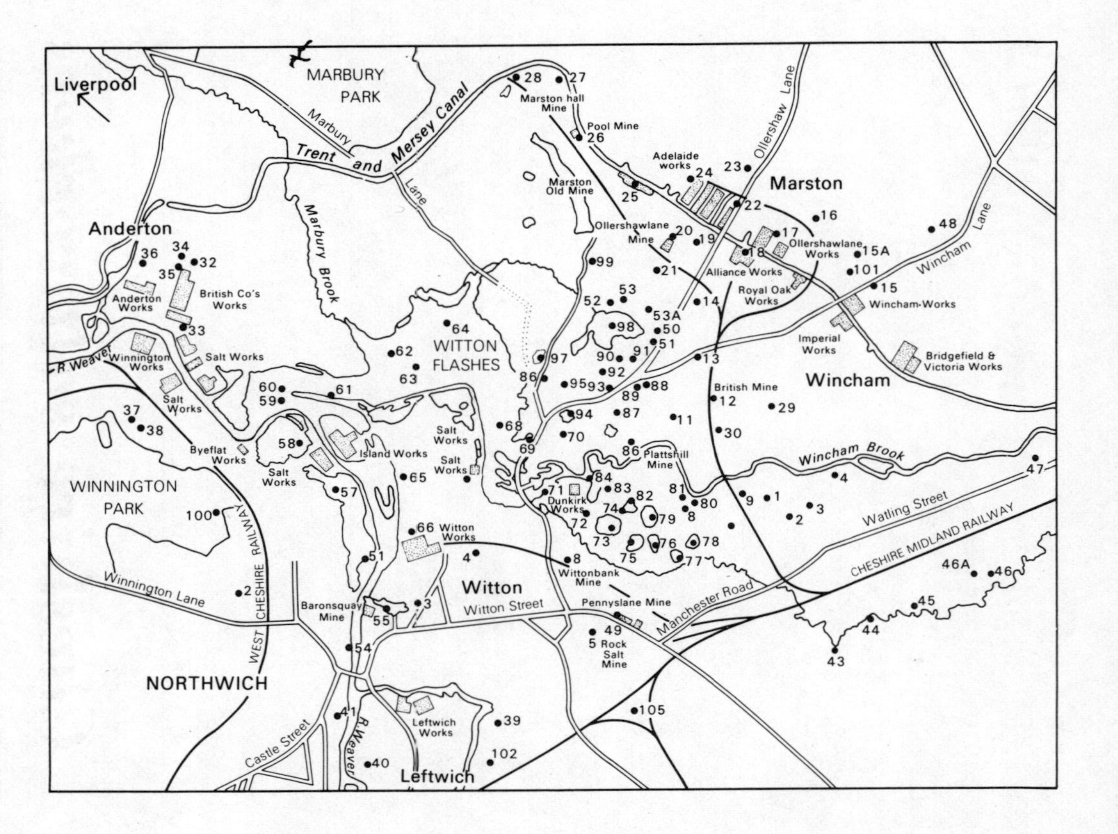

Figure 1 Salt borings in the Northwich district (adapted from A. F. Calvert, *Salt in Cheshire*, London, 1915).

I

The saltlands of south-east Cheshire occupy an area roughly thirty miles by ten along the Weaver valley. Towards the centre, where the waters of the rivers Dane and Weaver converge, is the district generally known as Northwich, consisting of Hartford, Castle, Winnington, Leftwich, Witton and Rudheath. A mile or two to the north-east are the villages of Marston and Wincham. Some six miles to the south, created by an amalgamation of the ancient townships of Wharton and Over, is Winsford. Throughout the nineteenth century most of the salt industry of the United Kingdom was based on these three districts.

The salt towns are towns literally built on salt. Extending beneath the whole district are two beds of salt: at Northwich the 'top bed' is about 100–150 feet below the surface and the 'bottom bed' about 330 feet. This salt was extracted not so much by direct mining (except in Marston and the Dunkirk area of Witton), as by tapping the underground brine streams.[1] Over many centuries (it is said that saltmaking in Cheshire was started by the Romans) countless borings were made throughout the district (see Fig. 1), and vast quantities of brine pumped up to the surface for the salt to be boiled out. The result of extracting the brine (and of the steady washing away of the salt itself) was the subsidence, gradual or sudden, of the land above.[2] The effects were perhaps most marked along the waterways, the river Weaver and its many tributaries, where the sinking of the land caused immense lakes to form which are locally called 'flashes'. What had once been green and pleasant meadows was submerged by water whose depth varied from a few feet to forty or fifty.[3] The process was described by a visitor to Northwich in 1879, who was shown the effects near Witton Brook, a tributary of the Weaver:[4]

> Here the ground sinks bodily in immense masses to a great depth. A tiny brook or ditch that a child could skip across passed over flat fields some five years ago. Gradually the land began to sink, and cracks opened in the surface right across the course of the brook. The water went down the crevices. The land immediately sank more rapidly to a depth of forty to fifty feet in the centre, and was filled to a certain height with water which covered the hedges and trees.

Nor were only meadows lost. The great mass of water to the north of Northwich, known in my childhood as 'the ocean', covered land on which had once stood saltworks and houses. Destruction was not always immediate: a writer in 1850 described how the Leicester Arms public house had been gently sinking for some twenty years, so that the sitting-rooms and tap room had become cellars, and its former sleeping apartments were being used in their place.[5] But sometimes subsidence was dramatically sudden. James Cowley, Clerk to the Northwich Local Board, spoke in 1891 of 'main roads going down without intimation at all, horses disappearing', and he described the narrow escape of a company of Volunteers drilling in Wheat Sheaf yard: 'they had only just passed over a certain portion of the yard and it immediately went in and a cavity was formed; had they been two minutes later they would have fallen into it'.[6] The appearance of the town clearly reflected the strains that were put on its buildings by subterranean shiftings. A writer in *Chambers' Journal* who visited Northwich in 1879 described it as follows:[7]

> a number of miniature valleys seem to cross the road and in their immediate neighbourhood, the houses are, many of them, far out of the perpendicular. Some overhang the street as much as two feet,

> whilst others lean on their neighbours and push them over. Chimney-stacks lean and become dangerous; whilst doors and windows refuse to open and close properly. Many panes of glass are broken in the windows; the walls exhibit cracks from the smallest size up to a width of three or four inches; and in the case of brick arches over doors and passages the key brick has either fallen out or is about to do so, and in many cases short beams have been substituted for the usual arch. In the inside, things are not much better. The ceilings are cracked and the cornices fall down; whilst the plaster on the walls, and the paper covering it, exhibit manifold chinks and crevices. The doors either refuse to open without being continually altered by the joiner, or they swing back into the room the moment they are unlatched. The floors cannot be kept level; and frequently a billiard-table will require packing at one end, some two or three inches, to keep it level. Many of the houses are bolted and tied together, but even then they cannot be kept upright. This is not merely an odd case or here and there a house; but for sometimes twenty, sometimes fifty, and occasionally a hundred yards each way from the little valleys crossing the streets, the houses are affected in this manner. If it were no worse, it would be bad enough; but unfortunately the bolting and tying of the houses cannot prevent their destruction. The time comes when they are declared unfit for human habitation and must be taken down.

The salt towns were covered in soot. The salt was obtained by a process of evaporation, the brine being boiled in huge pans till the salt crystallized and about 500,000 tons of coal were needed to make a million tons of salt. In Northwich alone, it was calculated in 1887, the chimney effluent contained 29,000 tons of sulphur dioxide per annum.[8] Hundreds of low-built chimneys belched black poisonous smoke which mingled with the steam to create a perpetual evil-smelling fog. The blackness of the saltmaking districts was legendary. Even in 1698, when Celia Fiennes visited Northwich, she found it 'full of salt works' and 'full of smoak from the salterns on all sides'.[9] And a writer in the *Manchester Guardian* in 1888 deplored the effects: 'Every blade of grass in the vicinity is being killed off, and hillsides and fields are reduced to bare surfaces of baked clay.'[10]

The coal burned in the saltworks came from Wigan and St Helens, and was locally called 'burgey'.[11] It was of the cheapest grade 'and sold at the coal pits as fit for no other purposes'.[12] This accounted not only for the immense amounts of black smoke, but also for the quantities of clinkers, or 'basses', which it left after burning. (The word is still in use: when household coal does not produce a pleasing fire older people complain that it is 'full of bass'.) This clinker residue, along with ashes from the fire and scale from the salt pans, was dumped in great mounds known as 'cinder middens', which frequently caught fire and smouldered for months,[13] giving off fumes that were said to be even more dangerous than the chimney effluent. The mixture of sulphurous gases from the coal and hydrochloric acid vapour from the decomposing salt scale was not only very offensive, but also 'irritating to the eyes and organs of respiration at a distance of at least 80 yards', and inhaled in sufficient concentration caused 'a sensation of suffocation or oppression at the chest, and cough, with general malaise, headache, and loss of appetite'.[14] Cinders were put to every possible use in the effort to get rid of them, and a special class of boat – 'cinder boats' – carried them to be dumped in the 'flashes'.[15] In the early part of the century they were sold soaked in brine to local farmers as fertilizer. They were spread on the roads, on the river towpaths, and on farm tracks, and used extensively in the construction of railway embankments, or 'batters' as they were called. They were even used for building, for pigsties, for instance, and in some cases for saltwork buildings. One salt proprietor, H. E. Falk, used them in the construction of a hundred workmen's cottages, which, although not pleasant in appearance, stood for nearly a century.[16] Noting this practice a correspondent of the *Manchester Guardian* wrote in 1888: 'The flaked cinders are being used as bricks and with the aid of liberal supplies of leaden-coloured mortar walls, buildings and embankments of a sober hue are constructed.'[17]

The wych-houses, where the brine was boiled, were sheds with louvred roofs and often an open side; they provided basic shelter whilst leaving as much open as possible so that steam could escape. Inside them were huge iron pans 'varying from 30 feet to 60 feet long, and about 24 feet wide, and from 1ft 6ins to two feet deep'.[18] The pan was supported by iron columns, by the front and side walls, and by brickwork in which were three or four fire-holes through which the huge fire running along inside, under the pan, could be coaled. Above the fire-holes along the length of the pan ran a gallery, the 'hurdle', where the workers would stand to rake in the salt with long-handled rakes of wood or iron. Behind and below the pan were flues connecting the fire to the chimney; in a shed where lump salt was made these flues would pass under and heat the room called the 'stove', where the lumps would be stood to dry.

In spite of all their ventilation the wych-houses were hot and steamy.

According to Meade King, a local Factory Inspector, writing after a visit in 1876, 'the vapour rising from the pans . . . is generally so thick that you cannot see across the shed in which they are working'.[19] The fires were kept burning night and day, and besides the general heat of the place there were snug corners by the brick walled flues or in the drying room or the store where in the middle years of the nineteenth century the salt boilers would often snatch their hours of sleep. For it was common then for the boiler to stay at his work through the week, perhaps with his wife and children coming to help him in relays. Passing tramps would also find shelter there,[20] or homeless locals like Michael Gray – 'poor old Mick' – a labourer who 'had no regular work but carried parcels and did odd jobs' and slept rough 'in out-houses, stables or salt works'. The salt boiler who found him on the towpath of the Weaver dying of exposure on a freezing night in January 1875 had seen him the previous morning 'coming out of the hot house of Messrs. Deakin's salt works . . . He had been warming himself; he had no settled abode.'[21] According to Inspector Dutton, giving evidence before the Northwich magistrates in 1875, 'there were continual complaints made about men resorting to salt works to sleep and damaging the salt.'[22]

The warmth was also the cause of the 'immorality' of dress of which saltworkers (particularly women) were sometimes accused. Because of the stifling heat in which they toiled the women removed their outer garments and worked in chemise and petticoat – a practice which in the 1870s led to ill-founded attacks upon their morals and reports that they were in the habit of working 'in almost a naked condition'. Mr Whymper, in whose district, as H.M. Superintending Inspector, the saltworks at Droitwich fell, reported in 1878:[23]

> The supposed tendency to promote immorality arises in this way . . . In 'drawing' (i.e. taking the salt from the pan) the women wear a shift not always closely fastened at the neck, and a petticoat. Often men are moving about the same pan, perfectly naked from the waist upwards. Owing to the heat and moisture, such clothes as are worn, instead of concealing, cling to and emphasize the figure. That is, when the mist lets you see any figure at all; for individuals are invisible to each other until they actually come into contact.

The mining of rock salt did not involve smoke and cinders (though subsidence sometimes occurred in connection with it), but it was the smallest branch of the industry. It was begun soon after the accidental discovery of rock salt during an unsuccessful search for coal at Marbury Lane, Marston, in 1670.[24] The first mines were sunk to a depth of about 100 feet, but it proved impossible to prevent fresh water seeping in, and one by one they all collapsed.[25] Nevertheless, in 1781 at the Marston Old Mine a trial shaft was sunk through the existing workings and a second layer of rock salt was discovered, the 'bottom bed', about 330 feet down.[26] From that time onwards all mining was done at that level. For a short time – before the river Weaver was opened up for commercial traffic in 1721 as the Weaver Navigation – mining salt was more profitable than making it.[27] But cheaper transport and improvements in saltmaking technique swung the balance the other way, and by 1882 it was estimated that 'the cost of making white salt from brine was . . . about one-fifth of the cost of obtaining such salt by mining'.[28] Even so, rock salt had a remarkably steady export market, and for almost a century its extraction provided fairly regular employment for about 300 people, who mined between 150,000 and 170,000 tons a year.

The opening of the Weaver Navigation in 1721, and of the complementary Sankey Canal in 1755, was of great importance to the salt districts, providing much cheaper transport (pack horses had been used before),[29] access to a major export port, and an incentive to the development of a local ship-building industry to provide the 'flats' which carried the salt to Liverpool and brought back coal for the salt house furnaces. Of the 16,500 people estimated in 1817 to be dependent on the salt trade for their employment, 5,000 were employed on the river and canals, and a further 5,000 were 'indirectly employed as ironmongers, shipwrights, carpenters, bricklayers, ropemakers, sailmakers, and so on'.[30] Ship-building seems to have started in Northwich almost as soon as the Navigation was completed: a 'flat' of 50 tons, the 'Friends Goodwill', is said to have been built there in 1740.[31] But Winsford was the town which in the end was to prosper most from the Weaver Navigation. Only two saltworks – 'and those on a very small scale'[32] – were in operation there when small boats were first able to make a passage down the Weaver from Winsford, in 1671. Once the Navigation was opened, however, Winsford began to grow, and much of the new capital which the Liverpool merchants invested in the industry found its way there.[33] In the early nineteenth century, production surged ahead: Winsford sent 44,384 tons of white salt down the Weaver in 1800, and increased that output by more than 730 per cent in the next fifty years, sending 324,249 tons in 1850.[34] By that time, too, Winsford's share of the salt shipments was 72·49 per cent to Northwich's

27·51 per cent, and although the proportions varied, Northwich never again sent as much as Winsford, let alone more. This was mainly because of the increasingly serious effects of subsidence around Northwich which was forcing the salt proprietors to build further away from the river.[35]

The great period of expansion in the whole salt trade was the nineteenth century, although technological innovation and (as we have shown) improved transport had happened long before. The introduction of coal to fuel the burners and of iron pans instead of wooden ones took place towards the end of the seventeenth century, and made it possible to increase the range of salts being made. In the eighteenth century engines were introduced to pump up the brine, worked at first by wind or horse power, and later by steam. But, in the nineteenth century, new demands (like those of the growing chemical industry) and the growing foreign markets which Liverpool merchants were opening up, brought about a massive increase in production, as Table 3.1 shows.

TABLE 3.1

Shipments of White Salt down the Weaver at ten-yearly intervals, 1798–1878 (A. F. Calvert, *Salt in Cheshire*)

Year	*Tonnage*
1798	100,544
1808	196,433
1818	154,255
1828	239,864
1838	328,318
1848	357,546
1858	647,347
1868	868.679
1878	913,952

II

Generally speaking, saltworks could be divided into two types – those where fine-grained, or 'lump' salt was manufactured and those where coarse-grained, or 'common' salt was made. In the former the brine was heated to boiling point and produced dense, fine crystals of salt suitable for use in the home (i.e. table salt) and in the manufacture of cheese and butter. The coarse-grained salts were produced at lower temperatures; the crystals were cruder and took longer to form. A temperature of 100°F. produced a coarse, hard, slow-melting crystal considered ideal for use in the fisheries. 'Common salt', used in the chemical industry, the potteries and other manufacturing industries, was produced when the brine was maintained at a temperature of 160–170°F. Finally a temperature of 130–140°F. produced large-grained flaky salt suitable for the preserving of meat. The manufacture of the coarse-grained salts took longer; for example, that used in the fisheries might be 'a fortnight forming in the pan'.[1]

The making of fine salt was governed by an ancient technology. Saltmakers engaged in it were known as 'lumpers', or 'lumpmen', after the practice of moulding the salt during drying in tapered lumps or bars. After the pan was filled with brine and the fires lighted it was about twelve hours before the brine began to boil, and at some stage during this period the lumpman would, in all probability 'dope the pan'. This practice was observed in the early eighteenth century by John Ray, who noted:[2]

> When the liquor is more than lukewarm, they take strong ale, bullock's blood and whites of egg mixt together with brine in this proportion: of blood one egg-shell full, the white of one egg and a pint of ale, and put it into a pan of twenty-four gallons (of brine) or thereabouts.

Over a century later, in 1808, the additives used included: 'acids, animal jelly and gluten, vegetable mucilage, new or stale ale, resin, butter and alum'. Dr Henry Holland commented:[3]

> various additions are often made to the brine, with a view to promoting the separation of any earth mixture, (which rose to the surface as a scum and was skimmed off) or the ready crystallization of the salt. These additions vary in different works, and many of them seem to have been made from particular, and often ill-founded prejudices; and without any exact idea as to their probable effects.

As the brine began to boil the first crystals would be seen rising to the surface of the brine; there they floated for a few moments until they increased in density, when they would sink to the bottom of the pan. When the lumpman considered sufficient salt had been thrown out of the solution it was drawn to the side of the pan with long-handled rakes and ladled into tubs. The tubs were then drained to leave the salt in the solid form of bars.[4] The next process was known as 'happing' or squaring off, by means of a wooden pat or 'happer'.[5] The lumps were then wheeled into a hot-house at the end of the pan and placed on end for a week to drain. Finally they were lifted on to the flues, where they were allowed to dry for a further two weeks.[6] By repeating these operations three times a day and by working a six-day week, the lumper would produce about 35 tons of salt per week.[7]

The contract between the proprietor and the lumper was simply an arrangement whereby the owner of the pan, i.e. the proprietor, undertook to supply the saltmaker with brine and coal, also to keep the pan and furnaces in good repair and to purchase whatever quantity of salt was produced at the agreed price.[8] Having made the agreement, the proprietor, in effect, relinquished direct control over the process of production. He had no say in the type or number of people the 'lumper' chose to employ to assist him, nor any control over the day-by-day production or the division of labour. These things were the sole responsibility of the saltmaker, and if he chose to engage others as hired helps, or to work with members of his own family sharing the work with him, he became an employer, or 'small master' in his own right. This contractual arrangement, struck between the proprietor of the wych and the individual saltmaker, formed the original basis of employment in the salt industry.

Lumpmen worked to an agreed list of prices which varied, depending upon what type of salt they were making or upon how many lumps were required to the ton. For example, if 100 lumps were required to the ton they would be paid for at a higher rate than if, say, 160 or 180 lumps were required. This was because the heavier lumps required 'mundling' (i.e. a mundling stick or peg was used for panning the wet salt in the tubs to achieve the extra weight), and consequently took longer to make.

The lumpman was paid only for the salt produced. Everything, therefore, depended on the efficiency with which he fired the pan and the speed with which he and his assistants could get it cleared. If difficulties were encountered during the process of manufacture they would have a damaging effect on his earnings. A recurring obstacle was the formation, at the bottom and sides of the pan, of 'scale' – the sulphate deposits to which the fine-grained salt adhered – and as this impeded the pan's efficiency, each week it had to be taken off work, drained and 'picked' clean. The hardest scale formed at the front end of the pan immediately above the fire area, and when this happened there was a danger that the pan would become overheated and buckle, because the scale acted as an insulator. So the lumpman had to be alert for the glow of an overheated pan bottom. When such a glow was observed (referred to as 'seeing the moon') it was necessary, to prevent damage to the pan, to remove the scale by a process called 'dodging'. This usually consisted of chipping the scale away with a peculiar-shaped hammer, one end of which was tapered to a blunted spike. Like the rakes used to draw the salt to the sides of the pan, dodging hammers had a handle over one half the width of the pan in length, and must have been very unwieldy to use. Dodging could not begin until the 'moon' had disappeared, otherwise there was a danger that the hammer would penetrate the red-hot metal or that the sudden entry of brine under the scale would cause an explosion. It was therefore necessary for the fire doors to be opened and the fire damped down.[9] This of course reduced the yield of the pan and, in consequence, the lumpman's earnings. If the scale was too hard to be removed by the dodging hammer it would be necessary for the entire pan to be drained and the scale removed with a pickaxe. Since this involved the loss of an entire day's earnings, the lumpers would risk injury and even death in an attempt to remove the scale as quickly as possible. The Department Committee on Miscellaneous Dangerous Trades reported, in 1899, that members had observed this operation being undertaken when there was still some brine at 190°F. in the pan. The man who performed the duty had his feet in two buckets, in which they were strapped down.[10]

In between draughts the lumpman divided his time between tending his fires and raking the salt – fineness was obtained 'by keeping the brine not only boiling, but well agitated by raking'.[11] The preparation of each 'draught' of salt consumed about one ton of coal, which had to be shovelled into the fire-holes. In addition much of the lumpman's time was spent in 'ranging' – stirring the fires and declinkering the fire bars with a 9-foot long steel bar (his 'ranger') and removing and quenching the ashes. In this way the men spent their days and nights – 'working in a semi-nude state for a while, then clothing themselves in flannel and resting'.[12]

Because of the contractual nature of their employment, the

lumpmakers worked as they pleased, and regular hours, in the strict sense of factories or other trades, were unknown in the salt industry – the proprietors 'being quite indifferent to the hours, so long as the quantum of work was performed'.[13] Nevertheless, they were subject to the dictates of the process (to produce the maximum yield of salt of the correct crystal size required that the brine be kept at a vigorous boil between 'draws') and the exploitation of price. The employment of women and young children was said to be largely caused by the remuneration, for the maximum amount of salt the lumpmaker could produce in a week was 'too small . . . to be shared with another man'.[14] From this probably stemmed the practice of 'employing' the members of one's own family, which became 'generally the case'.[15] This practice, it was said, caused 'a great destruction of domestic comfort'.[16] Life for the whole family was regulated by the dictates of the process – the pan became the focal point of their lives. Always referred to as 'she' it was spoken of as if it were a living being – 'she' would 'chuckle' when the salt yield was good and be 'sick' or 'sour' when it was not – it was almost like one of the family. In reality the pan ruled their lives, demanding absolute loyalty – it could provide them with a comparatively good standard of living, or bring economic disaster and destitution, kill them or, very occasionally, bring them wealth.

Saltmakers engaged in the production of coarse salt were known as 'wallers'. Their work was even more arduous than that of lumpers (they claimed that they often had to shift 15 tons of salt at a draught), but in the hierarchy of wages they ranked lower because they did not have to work to such fine limits. They had no responsibility for firing the pan, a fireman being specifically employed for this task. The 'wallers', like the lumpers, were responsible for 'dodging' and 'picking' their pans but the scale, although harder than that in fine salt, took longer to form because of the lower temperature, and the pans required cleaning only every three or four weeks. The wallers used rakes in the same way as the lumpers, but they did not need to use them often, since the salt did not have to be so fine; their main job was digging the salt out, once it had been raked to the side, and humping it on to the racks to be dried.

In coarse salt manufacture the pans were much larger – some of them 140 feet long by 30 feet wide and 2 feet deep. The waller had to look after a number of them.[17] The brine was not boiled but kept at a steady temperature, and the salt took a long time to form. When sufficient pans were working to keep them fully employed the wallers – perhaps four men to a pan – 'drew' one each day. The salt was left to drain on the 'hurdles' (i.e. wooden platforms running along the floor on each side of the pan). There it would be left to drain and only be removed to the warehouse – and heaped into a huge mound – after the second pan had been drawn on the following day.

Wallers spent more of their time in the open air, loading and unloading, and their pans were much less covered in than those of the lumpers. This was because their pans did not have to come to boiling point, and their salt, when gathered, could be dried by a simple process of evaporation rather than by being heated in a stove house. The shed covering was mainly to stop the salt from getting wet.

In coarse salt manufacture the real saltmaker was – so far as preparation was concerned – the fireman. His was the responsibility for maintaining the brine at the correct temperature during the making period – work he did 'almost by rule of thumb'.[18] The price of coal accounted for over 50 per cent of production costs,[19] and the skill of the fireman – particularly when prices were low – often meant the difference between a profit and a loss. Nevertheless, the fires did not require the constant attention that lump pan fires required, and for a period of about four hours in the middle of the day and for a similar period during the night, they were left unattended. For this reason the firemen's hours were exceedingly irregular. The first shift started at two o'clock in the morning, and finished at about 10 a.m., the second shift, who also worked for eight hours, began work at 2 p.m.[20]

The wallers did not have the same power as the lumpers to increase their earnings by taking on extra hands. They worked not as individualists, contracting all the work of a pan, but in gangs.[21] They were paid by the day or 'drought'[22] (i.e. the day's draw of salt). They did not work to set hours, but they had specific tasks to perform in unloading slack coals and loading salt, which made them subject to the movements of the flatmen and the tides.

The most independent class of saltworkers were the watermen. Originally, opponents of the Weaver Navigation Act of 1721 warned that to render the river navigable would attract thousands of watermen who were all beggars and vagabonds, 'who, generally, were very lewd and ill-disposed people'.[23] Nevertheless, when they came (by 1817 an estimated 5,000 people[24] worked between 300–50 boats,[25] locally called flats, up and down the Navigation), they quickly secured the respect and affection of the saltmaking communities. For over two hundred years 'to work the boats' was an occupation of great prestige. In some cases the flats were individually owned, in others the flatmen were employed by the salt proprietors. But even where the flatman was nominally an employee he was, like the lumpman, a part contractor. He worked not for wages but on a 'shares' system of payment (i.e. the captain received a share of the freightage rate per ton out of which he paid himself and his crewmen).[26]

In the earlier part of the nineteenth century, working the flats was, like lumpmaking, a family occupation. Numbers of women and children lived and worked on the vessels – the wives and elder children assisting with the steering and setting through the locks and 'trimming', i.e. loading the boats.[27] It was not a comfortable existence. Flats were shallow-draft, broad-beamed vessels – originally of 50–80 tons;[28] they gradually increased in size to between 100–88 tons[29] – they were not as stoutly built as ships and therefore could not resist the damp so well.[30] Ideal for sailing the tranquil waters of the Weaver, the flats had a rough time in the treacherous waters of the Mersey estuary or in the Irish Sea when taking salt to Ireland or Scotland. Northwich, an inland town some thirty miles from the sea, knew the horror and heartbreak of shipwrecks when 'all hands were lost'.[31]

A by-law forbade the passage of boats on the Weaver on Sundays,[32] and therefore, generally on Saturday nights, the families would come ashore to spend a night or two in small houses or sheds on the river bank.[33] A visitor to Northwich in 1838, commenting upon the watermen, observed: 'They have a fancy for dressing up smart on Saturday evening, after they have cleaned their boats, when they stand sauntering away their time in idleness.'[34] However, for the most part, they would be astir and ready to sail by midnight on Sunday. Then it would, perhaps, be six days and nights with the entire family – sometimes said to be as many as eight in number – sleeping and eating in a cabin 'about 6 feet long, 5 feet wide, and 4½ feet high' before they came ashore again.[35] Adding to their discomfort was the operation of the Dock Act (1802) under which neither fire nor light was allowed on any vessel in the docks of Liverpool.[36] Sleeping aboard the damp flats was a nightmare – the cabins were cold and bedding became so damp that, sometimes, water could be wrung from it.[37] Risking a fine of £10 – which, at the discretion of the magistrate, was reducible to 10*s*., and 4*s*. 6*d*. costs – for as much as lighting their pipes, the men tied up in Liverpool faced a miserable ordeal. Some refused to use the port, others left their wives and children at home, and often they preferred to sail to Birkenhead, where fires were permitted, pay the 5*s*. docking fee and spend the night there, but most simply endured the misery. As one flatman explained: 'No place is pleasanter to a flatman than his own cabin, but when the fire is out his best friend is gone. It is all gloom and misery, and he has no longer any comfort in it.'[38]

As ever in the saltmaking communities the driving force which led men to accept the risks and discomforts – to exploit themselves and their families – was the desire to be totally independent of the salt proprietors. To own their own flat – or, at least, a share in one – was the driving ambition which made the men impervious to the endless toil, lack of comfort and general wretchedness of their everyday existence. Many achieved their goal and for a century a large body of independent watermen plied the Weaver, some of them, in time, becoming salt proprietors. Even those who failed generally became 'pretty well-to-do men'[39] who owned their own houses and commanded the highest-paid jobs in the salt districts.[40]

The freight charge of 3*s*. per ton was paid by the merchant and remained unchanged for many years. Out of it the Weaver Trustees extracted 1*s*. and the remainder was divided between the salt proprietor and the flatman. (The actual division of the charge was a matter of confrontation and bargaining between the watermen and the proprietors; it was a dispute on such a question which led to the formation of the Flatmen's Association in 1792.) By periodically demanding a bigger 'share' of the freightage charges the Weaver watermen had, by 1853, been able to secure 1*s*. 3*d*. per ton of cargo.[41] Given a flat of, say, 70 tons, the captain therefore stood to receive five guineas for each load of salt delivered to Liverpool. Out of this, if the flat was not worked by his family, he had to pay a 'hand' about 10*s*. or 11*s*. per week.[42] Also he had to pay the lads who loaded the salt – these were boys of about sixteen years of age who earned 1*s*. per day[43] – and pay them, in addition, the traditional bonus of 6*d*. for every 10 tons loaded. For the journey from Winsford or Northwich to Weston Point and back, the boat had to be hauled – from Weston Point to Northwich the charge for the round trip was 15*s*. for a horse and 5*s*. for a man[44] to assist, or 5*s*. each for from two to six 'haulers', depending on the state of wind and current[45] – and finally 'salt heavers' had to be paid to unload it. Added to this was the cost of delays, caused either through the state of the tide or weather, and the incidental expenses incurred at Liverpool, where because of the law the captain had to pay 'a penny a morning for a bottle of hot water for his breakfast, a penny for cooking his dinner at a baker's, and a penny for his hot water for tea in the evening'. All of these expenses fell upon the captain of the flat.[46] Therefore out of the five guineas he received for the trip the flatman only retained about twice as much as he paid his 'hand' – 22*s*. per week.[47] However, where members of the captain's family helped to work the flat – nine families out of every ten were said to do so[48] – the savings could be considerable. Working in this way the family could hope to save sufficient to buy a flat of their own, or a share in one. It was a mammoth task – in 1840 a flat of 80–100 tons cost about £800[49] – but a fair number succeeded. These independent watermen carried salt for proprietors who had no flats of their own or too few, or they carried for the merchant who purchased salt at the works and paid the Weaver dues. In busy times the watermen bought a cargo on their own account and took it to Liverpool as a speculation. As the freight of 3*s*. per ton was paid by the merchant as soon as the cargo was discharged, there was always ready money about.[50]

The Flatmen's Society, though called a Friendly Society, was very much concerned with regulating working conditions and securing for the watermen a fuller share of freight charges. We catch a glimpse of it at work in 1859 when one, James Southern, informed the salt proprietors that he had been instructed, presumably by a committee of watermen, 'to issue out (to the Northwich and Winsford captains of flats) another list of prices'.[51]

The independent watermen were detested by the large proprietors, who accused them of extortionate practices. In particular, H. E. Falk, the biggest exporter in mid-Victorian times, complained of 'a solid combination and most tyrannical rules' which enabled the watermen to keep up the cost of transport, and he determined to break their power. Accordingly, in 1863 he built and launched the 'Experiment' – the first iron steam barge to carry salt on the Weaver – and in the following year launched the 'Improvement'.[52] By doing so he changed the face of the industry and gradually power returned to the large proprietors.

The coming of steam barges altered the balance of power on the Weaver. They were much more expensive either to purchase or to maintain – only the merchants and the larger proprietors could afford to keep them up. Although some of the old sailing flats survived, steam had made possible the reduction of freightage charges and they were not economical. A train of one steamer and three 'dumb' barges could now take a load of 1,000 tons down the river.[53] As *Reynolds News* noted in 1888, 'where, for instance, the loading and carrying of 1,000 tons of salt used to occupy thirty-two men and eight horses, steam now allows the same work to be done by nine men only'.[54] The number of independent watermen decreased rapidly, though some took up employment on the steamers or dumb barges. At the same time there seems to have been a partial shift from a 'share' or 'tonnage' to a wages system for those who worked as captains on the salt proprietors' boats, though as late as 1891, when both are recorded in the Weaver watermen's agreement with the salt proprietors, both systems of payment seem to have been widespread.

The lumpman's scope as a small-scale labour contractor was restricted when women were forbidden to work between the hours of 6 p.m. and 6 a.m., and when younger children were withdrawn through the Education Act of 1870. Henceforth he was liable to be restricted to three draws a day, and helps were more likely to be employed as labourers, rather than coming from the family group. In some works, too, a part of the lumper's work was hived off and given to a new class of 'lofters', paid by the proprietors, who were given charge of the stoving part of the work.

III

There was no 'light' work in a saltworks; making salt demanded, for the most part, sheer physical strength. The work consisted of raking, shovelling, carrying and wheeling immense quantities of salt and coal. On the landing quays, too, everything had to be handled in bulk, with no mechanical hoists or chutes to ease the burden.

Working hours, though irregular, were incredibly long, and it was common for the day's work to begin in the middle of the night, either because of the state of the tides (in the case of the watermen), or because of loading and unloading, or (for the lumpmen) in order to increase the number of the day's droughts. In fine saltmaking, hours were determined by the state of the pan, 'not infrequently commencing at three in the morning'. 'Why do they begin so very early?' a Factory Inspector was asked in 1876. 'That I have never been able to find out. I take it that it is rather a habit which has grown upon them than anything else.' 'They are very odd people in fact? – Yes!'[1] According to Robert Baker, the Factory Inspector, it was quite normal, even at this time, for men never to see beds but on Saturday night ('They lie down in the saltworks from day to day'),[2] while earlier the same was said to be true of the whole family group: 'The men, women and children were kept in the pan shed the life-long week, getting their meals as they could within the walls, and sleeping in some cavity which they had managed to hollow out in the neighbourhood of the furnace.'[3]

In the middle years of the nineteenth century a good deal of labour was supplied by women and children. This was especially the case in fine saltmaking, where the lumper contracted for the work. 'In many instances', according to E. O. Meade King, the Factory Inspector who had charge of the district in 1876, 'women, young persons, and occasionally children are employed . . . in a proportion of 40 per cent of the whole number of persons employed.'[4]

Before the limitation of women's hours, under the Factories and Workshops Extension Act of 1867 – and still to some extent in later years, though illegally – the woman's day might stretch from the earliest hours of the morning to late at night, though with substantial intervals at home in between, while the 'droughts' were being prepared in the pan. Groups of weary women (some with babies wrapped in their shawls) and sleepy-eyed children made their way through the pitch darkness towards the wych houses. Many lived within a short distance of the saltworks, in cottages rented from the proprietor, but others had to walk up to two miles.[5] Often these others would have to approach the saltworks by way of the river bank,[6] where the boatmen and their families would be astir, preparing to catch the early morning tide.

The lumpman, their husband or employer, would already be at work, preparing for the first of the day's 'droughts' or boilings. The younger children would be left in the comparative safety of the warehouse, where they worked crushing the dried salt with hammers and axes prior to its being bagged by the older children;[7] while the older children would go with their mother to the pan side and begin their first task of the day, which was to place the tubs on the 'dogs' – two parallel steel bars, 2 feet by 1 foot, fitted as a shelf along the two sides of the pan – on which the tubs stood while being filled. Then, assisted by the lumpmen, they shovelled the salt into the tubs. Before the 'drawing' began the men reduced the fire (by pushing it well back in the fire holes), and so, until the salt was removed and fresh brine ran into the pan, no fresh salt was being formed. For this reason the accent was on speed; there could be no rest until the draw was finished and the pan 'going ahead' (i.e. making salt) again.

If the lumper worked a pan of average size he could produce about 1½ tons of salt at each 'drought'. Working at top speed, he, his wife and children or hired hands could shovel this amount of salt into the tubs in about an hour. (Because the salt was still saturated with brine when it came from the pan, it might be necessary to shovel 3 tons in weight for $1\frac{1}{2}$ tons of salt to be extracted.) If a heavier weight of salt were being made, the salt would have to be 'mundled' (pounded) as it was put into the tub, a 'mundling' stick being used to try to force it in.

After the tubs had drained for about half an hour the salt was dry enough to be handled.[8] Work now began on emptying the tubs and 'happing' the salt into shape (i.e. squaring off the lumps with wooden 'butter-pat' implements made from willow or beech, which, when not in use, were kept in water to prevent warping).[9]

Whilst the women were 'happing', the saltmaker was topping up the pan with fresh brine and tending his fires – raking the embers to the front of the fire holes and refuelling. Until the next 'drought' was ready his work was finished. Not so that of the women: 'happing' completed, each lump had to be carried to the stove and stood upright in the 'ditches' (channels between the raised rectangular flues running along the floor of the stove). Each lump weighed about 45 lbs. while the salt was still wet, and the women had about fifty to seventy of these to move in quick succession.[10] Only after the last lump had been carried to the stove could they take a rest, and then only for a short time.

As soon as the salt was safely in the stove the women would return home to prepare breakfast and begin their household tasks. In a little over two hours, working in almost total darkness, intense heat and strength-sapping humidity, each woman had shovelled, shaped and carried about a ton weight of wet salt. For doing so she earned about *2d.*,[11] unless she was working for her husband (to whom her labour would be worth rather more), and many women had to pay others something 'for looking after their children, washing etc.'[12] Each 'drought' was about three to four hours forming in the pan, and before noon the women would be making their way to the saltworks again. This time, and at the two later draws – afternoon and late evening – the women might have the additional task of raking the salt to the sides of the pan, though by 1876 it seems that this was less a woman's work in Cheshire than it was in the saltlands of Worcestershire.

Raking was potentially dangerous and probably the most unpleasant task in the works, because it had to be performed at high speed over the boiling brine. The worker stood at the side of an unguarded pan, whose rim seldom rose above knee height, and flung the long-handled rake towards the centre. At most works 'the workmen considered it essential that there should be a space (called the 'toe-room') between the floor . . . and the bottom of the pan,'[13] and they used this toe-room to keep their balance, pressing their legs against the side of the pan. Where there was more than one pan to a shed the space between each pan was only 3 or 4 feet, which made raking still more dangerous. As a witness told the Factory and Workshop Commissioners in 1876:[14]

> The margin of ground or rather of coagulated salt upon which they stand is at times very narrow, and there is nothing whatever to prevent them in the effort of 'drawing' from tumbling into the scalding water.

Some did, and even if rescued at once were not likely to survive. Anyone who fell in unobserved would probably, like William Coppack in August 1888, be found dead in the pan next morning.[15] The danger was perhaps greatest for the less experienced, like the seventeen-year-old labourer, Michael Carr, who, 'when any of the men were off work . . . worked as a salt-boiler', but fell into the pan and lost his life;[16] or for the old and tired. The saltmakers were not sufficiently anxious about it, however, to accept safety proposals which would make their work more arduous and slower (thus reducing their earnings); when the Departmental Committee on Miscellaneous Dangerous Trades seemed likely to recommend that salt pans should be fitted with a guard-rail or made with higher rims, the saltmakers protested vigorously.[17]

At some works, especially if brine were scarce, the hurdles, instead of being at floor level, were raised to a height sufficient to allow excess brine to drain from the salt and return to the pan through strategically placed pipes.[18] Understandably, these so-called 'high hurdles' were not popular, because they rendered slower the task of shovelling the salt from the pan to the hurdle. They also had guard-rails fitted to protect workers who had to walk along the high hurdles whilst wheeling away the salt, because there had been 'a considerable number of accidents from falling off high hurdles, some of which proved fatal', and these remained unchanged in spite of the saltmakers' plea that they should be of a removable type so as not to interfere with the walling.[19]

Accidents which appeared to have happened with depressing frequency involved 'dodging', warehousing and loading. Because the hurdles did not extend beyond the right angle formed by the corners of the pan, accidents occurred during dodging, when workers were 'obliged, in some cases, to assist themselves across the intervening space by the dangerous practice of holding on to the edge of the pan and swinging themselves around'.[20] 'Lofters', who worked the stove-houses (i.e. the hot-houses where the lumps were dried) and who derived their names from that part of their work which involved passing the dried lump salt through a trap-door in the stove-house roof into the 'loft' or warehouse, were often injured when the bottom lumps in a stack crumbled and the stack fell.[21] Finally, once again illustrating the detrimental effects of the system of payment upon the workers' safety consciousness, loaders, who tipped the salt from hand-carts into railway wagons or boats, often worked at such a pace, running with their carts to make up time, that they plunged over the edge of the off-loading platform. In this way numerous loaders, falling 8 to 10 feet into a railway wagon or about 16 feet into the hold of a vessel, suffered terrible injuries.[22]

One reason why the saltworkers worked so hard was that they lost a great deal of time. The salt trade was seasonal, with a period of very dull trade from Christmas to Easter. It was the custom in the salt towns that as soon as the warehouses were full the pans were stopped – except for a very small fire to keep the stoves aired and thus prevent the salt spoiling – and the proprietors declared a 'holiday'.[23] By working flat out during the rest of the year the saltworkers hoped to be able to put enough by to see them through the winter-spring period of unemployment. To ensure a little money at Christmas some works had a 'benefit club' run by the workers themselves: having made small payments to workers off sick, they 'divided every Christmas anything they had and started again.[24]

Another reason why men were often laid off was because conditions in the industry were very unstable. Prices fluctuated wildly (in 1856, for instance, the price of salt at the works varied in the course of the year from 5*s*. 6*d*. to 8*s*. per ton),[25] and when they slumped, proprietors stopped some or all of their pans from working. Bouts of frenzied activity – as in November 1874 when trade 'was unprecedentedly busy, owing to the East Indian shipments having been so very large'[26] – alternated with years of dull trade.

The industry was often working at less than capacity. In 1874, for instance, according to the trade report at the end of the year, the tendency of prices was downwards, 'owing to fuel being more easily obtainable and at reduced prices'.[27] During the year the number of pans stopped 'varied from one quarter to a half'. This was not an exceptional position, as Table 3.2 indicates.[28]

The proprietors needed to have spare pans in order to capture a large share of the extra trade when times were good. Coupled with this was the ease with which newcomers could enter the industry. The proletarian makers, particularly those who employed their families, could usually afford to sell their salt at prices well below what the large proprietors considered remunerative. Although output was on a tremendous scale and the trend was for the market to expand, there was an inherent instability which caused sales to fluctuate wildly. Therefore each new pan laid down was a potential threat to someone else's

TABLE 3.2

Number of salt companies and number of pans in use or idle: 16 March 1876 and 12 April 1879

						No. of pans						*No. of pans idle*	
	No. of	*Total*		*Stoved*		*Butter*		*Common*		*Coarse*			
Town	*firms*	*1876*	*1879*	*1876*	*1879*	*1876*	*1879*	*1876*	*1879*	*1876*	*1879*	*1876*	*1879*
Winsford	35	612	607	96	99	108	127	187	223	55	3	166	155
Northwich	28	457	466	52	54	33	28	143	153	89	77	140	155
Middlewich	4	12	13	8	6	—	—	2	3	—	—	4	4
Sandbach	2	67	68	16	14	3	—	24	33	4	4	20	27
Total	69	1,148	1,154	172	173	144	155	356	412	148	84	330	341

livelihood. As over-production caused prices to fall, each proprietor had to struggle to maintain his share of the market, and prices spiralled downwards. At a certain point many of the small proprietors would be forced out of business and in the big proprietors' yards many pans would be idle.

At this point the surviving proprietors usually attempted to come to some arrangement to restrict output – various attempts will be discussed in detail later – and thus create an artificial scarcity so that prices could be increased. However, no sooner had they restored stability to the trade, when, in consequence of better prices, small men came flooding in once more. Forcing their way into the market by underselling the big proprietors – as in April 1874, 'when orders became more numerous, non-members of the Salt Traders' Association secured many orders by taking lower prices'[29] – the small men started the cycle once again.

IV

Despite the individualistic basis of employment on the saltworks there was a strong tradition of combination. The earliest to organize themselves were the flatmen, who formed their first trade union – the Northwich Flatmen's Friendly Society – in 1792.[1] (It was still going strong in 1875, when a local newspaper reporting the annual dinner and procession recorded the membership as 451.)[2] Shortly after the founding of the flatmen's society, day labourers on the Weaver demanded (6 August 1795) increases 'in consequence of the present high price of provisions'. They were granted a special allowance of 1*s.* per week, which the Weaver Trustees withdrew on 1 October but were forced to reintroduce on 12 November; this was apparently maintained at least until the end of the century. The success of the labourers brought about increases for other groups of workers on the Weaver. A request from five carpenters and joiners employed at the Navigation timber-yard for an increase from 11*s.* to 12*s.* was 'considered reasonable' and granted. The workmen, probably surprised at the ease with which they achieved their increase, reviewed their claim and within four weeks had secured a wage of 15*s.* per week. Ann Latham, blacksmith, was, on 11 May 1796, 'on account of the advanced price of corn and workmen's wages', awarded 5*d.* a lb instead of 4*d.* for new iron work and 2½*d.* a lb instead of 2*d.* for 'working up old iron'. On 2 June 1796 stonemason Nicholas Bower was granted a rise from 2½*d.* to 3*d.* 'for all stone dressed and set in the foundation of any new lock', and 4¼*d.* instead of 3¾*d.* for 'all hand dressed stone set in such locks'.[3] In December 1796 lock-keepers had their wages increased by 1*s.* a week. For keeping Acton Bridge Lock Elizabeth Gerrard had her wages increased, in February 1797, to 12*s.* per week, but this was stated to be due to the amount of traffic through the lock rather than the general increase in prices.[4]

The first recorded union of the saltmakers was the Brine Boilers Association, formed at Winsford in 1845.[5] (This association went under a variety of names, and in the 1880s was known as the Winsford Saltmakers' Association.) At first this association bore many resemblances to the local trade clubs of an earlier age. The central figure of the organization was an annually elected president, whose duties included acting as chairman at the monthly 'club night' meetings and at the half-yearly general meetings. A representative committee was elected at half-yearly intervals, and seems for the first few years to have been composed of members selected on more or less a rota basis without regard for the members' place of work. Its function was to consider all matters relating to the trade. However, for the first few years its role appears to have been purely formal and almost all important decisions were made by those members present at the monthly 'club-night' meetings.

These were, for many years, held at the Golden Lion Hotel, Winsford, where the landlady, Betty Williams, acted as custodian of the association's funds and records. Apparently slow to prosper, the association had on 26 March 1853 only £8 18*s.* 8*d.*, out of which Betty Williams was instructed to pay £2 10*s.* to five members who were out of work.[6] On 'club-nights' members paid their contribution of 1*s.*, failure to do so leading to a fine of 6*d.*[7]

Whilst in times of peace the organization could be governed by autonomous general meetings, conflict demanded an entirely different constitution. The Winsford saltmakers, metaphorically speaking, placed their constitution on a war footing in 1853. A general meeting, on 2 May 1853, resolved that 'the Committee consist of a qualified member from each Bank (i.e. saltworks), and shall be competent to act for all' and that 'the Committee sit in a private room, and no New Member to be admitted without their sanction'.[8] Henceforth, the meetings of the committee, unless they themselves decided otherwise, were to be closed to the ordinary members and to ensure secrecy it was resolved 'that any Member of the Committee, divulging any business or proceedings that has been transacted shall be expelled and shall not be eligible to re-enter until he has passed the Committee and paid an entrance fee of 20/-'. As there would be little point in the average member attending on 'club-nights', stewards were appointed to 'pay and receive all monies'.[9]

The 1850s saw a general revival of trade unionism, and under its new constitution the Winsford Saltmakers' Association prospered. By May 1857 funds were sufficient for the committee to order: 'That money belonging to this Society be deposited in the Over and Winsford Savings Bank, by the President and Treasurer accompanied by two Members appointed by the Committee and that all papers respecting these monies shall be signed by the Secretaries.'[10] In 1858 the union was successful in securing for labourers 'a small rise', which brought their wages to 10*s.* per week,[11] but their first major conflict with the employers was not to occur until 1868.

The union was not a craft union in the traditional sense. It paid few 'friendly benefits', it was not the preserve of an 'aristocracy of labour' and, according to its president, in the first forty years of its existence it fought and won 'a good many strikes'.[12] Although women were, apparently, excluded, membership extended to every class of male saltmaker – lumpers, wallers and labourers – and by 1878 it included in its roll of members nearly every saltworker in Winsford and Middlewich.[13] It also recruited watermen and craft workers from the saltworks and shipyards, and would even admit workers not connected with the industry. Some, like the railway workers, were actively encouraged to join, and their entrance fee was set at *2s. 9d.* Others were greeted with less enthusiasm, 'Farming Labourers', for example having to pay two guineas.[14]

Many of the saltworks craft workers belonged to their appropriate craft union, the pansmiths for example being members of the Boilermakers' and Iron Shipbuilders' Association, a branch of which had been established in Northwich in 1848.[15] In cases of dispute those who belonged to the Saltmakers' Association were invariably instructed to 'go with the majority' and informed that if strike action was forthcoming they would 'be supported by the Society'. On one such occasion in May 1889, it was recorded that: 'Three Members of the Bricksetters attended about one of our members doing work at night that one of their Members was on strike from.'[16] However, at the next meeting, after the offending bricklayer Amos Taylor had appeared before them to offer his explanations, the Committee 'believing he had been misled by an employer's son swept the table of it, asking him in the future to be more careful'.[17]

Once the union was considerably embarrassed, when James Woodier of Runcorn was called out by the Dockers' Union. He had failed to ask the committee for permission to strike and they refused his strike pay.[18] Two weeks later a red-faced committee instructed the secretary to write 'to the Secretary of the Dockers' Union thanking that Society for their courtesy to our Member (James Woodier) and asking them if they will allow us to refund the money that they have paid him . . . for the week's strike pay'.[19]

Understandably, because the lumpmen were employers as well as workers, occasional differences arose between them and their helpers. Such differences threatened the solidarity of the Saltmakers' Association and naturally warranted particular attention. A case of this kind occurred in January 1900. One Peter Boyle came to the committee with a complaint about his treatment by J. Burrows, the lumpman who was employing him at 'happing' (shaping the lumps). Burrows was also present, and said he thought he was 'using him fairly as the other happers were used and paying him equal to what the others were doing'. The committee took the matter seriously enough to send a deputation, consisting of the secretary and a member of the committee, to visit the works and determine whether the complaint was justified.[20]

Most of the strikes involving saltworkers – to judge by the 1890s when records are more complete – were confined to small groups of men: two, three, four or five – at a particular place or works. Sometimes these strikes went on for a fairly long time; and that no-one would blackleg in an area of such high unemployment is testament to the strength of the unions and to the solidarity of the saltmaking Communities.[21]

The Winsford union was particularly militant in fighting dismissals, and where they failed to get immediate reinstatement they declared the member(s) either 'locked-out' or, 'on strike'. This ensured that the aggrieved member could be supported on the funds, and also that no other worker would take his place. In one such case, in 1897 when three 'baggers' were dismissed for allegedly 'sending out bags which were found short of weight', the dispute dragged on from 29 July until 25 September, when, under threat that 'the rest of the baggers would be stopped', the salt proprietors gave way and the men were put back to their places.[22]

The leaders of the Winsford Saltmakers' Union had a tremendous task in holding their union together in the face of persistent unemployment. In an interview with a correspondent of the *Manchester Umpire* in 1889, its president, John Newall, explained that the union had been formed solely to prevent its members being overworked and not as a benefit society.[23] The rules did not provide for sickness or unemployment benefit, but at times the saltmakers' committee were moved to transcend their own rules. One such occasion occurred in March 1878, when half the pans were still idle after the Christmas 'holiday'. Under the headline 'Scarcity of Work at Winsford' the *Winsford and Middlewich Guardian* reported:[24]

> there are many families in the place who would be in a sorry plight but for the good work that is being done by a purely local agency, the Winsford Saltmakers' Association. The primary object of the association is to sustain its members in periods of strike; but fortunately its assistance for this purpose has not been needed for many years, and as the funds have been in a healthy state, a portion of them has been applied more than once to the relief of members absolutely destitute through depression of trade and scarcity of employment. Thus at the present time the Association is granting relief to about 100 persons . . . 5/– a week being given to each man and his wife, with an extra shilling for each child. (This takes the form of tickets to exchange at food shops . . . School Board remitting fees. . . .) The attendance officer reports that in many cases children are kept from school through being without shoes or clogs to turn out in; and in others the

parents having no money to buy fuel with send their children to pick cinders from among the refuse at the salt works . . .

Evidence of direct interventions by the union to limit the amount of unemployment, by pressurizing the proprietors, is difficult to come by, because entries in the first of the only two surviving minute books (1853–1900) are very sparse before 1889 (for the most part simply recording the names of those who served on the committee); but one rare entry, in 1861, shows that pressure was being put on proprietors to keep the men on during the dull season:[25]

Resolved: 'That the Society give to Mr. George Deakin a promise that if, he gives his Men one pound per week throughout the winter and if his men shall turn upon him next Summer for more than others are giving this Society will not support them.'

The records for the 1890s are much more informative and show that the union was prepared to challenge the proprietors whenever unemployment (euphemistically referred to as 'holidays') threatened. Under the standing orders of the Winsford Saltmakers' Association at this time, the president was 'empowered to call the Committee together when it becomes known that the Saltmen are to have holidays',[26] and almost as a ritual the secretary was instructed to write to the employers asking 'as a favour' if they would 'give as little holidays as possible'.[27] Whenever these appeals went unheeded employers were reminded that it would be to their advantage to 'see if the pans can in some way be left at work as it would be difficult to get the men to load (salt) if there is no pans working'.[28] Occasionally, especially when they considered the 'holiday' overlong, the committee instructed members to refuse to load further salt until the pans were re-started. This tactic appears to have been particularly successful during periods of deliberately restricted output, when, it can be assumed, prices were usually good. One such occasion occurred in March 1899 at the Little Meadow Bank where 'the men were only making one day and a half per week for the last 4 months'.[29] The intervention of the committee resulted in an immediate return to full-time working, the manager, Mr H. Stubbs, explaining that the stoppage of pans had been a mistake caused 'through his absence from illness'.[30]

Another way in which the union tried to limit the effects of unemployment was by preventing members poaching on other's places, limiting competition between them and trying to ensure that in slack periods work was shared rather than men being stood off. An entry in the minute book for 1853 designed to protect members' places says that:[31]

> Any Member going on a Bank (i.e. Saltwork) to apply for work knowing any of the Members have been turned away for being in the Union or anything prejudicial to the same interests of the same shall be expelled, but shall be eligible to re-enter by paying an entrance fee of 20/-.

In order to preserve the solidarity of the work groups, and avoid a disrupting scramble for jobs to be followed by an inevitable reduction in wage rates, it was the custom, whenever the proprietors did stop a proportion of their pans, for the men to share whatever work was available. Once again the scantiness of the early minutes precludes detailed examples of work sharing before 1890, but it was certainly already regarded as a custom of the trade in 1889 when the salt proprietors (as part of an attempted rationalization of the trade which is dealt with later) put it into question. At the heart of this confrontation, which spread over the entire saltfield and involved the right of unemployed wallers to have a place, were the demands that 'all the works shall have a fair share of the work' and that 'every man out of work shall have a fair share of work'.[32] (It is significant that the saltmakers of Northwich, whose union was apparently disbanded in the 1850s, and who were in the process of forming a new one in 1889, combined with the Winsford men to fight in defence of the custom of work sharing).[33] In the fuller minutes from 1890 onwards references to sharing are very frequent. Typical was the curt reminder to the manager of the Weston Point Salt Works, in 1900, that it was 'not customary to engage men to pans and then turn them adrift when the pan stops', accompanied by a demand that the men concerned 'should have a share with the other men'.[34] The degree to which the saltmakers would take their work sharing is typified by this case of the Newbridge wallers from the minutes of the Northwich union for 1908. Fourteen wallers who were employed at Newbridge Salt Works, Winsford 'for some time had only been "drawing" 6 pans making "14 days salt"' (i.e. fisheries salt two weeks forming in the pan), for which they were paid £6 per fortnight. Rather than see any of their number completely unemployed, the fourteen wallers shared, as equally as possible, the total amount of wages between them, so that twelve received 4*s*. 3½*d*. per week and two received 4*s*. 3*d*.[35]

V

Ownership in the Cheshire saltlands was divided between a small number of large proprietors and a much larger – though fluctuating – number of small ones. No sooner did trade enter a favourable phase than the numbers of proletarian proprietors increased. Many of these were lumpmen who had managed to accumulate sufficient capital to be able to rent a pan from a proprietor. Little capital was required to do this and it was claimed that the small firms were 'worked almost exclusively by families on a sort of patriarchal system'.[1] The ambition of any careful and pushing saltmaker was to rent not only a pan but an entire wych-house, to purchase his own brine and fuel and to be responsible for the sale of his own salt. To take such a step constituted a considerable risk – an entire lifetime's savings might be lost – at best it meant years of struggle to survive the onslaught of the giant concerns. Usually these small proprietors began by renting a wych that contained a single pan and working it themselves.[2] (This was possible, since many of the large firms had their pans 'distributed over the town' rather than in one large works.)[3] In time they might put down more pans, bringing in more members of the family to help them – one such works was said to provide employment for forty-three relatives of the proprietor[4] – until they resembled small collectives or co-operative concerns.[5] Throughout much of the nineteenth century this type of firm proved tenacious – some of them grew into very big concerns owning steam barges, flats, shipyards and waggon-building shops. By the 1870s it was being lamented that the salt trade did not hold out any strong inducements for capitalists to enter, being 'somewhat precarious and easily affected by external circumstances' – preventing companies paying 'dividends of an attractive character'.[6] Although there were large proprietors, the majority of pans were in the hands of proletarian saltmakers. When Ludwig Mond, the founder of ICI, moved to Winnington in 1873, his mother-in-law wrote with dismay of her daughter's social isolation: 'The salt manufacturers are not like us. They are not educated or sociable people. Like most people in England, they are artisans who have worked their way up . . .'[7]

There was a similar situation in the Worcestershire saltfield, Droitwich. A writer in 1865 blamed it for the backward state of the trade:[8]

> the number of small masters in the trade renders any effective combination among the manufacturers as difficult as the introduction of any substantial reform in the habits and conditions of the workpeople. A large portion of the trade in the aggregate at Droitwich is in the hands of such masters, who, not possessed of sufficient capital to undertake the manufacture on a scale large enough to permit the adoption of expensive improvements in machinery, or in the system of working and payment of wages, are still able, by employing their own family as workpeople, to produce salt at a rate so low as to render it difficult for those able to initiate the necessary reforms to realise an adequate remuneration from them. On the whole, therefore, although the experiment tried at Stoke, has we believe, been commercially as well as socially successful, and must in the long run materially affect the condition of the saltworkers, there is but little immediate prospect of their making any great general advance in the social advance. For the present the public is able to purchase salt at a price possibly lower by some infinitesimal fraction than it would be if the manufacture was entirely in the hands of large proprietors and the employment of women and children strictly prohibited; but the advantage is dearly purchased by the continued existence amongst us of a class of labourers which ranks below that even of the working colliers.

Unfortunately, we have no way of knowing how many salt manufacturers entered and left the trade during the period under consideration. The alternative is to consider the number of people who shipped salt down the Weaver. However, it must be remembered that while some shippers were not manufacturers in their own right, some manufacturers sold their salt at the works and thus do not appear on the lists of shippers. Nevertheless, we are able to obtain some idea of the numbers of people coming into the trade, and we can see the extremely high rate of turnover. It would appear that taking advantage of the periods when prices were relatively high, and prospects for trade good, new manufacturers flocked into the industry, but when prices fell many were forced out again.

We can see from Table 3.3 that of those trading out of Winsford more than 50 per cent were unable to sustain their position in the industry for longer than five years. Likewise, when we consider the number of people trading out of Northwich we find that out of a total of 407 firms which entered the trade between 1734 and 1894, 205 traded for five years or less and only sixty-six remained in the industry for twenty years or more.[9]

Calvert suggests that typical of those who entered the industry after 1840 would be the independent owner of a flat who had accumulated enough money to start a pan. If he was a pushing man, he would take a little set of works of two or three pans and make enough salt to keep his flat working steadily. As trade grew he put down more pans and got

TABLE 3.3

Number of firms trading out of Winsford, 1766–1897[10]

5 years and under	158
5–10 years	64
10–15 years	31
15–20 years	12
20–25 years	14
25–30 years	8
30–35 years	3
35–40 years	5
Over 40 years	2
	297

more flats. Two or three often combined in both works and flats, and the salt brokers of Liverpool sold their salt for them and became practically their bankers, while the more successful amongst them opened offices of their own.[11] By doing this they naturally aroused the ire of the larger firms. Herman J. Falk complained bitterly: 'In one firm . . . there are no less than 43 members of the proprietor's families employed in the works themselves, so that they are, as it were, paying themselves in paying their wages. It is almost a gigantic co-operative concern.'[12]

Two families of watermen who began in this way and became very successful salt proprietors were the Starkeys and the Deakins. In 1840 the three Deakin brothers were working a flat on the Weaver. Seven years later they established the firm of Deakin Brothers, which, for thirty-eight years, manufactured salt in Northwich. Likewise, George, Thomas and Joseph Starkey, who also owned a flat, established the firm of Starkey Brothers in 1850 and traded out of Northwich for twenty-five years.[13]

Small proprietors also rose from the ranks of the salt boilers. Encouraged by the system of contractual employment, they worked themselves and their families hard in the hope of accumulating sufficient savings to be able to rent a pan. Presumably they would start by renting a dilapidated pan shed – containing a single pan – one which had been standing idle, perhaps, during a period of trade restriction. It would be a tremendous gamble – the whole family's savings accumulated over years of unremitting toil would be at risk. To stay in business the family would have to work harder than ever before – their only hope being to sell their salt more cheaply than others were doing, in the hope of being able to make a steady income. Illness would ruin them and they would have to return to being lumpmen and start again.

The independent waterman who ventured into saltmaking was in a far less precarious position than the salt boiler trying to establish his own business. At least the waterman could earn money by working his flat even if his saltworks was doing badly. Presumably some of the independent watermen – or anyone else with a little money who wished to speculate in salt – first began accumulating capital by buying salt cheaply from these lumpmen proprietors – who always sold their salt at the works – selling it at a higher price to the Liverpool salt brokers. Provided the waterman could keep his flat fully employed it would seem that it mattered little if at first his pan made little profit, because the merchants still paid him for freightage. The lumpman who failed might find himself working for the man who had been buying his salt.

Partly because of the number of small proprietors, partly because of the ease with which output could be increased, saltmaking in Cheshire was haunted by the spectre of over-production.

The answer of the big proprietors was to form an 'association' or ring, restricting production where prices fell, and tailoring output to what the market would take. The most successful attempt was during the years from 1805 to 1825, when they were helped by the salt duties, which made it difficult for the small proprietor to survive.[14]

Between 1818 and 1825 the excise duty was removed and the saltmakers' association began to break up. Relieved of the necessity of carrying a large capital reserve for duties, small proprietors were entering the trade. For a time the society tried to fight the newcomers by reducing its prices, but on 5 January 1825 it collapsed.[15] During 1824, the last year when the manufacturers' association had control over output, 162,365 tons of white salt were shipped down the Weaver; ten years later shipments totalled 376,000 tons, an increase of 132 per cent. This huge increase in trade was accompanied by very unsteady prices, and in July 1833 a general meeting of white salt proprietors, convened at Northwich, decided upon the formation of a new society, to try to hold them steady: 'With slight changes the 1813 rules were readopted and handsome tributes were paid to Thomas and John Marshall, both recently deceased.'[16] This society seems to have continued in existence until about 1841, when a new association was formed.[17] Henry Waterman and John Cheshire, two members of the committee, were appointed to purchase all the salt manufactured by the trade at certain fixed prices.[18] Makers who exceeded their allotted quotas were to pay a fine of 5*s*. per ton to the funds of the association.[19]

Despite the efforts of the society, the price of salt continued on its erratic course. At times, in the 1840s, it rose to 12*s*. and at others it was as low as 5*s*. 3*d*.[20] Even in 1830, when the society managed to recruit all the salt proprietors of Northwich and Winsford, things continued much the same as before.[21] Apparently some manufacturers remained aloof, and frequently the committee were unable to maintain the agreed prices or even to restrict output. At various times the society appeared on the point of breaking up, only to pull itself together for a short time.

The reason for these wild fluctuations in prices was partly the ease with which production could be increased, partly the low capital cost of entering the trade, and partly the attempt, by restriction, to attain the unrealistically high prices of 1805–25. Whenever the majority of manufacturers, by agreement, managed, for a short time, to maintain falsely inflated prices, newcomers came into the trade to share in the bonanza and, as a result, prices were soon depressed. The struggles which ensued between the established proprietors and the newcomers caused prices to fall well below what would have been their normal level, and the gains from the short-lived price agreements were thus offset by the losses incurred as a result of price cutting. Each time the trade entered a period of stability the idea of a society to regulate output and inflate prices again seemed feasible. But the restrictions in fact created conditions which enticed fresh capital into the industry and so the societies sowed the seeds of their own destruction.

A renewed attempt to bring the larger proprietors together was made in the 1850s by Hermann Eugen Falk, a German by birth, who had first taken an interest in the salt industry in 1842, acquired his own works in 1846 and rapidly became a very influential figure in the industry.[22] On 30 August 1858, he convened, at Northwich, 'a numerous meeting of [salt] proprietors', who formally installed the Salt Chamber of

Commerce, 'its fundamental principles being the formation of an efficient representative body for the extension, general advancement, and protection of the trade'.[23] The smaller proprietors, including the watermen salt manufacturers, were deliberately excluded. The Chamber disclaimed any intention of interfering with prices,[24] but by 1865 it was openly admitted that members had 'wisely considered . . . the all-important subject of supply and demand', and that the Chamber would exercise its influence in asking for a remunerative scale of prices.[25] A meeting held on 16 January 1866, 'Resolved that the trade from time to time shall fix whom they will recognize as their brokers, and that the Salt Trade Committee be a Court of Appeal, to decide questions in dispute which the brokers themselves may refer; that every party wishing to become a broker must be proposed by a consultant proprietor . . . that the prices last recommended by the Salt Trade Committee namely 8*s*. per ton for common, 9*s*. 6*d*. for butter (salt), and 11*s*. for stoved, be strictly adhered to . . . that one-half the common pannage, except that on the banks of the canal, shall stop on Monday next.'[26]

During its early years the Chamber seems to have had little success in maintaining prices; in 1865, for example, the average price of Common was only 6*s*. per ton.[27] But they did manage to open new foreign markets; and their sustained agitation pressured the British Government into forcing the Government of India in 1863 to agree to abolish the State manufacture of salt in the Bengal Presidency.[28] The new markets in the East were instrumental in saving the industry from heavy losses when the American Civil War affected prices to such a degree that 'some lots' had to be sold for as little as '3*s*. 6*d*. per ton . . . the lowest prices ever known'.[29]

By the middle of 1870, in the face of falling prices and a consequent return to overproduction, the Salt Chamber was in disarray, but a newly formed Trade Committee, working closely with the Liverpool merchants, was able to steady prices, and the salt industry entered a short but highly profitable phase. Falk noted: 'the Salt Trade Committee had the extreme satisfaction of building the prices of salt up to figures not known by this generation. From 8*s*. to 10*s*. and 12*s*. to 15*s*., the prices of common salt were officially raised to 20*s*. per ton . . . , and 27*s*. 6*d*., per ton for Calcutta salt.'[30] Output grew to unprecedented proportions and for a while prices were maintained; but during 1876 over-production caused prices to plunge to as little as 5*s*. per ton and, as so often before, manufacturers responded by producing even more. Not once, during the remainder of the 1870s, did prices exceed 7*s*. per ton, and for a time they were down to 4*s*.[31] Falk had no hesitation in blaming the collapse upon the proletarian element amongst the manufacturers, insisting that they had increased production 'quite beyond the necessities of the demand', and that although established traders were not 'carried away by the spirit of extension', over-production had again led the way to very disastrous results.[32]

A new association formed in 1877 managed to raise the price of Common to 7*s*. per ton and maintain it there until late 1878, when it fell again to 5*s*. 6*d*.[33] Once again the major cause was over-production, induced by manufacturers attempting to maintain falling profit margins by increased turnover. It is estimated that during 1878 Cheshire manufacturers produced 2,055,000 tons of white salt. 'Winsford and District, 1,036,000 tons; Northwich and District, 880,000 tons; Middlewich and District, 21,000 tons; and the newly developed Sandbach District, 118,000 tons.'[34] Yet another association was formed in March 1881, but demand was already falling and no impression could be made on prices which in 1882 were down to 4*s*. or 4*s*. 6*d*.[35]

When the boom was at its peak in 1872, a group of Manchester capitalists had proposed that the salt proprietors form a limited company and secure 'all existing salt works . . . all salt lands in the neighbourhood of the Weaver and the railways', thus creating a virtual monopoly.[36] Then, trade had been so good that no-one wanted to listen: in 1884, when H. E. Falk again raised the idea, trade was bad, and the clamour of recriminations and bitterness was too loud for anyone to hear him.[37] In January 1888 he was more successful.[38]

> The salt trade [he told his erstwhile colleagues] is at the most deadly crisis. Implacable competition among a small section of the largest makers has brought prices below all records, salt being freely offered at 50 per cent below cost. All the large chemical contracts for 1888 have been taken at ruinous prices. Nor has there been any more extensive demand for the article below cost. The total export . . . proves a considerable decrease on average. The principle of association has been violated again, and with more disastrous results than ever yet known. Nothing but a new form of general consolidation can resuscitate the trade.

Having secured the conflict-weary proprietors' attention, Falk pressed home the attack. Various interviews and deliberations prepared the way for a meeting of salt proprietors, traders and manufacturers to consider proposals, similar to those made by the Manchester group ten years earlier, put forward by a group of London solicitors who had the backing of large finance houses. On Thursday, 5 July 1888, at the Adelphi Hotel, Liverpool, the saltmakers approved a momentous resolution: 'that each proprietor and manufacturer send to Messrs. Fowler & Co., within one week from this date, the price which he binds himself to accept for his works, including land and leases, buildings, railway wagons, steamers, barges, flats, boats and all other effects, goodwill and business, specifying the sum in a Schedule, and that Messrs. Fowler and Co., appoint a valuer on the part of the purchaser to check with the vendor'.[39] Consequently, on 6 October 1888, the Salt Union Ltd was registered as a company with a capital of £3,000,000. Having acquired its purchases with two-thirds cash and one-third shares, the Salt Union comprised sixty-four firms, including all those in Cheshire, 'all the really proved brine lands of Cheshire and Worcestershire, as well as the Irish salt lands, mines and works, together with the most important works at Middlesbrough'.[40] The new company controlled 90 per cent of the industry's output, 2,000,000 tons per year.[41]

For sixty years after the removal of the salt taxes, repeated attempts by the salt proprietors to maintain high prices by curtailing supply had failed because numerous 'small fry' were attracted into the trade and maintained their position by underselling. The Salt Union of 1888 was no more successful in bringing order to the trade than its predecessors had been, and soon found itself faced with undercutting. As usual, this led to divisions in the ranks of the 'old' proprietors; and in the case of the Salt Union, which had brought them together in the form of a limited company, this proved disastrous.

In the new company pricing policy had been entrusted to the Cheshire proprietors. These men – who had never advanced beyond the notion of creating scarcity by restricting output – managed, in the first year, to double salt prices and produced a net profit of £320,388.[42] The high prices inevitably encouraged new competitors to enter the market: only in Worcestershire, where they had a virtual monopoly of the salt-bearing land, were the Salt Union able to prevent others manufacturing salt.[43] In Cheshire, although they contested bitterly the right of anyone to set up in competition against them, the saltfield was too vast for the Salt Union to control it all. Henry Seddon of Middlewich, for instance, succeeded in establishing a small works despite the 'old' proprietors' attempts to destroy him (on one occasion they sent a group of men to smash his brine pipes, 'but Messrs. Seddon's men met and dispersed the intruders'), and eventually grew into the town's biggest proprietor.[44]

Although the Cheshire directors were forced to resign in 1891, the bitterness and internal division persisted.[45] Many of the proprietors were entitled to commission on salt made at those works which they had formerly owned or on salt sold under their 'brand' names, and each wanted the lion's share of the market. Many thousands of pounds were squandered in costly litigation over the interpretation of agreements, and rivalries were intensified. In vain shareholders besought the directors: 'Do for one instant all pull together, do for one instant sink your petty differences and petty rivalries and pull together for the benefit of the Salt Union.'[46] But after a century of bickering the salt proprietors were not disposed to listen, and the great salt industry of Cheshire was thrown into a decline from which it was never to recover.

The most serious blow of all was the loss of exports through the rise of foreign competition, notably in Germany and the United States.[47]

VI

From its very inception, the Salt Union – the coalition of proprietors established in 1888 to hold a monopoly of the trade – attempted to 'rationalize' the industry by reducing the number of men both at the works and on the waterways. This brought them into head-on conflict with their workers, and in particular with the collectivist tradition of work sharing. Never before had entire works been systematically closed down whilst others had all their pans working. Although the transfer of the displaced workers to the larger works prevented widespread unemployment, the saltworkers were alarmed.[1]

Matters came to a head in 1889 with a strike on behalf of wallers whose pans had been stopped and who had been refused their traditional right of taking a share at other pans. Twenty-eight men were involved, but the proprietors refused either to give in to their demands or even to acknowledge the fact that they were on strike. After six weeks of the strike the saltworkers of Northwich and Winsford met to formulate demands which aimed at reversing the rationalization programme as well as reinstating the twenty-eight men. These demands, which were accompanied by a warning that the workers intended to publish a list of their grievances, demonstrate convincingly their determination to honour the customs of the trade and defend the tradition of sharing:[2]

1. That [wallers'] work shall commence at 5 o'clock a.m. and finish at 4 p.m., and that early and late orders be abolished.[3]
2. That sixpence shall be paid to the workmen for every ten tons by rail or river.[4]
3. That in the opinion of the men a general strike should take place unless the men on strike are taken on again.
4. That where three men are now employed there shall be four.
5. That all the works shall have a fair share of the work.
6. That firemen finish work at nine o'clock on Saturday night, instead of 10.30 p.m. and commence on Monday morning at two o'clock instead of one o'clock; also on other mornings finish at two o'clock, instead of 2.45.

The men's demands were considered by the Salt Union's central authority, which consisted mainly of London businessmen and, in an unfortunately worded reply, they announced that if the wallers would not countenance 'early' and 'late' loading the directors would be compelled to take the work away from them entirely and give it to other men.[5] This was too much for a meeting already angry that their other demands for an equalization of work had been refused. Even offers to reinstate the twenty-eight strikers and reintroduce the 'sixpence tonnage allowance', did not abate their wrath. Angrily the wallers declared: 'They could employ Irishmen and Foreigners if they wished but it would cost them three times as much money. They would soon be taught a lesson if they brought in other men to do the work.'[6] Then the meeting reaffirmed their demand that the unemployed wallers have a share of the work available, and threatened that if the Salt Union had not granted their demands within one week they would give notice to strike.

Thoroughly alarmed, the Salt Union selected Thomas Ward, a much-respected works manager, to continue the negotiations.[7] He managed to persuade the Saltmakers' Committee that discharging and loading should be made into separate jobs, which could give employment to the out-of-work. To accept these proposals would mean that wallers' rates would be permanently reduced, whilst their unemployed brethren would be permanently degraded to taking on labouring jobs. As soon as the wallers heard of them, uproar broke out.

At a special meeting of the Winsford Saltmakers' Association, the wallers refused to listen to the chairman's attempts to explain what the proposals entailed. Some idea of the ensuing scenes can be gathered from the *Chester Chronicle*'s delicately worded report: 'One irate debater hurled an unparliamentary epithet at the chairman's ruling, and doubted the veracity of his statement with a vulgar severity which produced a general "sh" from the rest of the wallers.'[8] The president's threats to resign if order were not restored were met with cheers, and to cries of 'traitors', he and the committee left the room. Whereupon one member suggested: 'This is a waller's question and let them settle it'[9] and an ad hoc committee of wallers drew up fresh proposals which the president and secretary, having been invited back into the meeting, were ordered to take to the Salt Union. The wallers were now demanding that five men should in future do the work that four were now asked to do, that there should be no reduction in wages and that the wallers should be relieved of their loading and discharging work.[10]

Despite the strong expressions of feeling, a compromise on the issue was eventually reached on the lines of Thomas Ward's proposals.[11] The unemployed men were taken back, but the wallers' work was henceforth sub-divided, with two new classes of men – salt loaders and slack dischargers – taking on a portion of their work.[12] The amount of employment was increased, and the wallers gained a greater freedom in the management of their working hours, since they no longer had to attend to the boats.[13] At the same time they lost overall control of their job, since the new class of loaders were responsible not to them but to their banksmen. Another new class created at this time was the slack dischargers, who unloaded the coal boats and railway wagons. This was a less skilled job than loading – it was just a matter of shovelling the stuff out – but the work was enormously hard. Presiding over both these groups was the banksman, who, in the 1890s, was increasingly taking on a foreman's role.[14]

The second group of workers to contest the Salt Union's attempted 'rationalization' were the Weaver watermen. They had suffered immediately from the company's policies. With the formation of the union, the entire fleet of Weaver rivercraft came under their control. Vessels were sold off, sailing flats replaced by caravans of dumb barges, and crews thrown out of work. At the same time – according to the watermen – hours of work for those in employment had been increased.[15]

The watermen and their families endured the pressures which must attend any caring community when some of its members are without employment whilst others labour like slaves. But in 1891, when the Salt Union tied up a further twenty lighters and barges and threw the crews out of work, the watermen's patience was at an end. Granting the

Plate I Screen women in Whitehaven. Women and girls were widely employed as sorters and screeners in many classes of mineral work—'bal-maidens' in the Cornish clayfields, 'limestone girls' at the Welsh iron works, ore-dressers in the Cornish tin and copper mines, 'pit brow lasses' in the Wigan and Shropshire coalfields. This photograph was taken at the 'Grub House' of a Whitehaven coal-mine, *circa* 1900, where the women picked stone and metal out of the coal. 'It was a dirty job, but many of the women, among them Elizabeth Jane McCourt, third from the left, prided themselves on arriving at work each day with aprons and shawls clean, and their clogs shining like armour.'

Plate 2 Gravel pits at Hampstead Heath, *circa* 1867. Pits like this would have been found on the outer fringe of other towns besides London, wherever there were accessible building materials to be dug. In the case of Hampstead Heath the depredations of the sand-drawers and gravel-getters were one of the leading arguments used in the 1870s when the Heath was enclosed as a public park.

Plate 3 Sand crushing at Gornal. This photograph is taken from the heart of the Black Country where sand was extensively used in furnace work, as well as being employed by the brickmakers, and exported to the glassworks by way of the canals. Brickmaking at Gornal was women's work as late as 1934 when J. B. Priestley (*English Journey*, p. 113) visited 'this end of the earth . . . called Gornal' and noted 'women returning home from the brickworks' who, dressed in caps and shawls, 'looked as outlandish as the place they lived in'. In Gornal, as in other parts of the Black Country, women worked as moulders at the brickworks, as well as at carting—the work being done by the women in the photograph.

Plate 4 *above* Jet workers in Whitby. The manufacture of jet ornaments was a major industry in mid-Victorian Whitby, the jet being mined and quarried in the nearby cliffs by free-lance diggers, and then dressed and shaped in local work-rooms. This photograph, taken by Sutcliffe in 1890, shows a workshop in Haggersgate, the only one in the town to be worked by gas engines. But note the boy in the foreground working at the stove.

Opposite

Plate 5 *above* Slatemaker's shanty at Stonesfield, Oxfordshire, *circa* 1900. Slatemakers in Stonesfield, a mineral-working village ten miles north-west of Oxford, were divided into two main groups of workers: the diggers, who were largely part-time workers, engaged at agricultural work during the spring and summer, and at slatemaking in the autumn and winter, and the makers, or 'crappers', who worked full time. The 'crappers' usually worked in the open air, with a makeshift booth to shelter them from the wind and rain. Nineteenth-century brickmakers often worked in shanties like this, but standing up and moulding at a bench.

Plate 6 *centre* Lead miners of the Yorkshire dales, *circa* 1904. This photograph was taken at Lofthouse, Nidderdale, and shows the very narrow entrance to the levels, barely wide enough for the trucks to get in and out.

Plate 7 *below* Clay workers on Dartmoor. This photograph was taken at Redlake in the twentieth century but the method of traction—by horse and pulley—had still not changed from that of the nineteenth.

Plate 8 Cornish tinners. The photograph at the top of the page shows miners working by hand at an overhand stope in the Carn Brea tin mine, and was taken *circa* 1900. That at the bottom shows miners working by the rock-drill. Despite the fact that the first rock-drill had been patented in 1812, it was still comparatively little used in the Cornish mines eighty years later (at the Carn Brea tin mine three boring machines were in use in 1890).

Plate 9 Explosion alert, Penrhyn Quarry. A bell or hooter was usually sounded before explosive charges went off; here a bugle is being blown to warn men of an impending explosion, and to send them hurrying for shelter to small huts on the galleries. The lines of the giant galleries can be seen faintly in the background. The placing of explosives was a skilful and dangerous job performed by the rockmen themselves.

Plate 10 *left* Rockmen working their 'bargain' in the Penrhyn Quarry. Skill, strength and agility were needed to work in such conditions. The thick hemp rope was looped loosely around the men's thighs as a safety measure; but danger remained, most obviously from falls of loose rock. There were usually four partners in a 'crew', two working on the rock face and two in the sheds.

Plate 11 *above* In the shed or slate mill, Penrhyn Quarry. The slab arrives from the men on the galleries for their partners in the sheds. Here it is sawn into more convenient blocks on the mechanically operated sawing tables before being split into slates. Stacks of finished slates can be seen in the background. The major hazard of work in the sheds was the continual presence in the air of fine slate dust which ravaged the men's lungs.

Plate 12 *left* The final processes in transforming rock into slates. Despite the fact that these men worked in the sheds they considered themselves, and were considered, to be the 'true' quarrymen. Here they are working on three separate operations and adopting a different pose for each task. The man on the right wields an iron hammer, the man in the middle raises a wooden mallet while the man on the left holds a (barely seen) cutting knife for dressing the split slates. (Later on this knife was mechanically driven.) All the jobs demanded extraordinary skill and precision—the slate dresser on the left had to judge size while the splitter in the middle assessed width and the man on the right had to cut against the grain of the rock. All these jobs were interchangeable and the skilled quarryman would be able to execute all three.

Plate 13 *below left* Making slates: slate splitters in the Penrhyn Quarry. A task which right up to today has always had to be performed by hand. No machine could assess the potential number of slates which each block would yield, nor could it split the block without breaking it.

Plate 14 *below* Outside the office on pay-day at Llechwedd slate mine, Blaenau Ffestiniog, in the 1890s. A 'bargain' was struck for a month between the management and a 'crew' of rockmen and splitters, and payment made on the basis of the month's produce. The money was shared between the members of each crew and payment made to any journeymen who might have helped them. The haggling would then begin for a 'setting' price for the next month's bargain. In the Blaenau Ffestiniog area slate was mined in huge underground caverns rather than quarried from the surface, but though the techniques employed were different the skills were essentially the same as those needed in other quarries.

Ancient Salt-works.

A. Wooden Ladle. *B.* Cask. *C.* Tub. *D.* The Master. *E.* Assistant. *F.* The Master's Wife. *G.* Wooden Spade. *H.* Boards. *I.* Salt-baskets. *K.* Hoe. *L.* Rake. *M.* Straw. *N.* Bowls. *O.* Bucket for Blood. *P.* Beer Tankard.

(From an Old Print published in 1556.)

Plate 15 *left* Sixteenth-century saltworks. The process and even the equipment shown in this old print changed little in following centuries. The pan at this time was made of lead, and the straw or twigs used to fire it caused such damage that after each use it had to be repaired. When wood became scarce (which the profusion of wooden implements and containers in this print shows to have been at a later date), coal fuel was introduced and iron pans substituted for the leaden ones. In the background of this print the master can be seen loading salt into drying baskets, an early version of the process of stoving. The 'bucket for blood' identified by a letter 'O' shows the antiquity of the practice of 'doping' the brine with blood and other additives intended to clear it of impurities. The wife in the foreground snatching a bite at the workplace rather than preparing a household meal is clearly a worker with her husband. When this print was published in 1556, Northwich was already a saltmaking centre, with 300 or so wych-houses, whose smoke according to John Leland made the town unpleasant and dirty. The industry was still at this time under strict guild control by the 'Rulers of the Brine'; later the saltmakers moved outside the town to exploit new brine springs, and the guild's power was broken.

Plate 16 *above* The Old Bridge at Northwich. The towing horse in the foreground dates this print sometime after 1792 when the first tow-path suitable for horses was constructed. Until the 1860s when steam barges made both redundant the waterman had the choice of hiring a horse at 15*s*. or men at 5*s*. each per round trip to haul his barge to Weston Point, after which it entered the Mersey and proceeded under sail.

Plate 17 Firing the pans. These men are fine saltmakers, or lumpmen, tending their fires. Each man had a single pan to attend to which he contracted to work being paid a fixed price for the salt he produced. He had to ensure that his brine was kept boiling vigorously and two tons of coal had to be shovelled into the fire holes for every three tons of salt produced. On the other hand the firemen who made coarse salt each had to attend to three or more pans, burning one ton of coal for every two tons of salt produced. Note that the men are unprotected should any of the boiling brine splash over into the fire area. Later they were protected by a 'caboose' which projected beyond the bottom of the pan to form a cover.

Plate 18 Making the squares. This photograph shows the 'draught', when the salt crystallized out of the boiling brine was removed from the pan. Several jobs were done in quick succession and close proximity at this stage in the making of fine salt. On the right here can be seen the saltmaker with his long rake drawing the salt in to the side of the pan. Behind him his assistant is shovelling the salt into a 'tub' which stands in the steaming brine. In the foreground is the 'happer' squaring off the lump or square, and behind him his mate with the mould. All this work was heavy, and the atmosphere hot and steamy, but these jobs were done at top speed so that the pan could be refilled and put to making more salt in the shortest possible time. Until legislation in 1867 began to affect the industry, women as well as men did all these jobs (especially 'happing' and loading), and work started very early (2 or 3 a.m.) so as to fit in 4 draughts a day. Women's employment in saltworks was deplored by outsiders, not so much because of the heavy exertion, or the long hours, or in some cases the risks involved; but because men and women were working close together without their top garments—on account of the heat men (as here) took off their shirts, and women were in chemises—which was considered immoral.

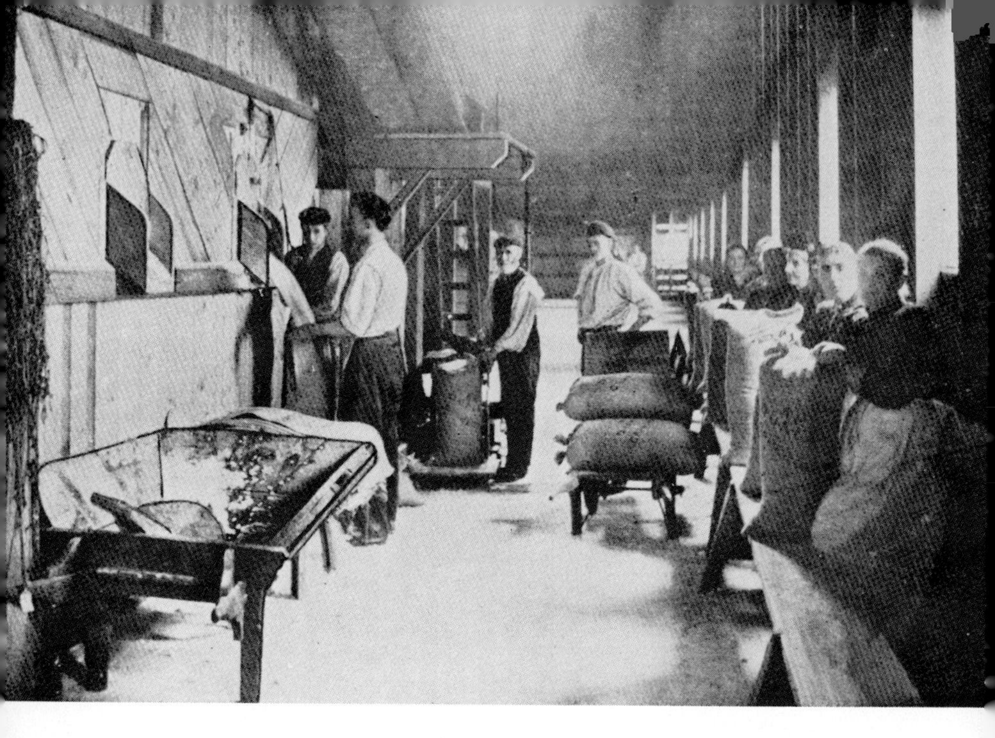

Plate 19 Bagging the salt. Fine salt, which was made into 'lumps' or 'squares' when drying, was usually broken up before it was sold. Until the 1860s, while saltmaking was usually a family undertaking, this stage was the responsibility of the younger members. Boys and girls working for their father or another saltmaker would crush the salt with hammers and axes, then older children (13 to 16 years old) would bag and load it. If paid they would earn 1*s*. per day. In the 1870s, after the exclusion of children under the Factory Acts, milling machines were introduced and the operation was reduced to two stages: filling the bags, and sewing them up. This photograph shows two youths and an old man doing the first job, and half a dozen women the other, which probably represents the usual character of workers engaged on this work. The sacks when filled could weigh 2 cwt., and had to be lifted on to the stitching bench. As the women could stitch and tie a bag in seconds, the men were working flat out to keep them fully occupied. Bagging went on being done like this for perhaps another fifty years in some saltworks. Some of the older women pictured here, and almost certainly the old man, had probably spent their youth at the pan-side toiling in the heat and semi-darkness.

Plate 20 Drawing broad salt. Broad (or coarse) salt was formed by a slow process of evaporation, and took from three to four days to form in the pan, according to the temperature at which the brine was maintained. The coarse salt pans—like the one in the photograph—were extremely large, some measuring 140 feet in length. Their width was governed by the size of the rake (the man in the background on the left is raking) which had to be flung out into the centre without the user overbalancing and falling into the pan. The workmen were called wallers and they usually started work at 5 a.m. and could be finished by 10 or 11 p.m. In this time, it was claimed, a waller might have to shovel 15 tons of salt from the pan and pile it as a 'draught' on the 'hurdles' (i.e. the standing platforms running the length on each side of the pan). After lying on the 'hurdle' for a time whilst the excess brine drained from it the salt was usually wheeled to the warehouse. Sometimes, though, it might be loaded directly into a river boat and then the waller would get an extra payment of about 2*d.* per ton.

Plate 21 Loading broad salt. Until 1889 all broad or coarse salt was loaded by the wallers as part of their responsibility. However, when they campaigned to get more men employed, a new category of workers—the loaders—came into being. They were paid by the ton and if there was no loading to do—during neap tides, for instance, when no boats could enter the Mersey from the Weaver—they got no pay.

From the photograph we can see that it was a job for older men; but the weight of the huge hand-cart and the pace of work often resulted in terrible injuries. The salt was tipped off loading platforms ten or twelve feet above the river craft or rail waggon being loaded; and occasionally the unfortunate loaders plunged over the edge, cart and all crashing sixteen or twenty feet into the hold of the barge. Above the loaders are two wallers warehousing their salt; note the absence of a safety rail along their pathway.

Plate 22 *above right* Subsidence at Northwich. The pumping of brine caused widespread subsidence in and around Northwich. Houses tilted and many either collapsed or were dismantled because they constituted a public danger. Least affected were the old half-timbered buildings; so during the nineteenth century it became the recognized practice to construct all town buildings in timber frames, and, in later years, in steel frames. A lifting system involving the use of hydraulic jacks was also devised and it became commonplace to see a complete building being raised—like this one in the photograph—to a new level.

Plate 23 *right* Percy Pit Colliery, 1844. The 'heap' or 'bank' as miners refer to the surface buildings. In early times—as in this picture—the shafts and buildings were made of wood, and the lighting at night was by open braziers. To a stranger it must have looked like a gateway to the inferno.

Plate 24 Drops at Wallsend. Drops were piers at the end of the dillyways which brought the coal from the pits to the rivers. Here you can see the colliers being loaded with corves. Another method of loading the boats was by funnelling the loose coal from a staith.

Plate 25 *below left* At bank, *circa* 1913. This looks like a one-deck cage (some cages have three or four decks). The man on the left — the banksman — raps to the winder, the cage is lifted slightly, the keps (which hold the cage while it is stationary at bank) are withdrawn and the cage drops. The banksman is usually a stern-faced man who doesn't like horseplay and tries to keep the men orderly. As the men come up to the banksman they give him their brass token — the last remaining token still in use (in the old days men had tokens for their tubs).

Plate 26 *below* Hewers at Cannock Chase, *circa* 1920. A good seam compared with county Durham, where an average seam height would be 2–2½ feet. In the days of bord and pillars hewers generally worked in pairs, with their own working place to themselves. The man on the right, loading the big coals into the tub, may be a putter, helping out (the putter's main job was to keep the men supplied with empty tubs, or 'chummins' and take their 'fullins' out; he might have five or six pairs of hewers to supply).

Plate 27 *below* Picking coals on the screens. A job for old or disabled miners (in earlier days for women, too), and also young boys. A song I vaguely remember about this is 'Geordie Black':

I'm gannin' doon the hill
I cannot use the pick
The masters habe ney pity on old bones
And now in latter days
I pass me time away
Among the bits o' lads a-picking stones

Plate 28 *right* The Washington disaster of 1908. This was a postcard printed at the time, probably to raise money (this one was given to me by my father). In the Washington area there were three collieries — Usworth, Washington Fanny pit, and Washington Glebe (the two latter have closed in the last five years). The pits in this district were highly explosive because of the very gassy coal. Amongst some earlier explosions at Washington were those of 1798 (7 killed), 1828 (14 killed) and 1851 (35 killed).

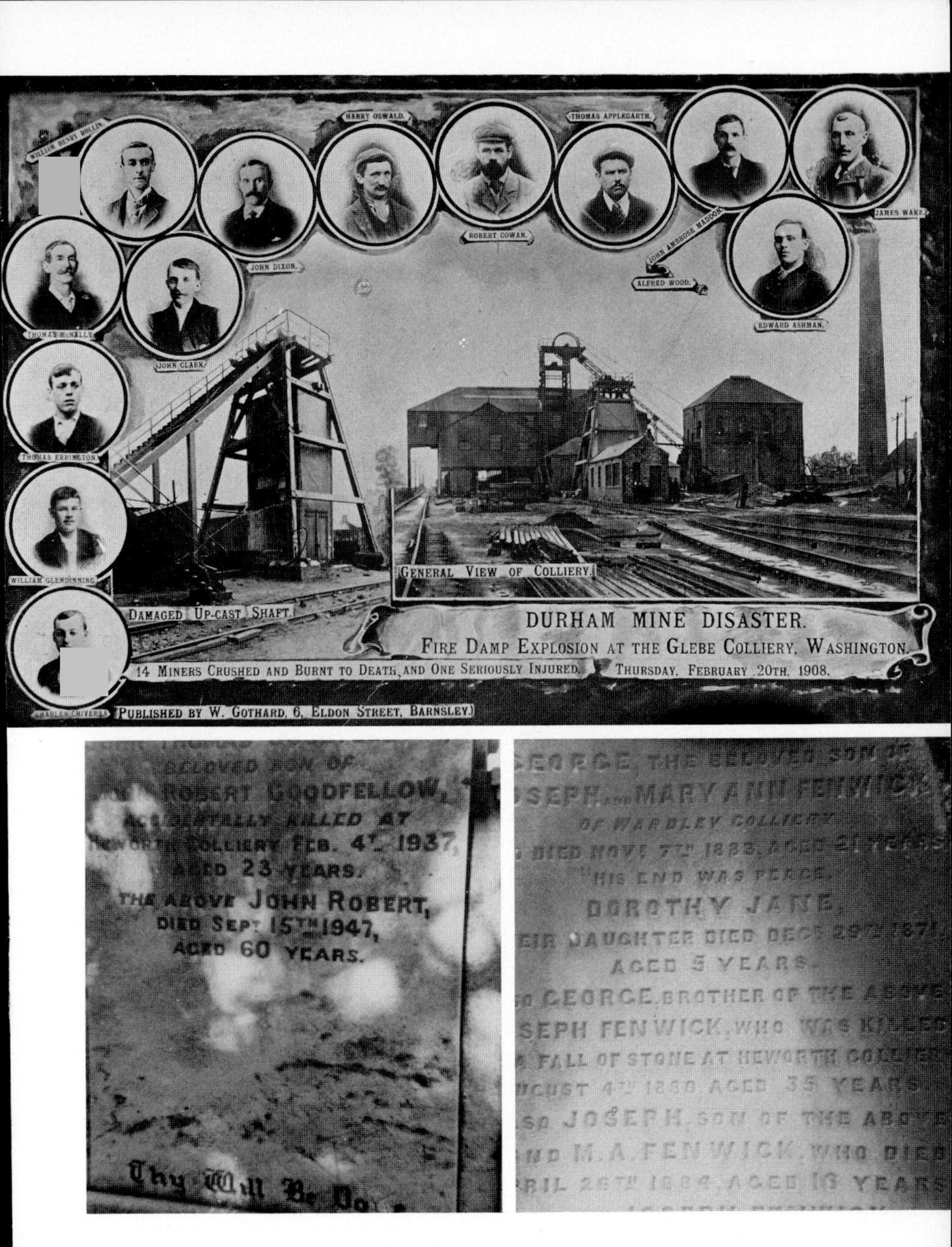

Plate 29 Gravestones at Heworth. The cemetery contains the obelisk to the 92 miners, men and boys, killed in the Felling disaster of 1812, and often one will find gravestones like the ones above, a record in stone of the numberless disasters of mining life in Wardley, Heworth and Felling.

Plate 30 Cover of a pamphlet by George Harvey. Harvey was a lifelong miner, a rank-and-file agitator, and a thorn in the flesh of full-time officials of the Durham Miners' Association. In my own village, in which he lived for most of his life, he is remembered as a 'Bolshevik' ('ney Communist, ye knaa, a Bolshevik'). In earlier days he was a follower of Daniel de Leon, a supporter of industrial unionism, and of the Socialist Labour Party.

redundant men £1 per week strike pay, the Watermen's Association (founded in 1888) gave notice to strike. The Salt Union apparently capitulated, and not only agreed to take back the discharged crews, but also entered into a most detailed agreement about wages, payments and working conditions, which set limits to over-work, and codified the customs and craft privileges of the river[16] (extracts from this agreement are printed in the appendix).

The agreement was intended to serve as a restriction on the number of hours that the waterman could be made to work and the number of trips he could undertake in a given period. According to this agreement, the men could not normally be employed actually sailing, loading or discharging their craft for longer than forty-eight hours. However, the watermen remained unsatisfied and, as can be seen from the evidence of William Hough, Secretary, Weaver Watermen's Association, before the Royal Commission on Labour, in March 1892, much of this dissatisfaction stemmed from the fact that the watermen felt the trade was being mismanaged. This feeling can easily be understood when we consider that the watermen until recently had been largely responsible for managing their own affairs and now found themselves under the direction of employers who had no traditional connection with the river.

The feeling of frustration must have been particularly acute for men, who, like William Hough, had been born to be watermen. His father, grandfather and great-grandfather had all been watermen before him, and he had only left the river to take up his position as full-time secretary to the association.[17] Before the commission, he complained of the long hours the watermen were still being forced to work, and, dismissing any suggestion that this might be due to the 'uncertainty of trade', he declared it was entirely the result of the incapacity or culpable neglect of the new managers, the officers of the various firms. The heads of the firm would not listen to information tendered by practical men. Large orders found only half the craft available while the rest were bound fast at Birkenhead, Liverpool and other points, 'carrying things they ought not to have in them'. What was needed was better management.[18]

William Hough had insisted to the commission that sometimes the watermen were being forced to stay aboard their craft 'from 12 o'clock on Sunday night till 12 o'clock on the Saturday night following, or every hour of the week, day and night.'[19] In the watermen's opinion this was a breach of the October 1891 agreement, but the Salt Union insisted that when the vessel was moored and not being either loaded or discharged, the crew were not actually working, even though they had to remain on duty.

There was a series of unsatisfactory meetings and on 17 August 1892 the watermen, ignoring a last minute offer by the Salt Union 'to frame and consider with the representatives . . . a scheme by which payment will be entirely on the basis of the tonnage carried', struck work. The Watermen's Executive issued the following notice:

> To flatmen, engine drivers, and all men connected with river craft. Fellow-workmen, please oblige by keeping away from the Salt Union offices, at Winsford and Liverpool, as their men are out on strike. It is not a question of wages, but excessive long hours. After 36 or 40 hours' work, the men ask for a few hours' rest. This the Salt Union directors will not agree to, but insist on the men working three or four days and nights in succession without rest. The men feel that this is too much for them, and cannot stand it any longer. They were promised last year that they would not be compelled to work a second night but this the directors have not adhered to. All that the men ask is that you will not interfere in this dispute by working their craft or carrying their salt, but allow the directors and their men to settle this matter between themselves.

The *Northwich Guardian* commented: 'The strike shows signs of becoming formidable, as the Salt Makers' Union – a very strong organisation – is standing by the watermen on sympathetic grounds. . . .'[20]

The first few days of the strike passed quietly, each side denouncing the other's claims but without, as yet, real anger. Meetings were held at Winsford and Northwich, presided over by William Chadwick of the Mersey Flatmen's Association, the object being 'to put the cause of the watermen's strike clearly before the public, to prevent any wrong impression being made by matter now being circulated by Salt Union Officials'. Various watermen spoke, giving precise details of the long hours they had been working. One of them, Henry Deakin, informed a meeting, that the Birkenhead salt heavers had gone so far as to threaten not to work at their craft because the men, who had been up night and day, were unable to keep their eyes open. 'They would rather work with drunken men than for men who were in want of sleep.'

Meanwhile, the Salt Union issued a statement to the effect that they were 'gradually filling the places of the men on strike'.[21]

Originally 230 watermen had given the seven days' notice required by law, but it had been subsequently discovered that for various reasons the notices of thirty men were invalid and they had had to re-submit them. Therefore, when the strike commenced, thirty watermen were still on duty and remained aboard their craft. Of these thirty, Thomas Wilkinson, Thomas Atherton, Thomas Holland, Solomon Taylor, Arthur Walker, William Holland and James Wood, all of Winsford and Wharton, all refused to move from the particular craft upon which they normally worked, and which, being undermanned, they refused to sail, to make up crews on other vessels. They were summoned under the Merchant Shipping Act by the salt proprietors, who claimed £10 damages from each. They appeared before the magistrates (all of whom were shareholders in the Salt Union) at the Middlewich Petty Sessions, but half-way through the hearing the Clerk informed the Bench that the Act only applied to sea-going vessels. Therefore all the summonses (except that against James Wood, which was dismissed) were adjourned *sine die*. Nevertheless, the salt proprietors announced that they were to take out summonses against those saltworkers who refused to load craft manned by non-Society watermen.

At Northwich a number of craft manned by strike breakers were stoned and 'while passing the town Bridge . . . where a considerable number of watermen had congregated, the men in charge of the craft were vehemently hooted and jeered. . . .'[22]

With the strike entering its second week the watermen and their families were beginning to anger at the sight of vessels manned by blacklegs sailing the river. The watermen crowded daily upon Northwich's Town Bridge to watch keen-eyed for the approach of any vessel.

On the evening of Thursday, 25 August 1892, the steamer 'Cynosure', towing the barge 'Antelope', was observed approaching Northwich. A crowd of men, women and children dashed to the river bank and, groaning and hooting, commenced hurling stones at the Liverpool blacklegs who were manning the craft. The police watched but made no move to interfere. The crowd kept pace with the craft for half a mile, showering it with missiles, until finally the engineer, 'declaring that he would proceed no further', leapt ashore. Within minutes the boats were secured, and the blacklegs, unmolested, set off for the railway station and Liverpool.[23] Heartened by their success, the people of Northwich awaited the appearance of the next blacklegs. These sailed into the town early on Saturday morning. Here, they picked up an escort of four policemen and proceeded towards Winnington Bridge. When the craft reached the Winnington Chemical Works the workmen hurried to the factory fence to hurl bricks at the blacklegs on board. The blacklegs, terrified, tried to leave their posts, and take shelter with the police escort on the prow, but the constables ordered them to return. When the boat reached Winnington Bridge, the outer limit of the Northwich police district, the police abandoned them, despite (or because of) the presence of a large and threatening crowd 'composed principally of labourers and women'. The crowd followed the craft along the towing path 'hooting and throwing stones'; bricks were thrown by the men who were watching the proceedings from the fences of a chemical works which overlooked the Weaver, and at 'Three Bridges', near Saltersford, 'the captain and men of the Albion deemed it the wisest policy to go below'. Finally the boat ran aground, after being damaged in Saltersford Locks, and the captain and his hands surrendered, some of them returning to Liverpool in the steamer 'Gladiator', and the remainder walking to Acton 'where they took the train to Liverpool'.[24]

Similar scenes occurred at Winsford, and when the Winsford Saltmakers' Committee advised the members that although they should not refuse to start loading craft, they should not finish the job, the salt proprietors brought in 200 men from Liverpool to load salt. The blacklegs, although under a heavy police guard, were attacked by huge crowds of saltmakers and their families, and 'the new hands, leaving their bedding behind them . . . were escorted to the railway station'. The next day the town of Winsford was occupied by a squadron of the 14th Hussars, but the strikers and their supporters were undeterred.[25] Eventually, after a strike which lasted from 17 August till the second week of September, the alliance of the saltworkers and watermen defeated the salt proprietors. The Bishop of Chester and the High Sheriff of the County interceded, and with the Bishop acting as arbitrator the dispute was settled. The watermen paid out in compensation £500 to the blacklegs for 'loss of Employment', but every Weaver waterman went back to his place, privileges intact and hours of work reduced.[26]

The saltworkers of the 1890s were strong enough to resist the 'rationalization' of the proprietors, but they could not prevent the ebbing of trade, and, moreover, they were under new threat, from the establishment in Northwich of an entirely new industry, using new technologies and with a system of industrial discipline entirely divorced from anything within the experience of the saltworkers. Within a little more than a quarter of a century Northwich became a company town almost completely reliant upon the good offices of two men, Dr Ludwig Mond, a German chemist, and his business partner, John Thomlinson Brunner, an ambitious ex-company clerk. Under their paternal dominance the working community of Northwich lost its proud independence and assumed an outward display of loyalty and co-partnership to cloak its enforced humiliating economic dependence.

In 1874 the partners Brunner and Mond began producing soda ash, by the recently patented Solvey process,[27] at their newly built chemical works at Winnington near Northwich. The location of the works had been chosen, primarily, because of the availability of a vital raw material, brine, and because it had, nearby, established transport links with areas producing limestone, an equally important raw material.

Ludwig Mond professed to have been a socialist in his youth, but his attitude towards his workers caused him to be feared.[28] Soon after setting up at Northwich he complained: 'I have over one hundred workmen. Among them many are foolish, obstinate and lazy; some even malcontent. To keep going effectively I have to keep them and my machines all in regular working order'.[29] Brunner was a firm believer in the Victorian gospel of work 'which he advocated with all the fervour of a Samuel Smiles'.[30] Together the two men carved out an industrial empire in the heart of the saltmaking district. In 1881 they sold their business to a limited company for £400,000 and became joint managing directors of the new company. The lot of the early workers who came under them was not envied: as the 'dangerous trades' inquiries got under way they came to be known as 'the White slaves of Winnington'.[31]

To manufacture soda ash by the method adopted by Brunner and Mond required continuous operation, and two shifts of men were employed working days and nights, turn and turn about. Each week, in order to facilitate the changeover from days to nights and vice versa, one shift had to be in attendance for almost twenty-four hours without relief. Condemning this practice the *Northwich Chronicle* complained:

> One intolerable effect of the two-shift system is the practical annihilation of Sunday as a clear rest day. Every alternate Sunday a shift labourer is at work from 7 a.m. throughout the day until 6 a.m. Monday – 23 hours. The other Sunday he leaves work at 7 a.m. after doing a 17 hours' shift. He is, of course, tired when he gets home. After refreshing he goes to bed, and when he rises after sufficient sleep his Sunday is gone.

Although the two-shift system was usually referred to as the twelve-hour shifts system, in reality men worked seventy hours one week and ninety-six hours the next, fifty-two weeks per year.[32]

Tom Mann came to work for a time at Winnington, to collect evidence about exploitation at first hand. The worst conditions he found were in the 'milky lime' department. A man working here had to push a heavy iron wheelbarrow to a furnace, where it was loaded with red-hot lime, hence along planks up to a big tank, into which the barrow load was tipped, then back to the furnace, and so on for a twelve-hour shift.[33] A writer in the *Northwich Chronicle* in 1889 also identified this as 'one of the worst places in the works', describing the work as 'horses' work'. The men on this job had no breaks for meals, but got their food 'the best way they can'. Only strong men could do it, and even they lasted 'two or three months at longest, sometimes not one day'. The gas from the hot lime and the heavy work made them queasy, but if they couldn't stomach their snatched mouthfuls they were likely to collapse. However, other jobs seem to have been little better. In the filtering rooms most of the men worked without shirts because of the heat, and the air was 'heavily loaded with ammonia'. At the grinding mills it was 'quite as bad as being in a snuff-box . . . one cloud of soda and dust'. Ammonia fumes in the distillation department were suffocating. Furnace men worked without food breaks in great heat, 'their faces, bodies, and arms . . . as a rule, badly burned'; they were 'poor looking objects after 12 months' work'.[34] At the lime kilns the work was outdoors 'in all weathers, night and day', without pause for meals, and, again, very heavy. Some men, the lime-pickers, had the task of removing partly decomposed pieces of limestone from the bottom of the kiln, while others were continuously engaged carrying up wicker baskets of coke and limestone to the top, and emptying them in. In all, 600 men were engaged at these various jobs.[35]

In its early years Brunner Mond had recruited much of its labour force from Widnes, while local saltworkers took up employment only for short periods. In the 1890s, however, the local salt industry went into a spectacular decline. The attempted monopoly of the salt proprietors was a disastrous failure, and in the following years many of the banks were closed down and left to decay. In ever-increasing numbers saltworkers from Northwich and Winsford were forced to take up employment with Brunner Mond. A typical example of the decline of salt was experienced in the parish of Over-Winsford: by 1913 there were only forty-eight pans working of the 244 which had been working in 1888.[36]

In the 1890s Brunner Mond gradually assumed a position of dominance as the major source of employment in the district, and Northwich, their original base, became, in effect, a company town. From their position of high dominance, Brunner Mond were able to command more loyalty from their work force. Gone was the need for the harsh discipline of the earlier years, with its brutal fines and summary dismissals. Economic dependence ensured obedience, and from 1889 onwards Brunner's undoubted mastery of the art of management came to the fore. The open and crude oppressive exploitation of the early years gave way to more oblique forms of control, all the more effective for their subliminal nature.

In 1880 Brunner Mond had 2,000 employees; by 1914 this had increased to 4,000 and by 1920 to 8,000. In addition, numerous small firms were dependent upon their patronage, so that altogether perhaps three-quarters of the local community were dependent on them for a livelihood. They were able to be extremely selective about the type of person they employed. A works rule forbade the employment of boys who had failed to achieve Standard IV at school, and from 1890 onwards it was a condition of employment that youths between fourteen and seventeen had to attend night school classes. By 1904, boys who failed to attend 90 per cent of such classes during the year could be dismissed.

In 1899 the firm decided that no man over thirty would in future be considered for employment. Brunner explained that the decision had been taken because the company had decided to introduce a pension scheme. 'Was it fair', he asked, 'that men should go to the Works at the age of fifty years, and after a few years' service ask for a pension? . . . He regretted most bitterly that they could not take on more men, and they could not take on men past their best in life.'[37] The object was to attach a man to the company from boyhood to retirement and, if possible, after him, his sons. In time it became accepted at Northwich that just as sons had followed their fathers to become saltmakers or watermen, so they now followed them into the service of Brunner Mond.

Politically, Northwich has reflected the dominance of Brunner Mond since 1885 when Brunner was first elected MP. As a Liberal and a proclaimed Radical, he laid claim to the working man's vote. In 1905, when the local Trades Council was canvassing support for Labour candidates at local elections, they were told by the Northwich Salt and Chemical Workers Union that 'in the opinion of the Committee, seeing that working men generally in Northwich and District have hitherto taken so little interest in supporting labour candidates at various elections it would be useless and foolish to repeat the attempt . . . and therefore they regret to say that they cannot see their way to take any part in the matter at the forthcoming election.' In 1909 John Fowler Brunner took over the seat from his father and held it until after the First World War.

With the decline of the Liberals it was the Conservatives who came to be identified, like their predecessors, with the prosperity of ICI. Today, Northwich may be regarded as a safe Tory seat, commanding a majority of from 4,000 to 6,000 votes. Apparently little effort is required to maintain this majority, despite Northwich being an overwhelmingly working-class constituency. Sir John Foster, who, until retiring before the 1974 election, represented Northwich for many years, encouraged his supporters from time to time with reminders that the Labour Party stood for the nationalization of ICI. His successor managed to show an increased majority in the 1974 election, and by reference not to 'Reds under the Bed' but to the twenty-five companies which the Labour Party allegedly wanted to take over.

To be born in Northwich during the twentieth century almost certainly meant, for a boy, that he had been born to the service of Brunner Mond. In 1926 Brunner Mond became part of ICI, but to the local lads it was still the 'Chimic'. Kids not yet out of school talked of the day when they would join the multitude hurrying to work 'across Golf' (many years ago a golf links; now covered with company houses) to beat the 'buzzer'. I speculated with the rest. Practically everyone's dad worked at the ICI, and if he didn't it might be hinted that it was because he was not very 'steady': he might be earning good money now but it was doubtful if he would ever get a pension. Children knew the names of the different process plants and characteristics of the products before they knew the names of trees and flowers. We'd almost all work at the 'Chimic', where, we'd heard it said, a five-year-old could do a manager's job, and a ten-year-old the foreman's. In truth there didn't seem anywhere else to go.

Those about to marry often put their names down for a company house. Probably one in five workers lived in company houses, and there was usually a waiting list. Living with in-laws until such a house became available was common. People took the first house offered and then looked around for another. These houses were let on a week's tenancy, and they were conditional upon keeping your job.

Young men present the company with a challenge: how to keep them until family responsibilities, the tied house, and fear of changing the known for the unknown work their own unseen influences. One of the things about the company, when you first go there, is that you will get fairly steady promotion. The opportunities for promotion seem endless – surely conscientiousness and keeping a 'clean sheet' must bring its rewards. The job structure plays a part in this. The practice of 'relieving upwards' – when men in more senior positions are off during holidays, or sick, for instance – means that you are constantly getting a short spell in a slightly higher status job. This may go on for years without you ever getting a permanent promotion. Likewise, it used to be the custom to hold some jobs open for a year or so, in order to ensure that the appointment, when made, would be 'correct', according to the official explanation, but also keeping the ambitious youngster on his toes, while the vacancy was filled on a temporary basis. It was the foreman's job to make nominations and he would tell a number of candidates that they were in with a chance, that he had put in a word for them. If they didn't get the job they would be given the impression that they would be in line for the next vacancy that occurred. A remarkable number of men seem to have spent a short time in their early years as relief foremen without ever getting the actual post.

People rise steadily through the ranks in their early years but towards middle age their chances of promotion fade, and if their health begins to decline they come down the tree again. Consequently, there are always a large number of older men who feel that they have been unfairly 'passed over' in face of younger men, and that their experience has been ignored. This, apparently, was felt in the early years of this century, when the first eager young men, fresh from their enforced night school, came in expecting promotion. In 1904 union records show complaints of 'younger less experienced men' being unduly favoured.

Disillusion creeps in about the age of forty. If the company can keep a man till that age then there is little chance of him leaving of his own accord. Men who have never experienced any other kind of working environment cannot suddenly face the challenge of change. 'The pension' is given as the reason for staying. It now assumes an awesome importance. For the next twenty to twenty-five years, the promise of youth squandered, retirement is the goal. When a man is nearing middle age and his strength begins to fail, he stops hoping for anything better. Perhaps for the first time in his life he may begin to lose time. Now he and the company may become involved in an elaborate charade. Perhaps he acquires a liking for losing time, or goes off sick. A visit to the company doctor follows. No examination, just a chat. Concern is shown and the worker acquires the reputation of being 'ailing' and he is invited to consider the taking of some less exacting job. It is treated as an act of kindness on the part of the company, 'finding a place for him', but in the newer post there will be less leniency for losing time.

Competition at Brunner Mond was encouraged from early days by the variety of ranks and rates. In 1888, for instance, shift jobs ranged from 3*s*. 4*d*. per shift to 6*s*., though there were 'considerably more men at 3*s*. 4*d*. than at any other price'.[38] In later years differentials became much more complicated with all kinds of groups earning an extra halfpenny or penny an hour. In this way competition was promoted within each group as well as between the groups themselves. Until recently there were no less than twenty-seven job ratings for process workers to compete for.

An early indication of the destructiveness of competition through differentials occurred in 1904, when, perhaps for the first time, all the different groups of shift workers submitted a joint claim. When the company rejected it, the distillermen directed their anger not at Brunner Mond but at the union. To include other groups in the claim, they alleged, had been injurious to themselves 'and caused the advance they asked for on their behalf not to be granted'.[39]

It is the company's ability to deflect resentments to other members of the workforce which is responsible, perhaps more than anything else, for the passivity of the workforce, and the weakness of trade union and shop steward organization. In the 1960s 100 per cent trade unionism arrived, but there are comparatively few committed trade unionists, and by comparison with other large factories shop steward organization is unadventurous in its aims. A few well-judged offers of promotion to the less committed has usually been enough to undermine solidarity and to destroy the shop steward's credibility in the eyes of his fellow workers. Ex-shop stewards spend miserable years as half-efficient supervisors resented by their colleagues and vilified by the many workers who like to feel that the job might otherwise have been theirs.

Shift workers, even without such management interventions, lack natural unity. Each shift group is made up of numerous sub-groups, usually not more than four to eight in number, who man the various plants. It is unusual for the whole shift ever to meet together. Not infrequently, the members of one sub-group will scarcely have any knowledge of the type of work being done by the others. Opinions based upon hearsay or half-remembered fact lead to an undue emphasis being placed upon the physical rather than the mental requirements of each task. Workers from different plants chide each other for having it 'easy' and fresh resentments are stirred up.

Appendix: Weaver Watermen's Association

Undated document, possibly 1891, in the possession of Lady Rochester of Northwich, and kindly lent to the author.

This document is almost certainly the agreement between the Weaver Watermen's Association and the salt proprietors in 1891; the rates given are the same as those quoted to the Royal Commission on Labour in 1892 by William Hough, Secretary of the Weaver Watermen's Association (cf. P.P. 1892 XXXVI Royal Commission on Labour (6795) Group B, Pt II, Q. 15, 596).

Agreement as to hours and tides for the working of Crafts, as agreed between the members of the Weaver Watermen's Association and their Employers.

1st Any steamer or barge loading on day and going for any tide and discharging the next day, and the tide is before a.m. 0.30 o'clock tide at night, she must go away and it shall not count a night.

2nd But if the tide flows between the hours of 8.30 p.m. and 6 a.m., she must not go away before the following tide. This has reference only to the Craft and men who have worked the night before.

3rd Also any Craft going away after the 8.30 o'clock tide at night and loading on arrival at Works, shall not go for any tide again flowing sooner than the 12.30 o'clock tide the next day.

4th Any Craft that has not worked the night before, and goes for a tide flowing before 8.30 o'clock tide at night at Liverpool, will not count a night.

5th Also any Craft that has to go for a tide flowing after 12.30 o'clock tide in the afternoon must not go for the level at Liverpool, without special orders.

6th And any Craft leaving Liverpool for the Works after the 5.30 o'clock and before the 8.30 o'clock tide at night, shall not be compelled to be at the Works before nine o'clock on the following morning, or this will count a night.

7th Further, no Craft shall be engaged shifting Docks more than two nights in succession, and if the Craft has already worked one night in going down to Liverpool, she has only to shift Docks one night.

It has been mutually agreed that, in all cases of emergency, any of these regulations may be suspended, but that a special order form on such occasions shall be given to the Captain of the Craft.

A. Burrows,
Secretary.

Explanations:

1st Further, no Craft shall be engaged shifting Docks more than two nights in succession.

Included in Dock shifting – to mean if a Craft works coming down in the night, when the tide flows after 8.30 p.m., she will be allowed to shift next night only, and on the following or third night she must not shift or go away.

But if a Craft has been laying down a few days or more, she shall shift Docks two nights in succession, but the third night she must not shift Docks or go away.

Further, when a Craft has been laying down and not worked the night before, she shall shift Docks one night, work up the second night, and, if required, work down again the following night if an emergency order is given.

2nd Any Craft having worked the night before, shall cease work at 5 o'clock, or at the times of ceasing the days at the place the Craft may then be fixed.

3rd A Craft which is not eligible to come away from Liverpool and District on a night's tide, shall not lock out after 6 a.m. if it is an ebb tide, but shall wait for the following flood.

4th Any Craft leaving Liverpool or Birkenhead on any tide and fails to get in Weston Point or Eastham until the following tide, and such tide flowing after 8.30 p.m., it shall count a night, and such Craft shall not work the following night except on being given an emergency order.

5th Any Craft trading to Weston Point and District must not leave Weston Point or Marsh Lock after 8.30 p.m. or before 6 a.m., or it will count a night.

6th Also any Craft crossing Canada Basin after 8.30 p.m., and before 6 a.m., shall be counted a night.

7th If a Craft comes up on a night's tide to the Works and loads the following day and goes down on the night's tide with an emergency order, such Craft shall neither come away or move Docks the third night.

8th When a Craft leaves Liverpool on a night's tide for Frodsham and is discharged the next day, and lying still all night, such Craft shall go down light the next day's tide without an emergency order, although the tide may flow before 12.30 noon.

9th Any Craft loading Staffordshire Goods shall take the last turn with Craft leaving Liverpool on the same tide, but shall be on turn to load before Craft leaving the following or any later tide.

Night Money at Weston Point and District –

Half night to commence after 5.30 p.m. and to finish at 12 midnight, the second half to commence at 12 midnight and finish at 5 a.m.

Overtime to commence on Saturday, at 2 p.m., and half-time to finish at 6 o'clock, full time at night to finish at 12 p.m., prompt; also full night to be paid from Sunday midnight (12 o'clock) to 6 a.m.

10th Any Member or Members accepting or receiving Clearance Notes or Guarantees from Employers, Skippers, Stevedores or Others (to pay fines for any offence against the rules), the person or persons receiving or benefiting by such notes shall be fined in any sum not exceeding £5 for each offence, together with the usual fines imposed.

11th Should a Craft leave Works for a night's tide and be prevented from going out at Eastham until a later tide, and leave Eastham on the next night's tide, i.e. any tide flowing between 8.30 at night and 12.30 the next day, such Craft shall be allowed to shift Docks the next night, but must not go away.

Sunday Rules:
Members are particularly requested to use their best judgement when working week-end tides, and work the tides which will involve them in the least Sunday work; but in no case shall they leave any 'Port Dock, Lock, or Tying up place' after 12 o'clock on Saturday night, or before 12 o'clock on Sunday night.

Any member or members Discharging or allowing anyone to discharge their Craft on Sunday, or in any way whatsoever assisting to do so, shall for this, and the foregoing rule be fined in any sum not exceeding £1.

Price List for Weekly Craft (Salt):

	pr wk	Loose Cargo	Sacks & Cases	Trip Money
Steamers:				
Captain	23/-	1d per ton	2d per ton	6/- per trip
Mate	20/-	½d ,, ,,	1½d ,, ,,	3/- ,, ,,
Engineer	24/-	½d ,, ,,	½d ,, ,,	3/- ,, ,,
Barges over 140 Tons:				
Captain	21/-	1d ,, ,,	2d ,, ,,	6/- ,, ,,
Mate	21/-	½d ,, ,,	1½d ,, ,,	3/- ,, ,,
Barges under 140 Tons:				
Captain	21/-	1d ,, ,,	2d ,, ,,	7/- ,, ,,
Mate	21/-	½d ,, ,,	1½d ,, ,,	4/- ,, ,,

No trimming or stowing to be done by the Crews.
Winding Money 1d per ton in and 1d per ton out.
Night Money to be paid as per card.
River Money 5/- per man.
Trimming (Salt)

Sacks: Three men to trim Sacks at 2½d per ton to be equally divided, together with an allowance of 1d on every 10 tons.
Bulk: Two men to trim Bulk at ½d per ton to be equally divided, but if 100 tons or any less quantity be loaded at any division or works, ¾d per ton shall be paid; over 100 tons and up to 156 tons 6/6d to be paid.
Squares: What men required at 3d per ton for 80s and 4d per ton for 120 squares and 160s.
Towings: For Barges over 140 tons 13/6 round.
Trip Money to Weston Point from Winsford and Northwich, 1/- per man for Craft over 140 tons, and 2/- per man for Craft of 140 tons and under.

Order of Trip Money from Weston Point:

	Steamers		Large Barges		Small Barges	
	s	d	s	d	s	d
Captain	4	4	4	4	5	4
Hand	2	4	2	4	3	4
Engineer	2	4	–	–	–	–
	9	0	6	8	8	8

Price List for Ton Crafts:

	From Winsford to Liverpool, Birkenhead, Garston, Ellesmere Port and Eastham.		from Winsford to Salt Port, Weston, Runcorn, and Widnes.	
Description	Captain's Sh.	Flat's Sh.	Captain's Sh.	Flat's Sh.
Common Salt	1/9 per ton	6d per ton	1/- per ton	3d per ton
Shute stoved	1/11 ,, ,,	6d ,, ,,	1/2 ,, ,,	3d ,, ,,
Sacks of all kinds	2/1 ,, ,,	6d ,, ,,	1/6 ,, ,,	3d ,, ,,
Handed squares	2/5 ,, ,,	6d ,, ,,	1/8 ,, ,,	3d ,, ,,

From Northwich to above places:

Common Salt	1/6 ,, ,,	5d ,, ,,	9d ,, ,,	3d ,, ,,
Shute stoved	1/8 ,, ,,	6d ,, ,,	11d ,, ,,	3d ,, ,,
Sacks of all kinds	1/10 ,, ,,	6d ,, ,,	1/4 ,, ,,	3d ,, ,,
Handed squares	2/2 ,, ,,	6d ,, ,,	1/5 ,, ,,	3d ,, ,,

No Allowance to be paid for Cargoes loaded for Salt Port, Weston, Runcorn, or Widnes, by the Captain.

A. Burrows,
Secretary.

Notes

I

1 In Cheshire the brine is formed by surface water infiltrating the rock-salt bed and forming underground streams. In some of the rock-salt mines it was said that the brine could be 'heard rushing like a torrent'. James Stonehouse, 'Salt and its manufacture in Cheshire', *Trans. of the Hist. Soc. of Lancashire and Cheshire*, vol. V, 1853, p. 115.

2 Cheshire brine contains practically 25 per cent, about 2 lb. 8 oz., of salt per gallon. In 1878 it was estimated that if to the total tonnage of salt produced in the year was added 30 per cent for brine wastage, the total amount drawn from beneath the surface of Northwich and Winsford was about 1,300,000 tons or, as one ton of salt measures one cubic yard, 1,300,000 cubic yards of the substratum which supported the towns. P.P. 1878–9 LVIII, Memorandum of Lt/Col. Ponsonby Cox R.E., Landslips in the Salt Districts, p. 560.

3 A. F. Calvert, *Salt in Cheshire*, London, 1915, p. 328.

4 Ibid.

5 S. Bagshaw, *Directory of Cheshire*, 1850, p. 458.

6 P.P. 1890–1 XI, S.C. on Brine Pumping (Compensation for Subsidence) Bill, p. 275.

7 Calvert, op. cit., p. 327.

8 P.P. 1887 XVII, p. 50, 23rd Annual Factory Inspectors' Report – Alkali works Regulation Act 1881.

9 C. Morris (ed.), *The Journeys of Celia Fiennes*, London, 1947.

10 Extract from the *Manchester Guardian* reprinted in the *Chester Chronicle*, 10 November 1888, p. 4.

11 Stonehouse, op. cit., p. 112.

12 J. T. Arlidge, *Diseases of Occupation*, London, 1892, p. 501.

13 *Widnes Guardian*, 16 January 1875, p. 6.

14 Arlidge, op. cit., pp. 501–2.

15 It was estimated that over a period of ten years 200,000 tons of cinders and other refuse, including dredgings from the Weaver, were dumped in the Winsford Flashes. P.P. 1890–1 XI, S.C. on Brine Pumping (Compensation for Subsidence) Bill, p. 108.

16 These cottages were built at Winsford in the 1860s and demolished in 1947.

17 Extract from *Manchester Guardian* reprinted in the *Chester Chronicle* 10 November 1888, p. 4.

18 P.P. 1873 LIII, Report on Landslips in the Salt districts, by Joseph Dickenson, p. 30.

19 P.P. 1876 (C 1443–1) XXX, S.C. on Factories and Workshops Act, vol. II, pp. 8728–9.
20 P.P. 1887, (C 5002) XVII, Chief Factory Inspector's Report, p. 87. See also *Widnes Guardian* 27 March 1875, p. 4, for prosecution of two men 'without visible means of subsistence' found sleeping on Aston's salt works, Witton.
21 *Widnes Guardian*, 9 January 1875, p. 5.
22 *Widnes Guardian*, 27 March 1875, p. 4.
23 P.P. 1887 (C 5002) XVII, Chief Factory Inspector's Report, p. 67.
24 P.P. 1882 (C 3241) XVIII, Report of Joseph Dickenson, Inspector of Mines, p. 303.
25 Calvert, op. cit., p. 201.
26 See note 24.
27 Daniel Defoe, *A Tour of the Whole Island of Great Britain*, Everyman edition, London, 1962, pp. 72–3. Defoe visited the town of Northwich to view the salt pits which he found to be 'odd indeed, but not so very strange as we are made to believe'.
28 P.P. 1882 (C 3241) XVIII, Report of Joseph Dickenson, Inspector of Mines, p. 338.
29 Willan suggests that in the later eighteenth century between 3,000 and 4,000 families were engaged in 'carrying'. Many were tenant farmers who, during the summer, carried the coal to their own homes for storing, and then sold it to the saltmakers during the winter when prices were higher. T. S. Willan, *The Navigation of the River Weaver in the Eighteenth Century*, Chetham Society, third series, vol. III, Manchester, 1951, p. 5.
30 P.P. 1817, III, S.C. on Rock-Salt in the Fisheries, p. 142.
31 Robert Craig and Rupert Jarvis, *Liverpool Registry of Merchant Ships*, Chetham Society, third series, vol. XV, Manchester, 1967, p. 34.
32 See note 24.
33 Francis E. Hyde, *Liverpool and the Mersey*, Newton Abbot, 1971, p. 30.
34 A. F. Calvert, *Salt and the Salt Industry*, London, 1919, p. 55.
35 K. L. Wallwork, 'The mid-Cheshire salt industry', *Geography*, 1959, p. 174.

II

1 J. J. Manley, *Salt and other Condiments*, London, 1884, pp. 43–4.
2 W. H. Chaloner, 'Salt in Cheshire 1600–1870', *Transactions of the Lancashire and Cheshire Antiquarian Society*, 1961, p. 63.
3 A. F. Calvert, *Salt in Cheshire*, London, 1915, p. 131.
4 Manley, op. cit., p. 44.
5 G. D. Twigg, 'Glossary of Open Pan Salt Terms', 1963. Typescript in Brunner Library, Northwich. Unless otherwise stated all the saltmakers' terms used in this passage are taken from this source.

6 G. T. Warren, 'The Salt Industry', in *Chemistry in Commerce*, E. Molloy (ed.), vol. III, London, 1934, pp. 1149–50.
7 Manley, op. cit., p. 76.
8 P.P. 1876 (C 1443) XXIX, Report on Factory and Workshops Act, vol. I. App. D, No. 94.
9 Twigg, op. cit., p. 70.
10 P.P. 1899 XII, Departmental Committee of the House of Commons on Miscellaneous Dangerous Trades (final report), p. 248, para. 51.
11 *Spon's Encyclopedia of Industries, Arts, Manufactures and Commercial Products*, vol. V, London 1882, p. 1732.
12 P.P. 1868–9 (C 4093–1) XIV, Factory Inspectors' half yearly report, 31 October 1868, p. 164.
13 Ibid., p. 163.
14 P.P. 1876 (C 1443) XXIX, Factory and Workshops Act, vol. I, App. D, no. 94.
15 P.P. 1868–9 (C 4093–1) XIV, Factory Inspectors' half yearly report, 31 October 1868, p. 163.
16 P.P. 1876 (C 1443–1) XXX, Factory and Workshops Act, vol. II, Q. 1017.
17 Manley, op. cit., p. 45.
18 A. F. Calvert, *Salt and the Salt Industry*, London, 1919, p. 126.
19 P.P. 1882 (C 3241) XVIII, Inspector of Mines Report, p. 338.
20 P.P. 1892 (C 6795) XXXVI, Pt II, R.C. on Labour, Q. 21,069.
21 For instance, in 1908 fourteen wallers were contracted at the Newbridge saltworks, Winsford to wall fourteen day Common (i.e. coarse salt that was fourteen days forming in the pan) at £1 per drought. Minutes of the Northwich and District Amalgamated Society of Saltworkers, Rock-salt Miners, Alkali Workers, Mechanics and General Labourers, 22 January 1908. This Union, founded in 1888, continued in existence with various changes of name until, as the Process and General Workers' Union, it amalgamated with the Transport and General Workers' Union in 1968. The author is indebted for the loan of this union's minute books to Mr R. M. Moss, former General Secretary, Process and General Workers' Union, now Cheshire District Secretary, TGWU.
22 P.P. 1892 (C 6795) XXVI Pt II, R.C. on Labour, Q. 21,069. In the 1880s, a time of fierce competition, it seems that the proprietors were surreptitiously increasing the intensity of the wallers' working day. 'Wages were not lowered', the *Manchester Examiner* reported in 1889, 'but the salt pans were made larger. The hands (i.e. wallers) had to draw more salt for the money they received, and when an example of this sort was set by one firm it was copied by others, in order that they might enter the market on terms of equality with their competitors.' *Manchester Examiner*, 29 April 1889, p. 3, col. 1.
23 T. S. Willan, *The Navigation of the River Weaver in the Eighteenth Century*, Chetham Society, third series, vol. III, Manchester, 1951, p. 50.
24 P.P. 1817 III, S.C. on Rock Salt in the Fisheries, p. 142.

25 Ibid., p. 144.
26 James Stonehouse, 'Salt and its manufacture in Cheshire', *Trans. Hist. Soc. of Lancashire and Cheshire*, vol. V, 1853, p. 113.
27 P.P. 1876 (C 1443) XXIX, Factory and Workshops Act, vol. I, App. no. 3, p. 116, para. xi.
28 P.P. 1818 V, S.C. on Salt Duties, p. 481.
29 P.P. 1892 (C 6795) XXXVI Pt II, R.C. on Labour, Group B. Q. 15,623.
30 *Morning Chronicle*, 3 June 1850, p. 5.
31 Robert Craig and Rupert Jarvis, *Liverpool Registry of Merchant Ships*, Chetham Society, third series vol. XV, Manchester, 1967, *passim*.
32 House of Lords Record Office, Committee Office, vol. 29, House of Commons Committee, 1840, Weaver Churches Bill, p. 88. The author is indebted to Mary Prior for this and subsequent references to the Weaver Churches Bill.
33 P.P. 1876 (C 1443) XXIX, Factories and Workshops Act, vol. I, App. no 3, p. 116.
34 Anon., *Osborne's Guide to the Grand Junction of Birmingham, Liverpool and Manchester Railway*, Birmingham, 1838, p. 226.
35 P.P. 1876 (C 1443) XXIX, Factories and Workshops Act, vol. I, App. no. 3, p. 116, para XI. Under the Weaver Navigation Act 1840, the Trustees of the Navigation undertook to build cottages for the watermen, and by the later nineteenth century it seems that the practice of families living on board the flats disappeared. Harry Wardale, 'The Weaver Navigation', *Trans. Lancashire and Cheshire Antiquarian Society*, 1933, p. 3.
36 *Morning Chronicle*, 3 June 1850, p. 5.
37 Ibid.
38 Ibid., pp. 5–6.
39 P.P. 1892 (C 6795) XXXVI, Pt II, R.C. on Labour, Group C. Q. 21,045.
40 Interview with Harry Healey of Winsford, May 1969.
41 Stonehouse, op. cit., p. 113.
42 *Morning Chronicle*, 3 June 1850, p. 6.
43 P.P. 1868–9 (C 4093–1) XIV, Factory Inspectors' half yearly report, 31 October 1868, p. 164.
44 Stonehouse, op. cit., p. 113.
45 Weaver Churches Bill, p. 83.
46 *Morning Chronicle*, 3 June 1850, p. 6.
47 Ibid.
48 Weaver Churches Bill, p. 42.
49 Ibid., pp. 23–4.
50 Calvert, *Salt in Cheshire*, p. 631.
51 W. H. Chaloner, 'William Furnival, H. E. Falk and the Salt Chamber of Commerce 1815–1889', *Trans. Hist. Soc. of Lancashire and Cheshire*, vol. 112, 1960, p. 123, n. 9.
52 Ibid., p. 123.

53 George Cadbury and S. P. Dobbs, *Canals and Inland Waterways*, London, 1929, p. 106.
54 *Reynolds News*, 21 October 1888, p. 5.

III

1 P.P. 1876 (C 1443–1) XXX, Factories and Workshops Act, vol. II, Qq. 8734–5.
2 Ibid., Q. 1017.
3 P.P. 1887 (C 5002) XVII, Factory Inspectors' yearly report to 31 October 1886, p. 69.
4 P.P. 1876 (C 1443) XXIX, Factories and Workshops Act, vol. I, App. D, No. 94.
5 P.P. 1876 (C 1443–1) XXX, Factories and Workshops Act, vol. II, Q. 8784.
6 In the dark these paths were a source of danger even to those familiar with them: Joseph Dutton of Castle, Northwich, giving evidence at the inquest on James Royle of Leicester Street, Northwich in March 1875, told the Court: 'They were (both) employed on upon Messrs Verdin Brothers' cinder boat. On Saturday night last deceased and witness went upon the towing path of the river Weaver by Winsford Bridge, at a quarter to eleven o'clock, to go to the hot-house in Messrs Verdin's salt works. It was very dark – they held each others' hand for protection. Deceased said he knew the road; witness did not, being a stranger, and having only been on the boat since Friday. The towing path is not protected from the river. They were both sober, each having had three gills of beer. When they got to the stage at Cross's works the path was narrower, and witness's foot slipped on the stone wall and he fell into the river. Deceased fell directly afterwards.' *Widnes Guardian*, 6 March 1875, p. 5, col. 3.
7 P.P. 1876 (C 1443–1) XXX, Factories and Workshops Act, vol. II, Q. 8733.
8 J. J. Manley, *Salt and other Condiments*, London, 1884, p. 44.
9 G. D. Twigg, 'Glossary of Open Pan Salt Terms', 1963. Typescript in Brunner Library, Northwich.
10 P.P. 1876 (C 1443) XXIX, Factories and Workshops Act, vol. I, App. E, nos 10, 11.
11 Ibid.
12 Ibid.
13 P.P. 1899 (C 9509) XII, Departmental Committee on Miscellaneous Dangerous Trades (final report) p. 250, para. 51.
14 P.P. 1876 (C 1443–1) XXX, Factory and Workshops Act, Q. 13,852.
15 P.P. 1899 (C 9509) XII, Departmental Committee on Miscellaneous Dangerous Trades (final report) p. 267, Table I.
16 *Widnes Guardian*, 27 March 1875, p. 4, col. 4.
17 Brunner, Mond and Co., whose salt pans were inside their chemical factory,

had altered the height of their pans from the floor to comply with the 'Special Rules for Chemical Works'. The committee used the following letter to counter the objections of the saltworkers:

Brunner, Mond and Co.
14th June 1899

> We have lowered the standing platform with three of our pans, which are all that we work at at present, so as to give 3 feet from the standing floor to the top edge of the salt pans. These have been working, since your visit, and we are glad to say that we have suffered no loss in the output of the pans in consequence, although the men at first, of course, grumbled slightly.
>
> We should point out that our men are working under very favourable circumstances on eight hours' shifts. As the labour will be certainly rather more, lifting the salt from a less advantageous position, men on twelve hour shifts may find rather more inconvenience but the extra labour required is not much, and we think men should soon accustom themselves to the altered position of the work.

P.P. 1899 (C 9509) XII, Departmental Committee on Miscellaneous Dangerous Trades (final report), pp. 249, para. 56.

18 James Stonehouse, 'Salt and its manufacture in Cheshire', *Trans. Hist. Soc. of Lancashire and Cheshire*, vol. V, 1853, p. 112.

19 P.P. 1899 (C 9509) XII, Departmental Committee on Miscellaneous Dangerous Trades (final report), pp. 249–50, para. 61.

20 Ibid., p. 250, para. 62.

21 Ibid., p. 250, para. 60.

22 Ibid., Appendix I, p. 267.

23 P.P. 1892 (C 6795) XXXVI, Pt II, R.C. on Labour, Q. 21,096.

24 When the Salt Union was formed in 1888 the proprietors offered to put £100 in the bank to provide the saltworkers with capital to start a permanent benefit club but the workers refused and carried on as before. Ibid., Q. 21,103.

25 *Mineral Statistics for 1856*, p. 21.

26 *Widnes Guardian*, 9 January 1875, p. 5, col. 7.

27 *Widnes Guardian*, 30 January 1875, p. 5, cols 4–6.

28 A. F. Calvert, *Salt in Cheshire*, London, 1915, pp. 527–35.

29 *Widnes Guardian*, 30 January 1875, p. 5, cols 4–6.

IV

1 P.P. 1892 (C 6795) XXXVI, Pt II, R.C. on Labour, Group C, Q. 21,061.

2 *Widnes Guardian*, 9 January 1875, p. 5, col. 5. For a report of the Flatmen's

Society's centenary celebrations see *Winsford and Middlewich Guardian*, 6 January 1892, p. 6.

3 T. S. Willan, *The Navigation of the River Weaver in the Eighteenth Century*, Chetham Society, third series, vol. III, 1951, p. 134.

4 Ibid., pp. 134–5.

5 The union was registered as the Winsford Saltmakers' Association on 20 June 1889, and continued in existence as the Winsford Saltworkers' Union until about 1969–70, when it amalgamated with the National Union of General and Municipal Workers. The author is indebted for the loan of these records which survive to Mr Harry Healey, of Winsford, former General Secretary, Winsford Saltworkers' Union, now Regional Organizer, NUGMW.

6 Winsford Saltmakers' Association Minutes, 26 March and 9 April 1853.

7 Ibid., 2 May 1853.

8 Ibid.

9 Ibid.

10 Ibid., 2 May 1857.

11 P.P. 1888 XI, S.C. on Emigration and Immigration (Foreigners), Q. 3361.

12 *Manchester Umpire*, 10 March 1889, p. 6, col. 1.

13 *Winsford and Middlewich Guardian*, 13 March 1878, p. 5, col. 2.

14 Winsford Saltmakers' Association Minutes, 26 August 1854.

15 *Chester Chronicle*, 27 April 1889, p. 6.

16 Winsford Saltmakers' Association Minutes, 7 May 1898.

17 Ibid., 21 May 1889.

18 Ibid., 30 June 1900.

19 Ibid., 14 July 1900.

20 Ibid., 27 January 1900.

21 One such strike occurred at the Jubilee Works, Winsford, during the annual 'holiday'. Twelve lumpers whose pans were stopped were offered work as wallers, but although they at first agreed, when they found they were working a man short they struck work. This particular strike lasted for three months and only ended when the workers' own pans were restarted. P.P. 1892 (C 6795) XXXVI Pt II, R.C. on Labour, Group C, Q. 21,096.

22 Winsford Saltmakers' Association Minutes, 29 July–25 September 1897.

23 *Manchester Umpire*, 10 March 1889, p. 6. col. 1.

24 *Winsford and Middlewich Guardian*, 15 March 1878, p. 5. col. 2.

25 Winsford Saltmakers' Association Minutes, 26 November 1861.

26 Ibid., 30 July 1898.

27 Ibid., 13 December 1902 and *passim*.

28 Ibid., 7 April 1900 and *passim*.

29 Ibid., 25 March 1899.

30 Ibid., 30 March 1899.

31 Ibid., 2 May 1853.

32 *Chester Chronicle*, 4 May 1889.

33 On 2 May 1853 the Winsford Saltmakers' Association resolved that members could transfer from the Winsford Union to either the one at Northwich or the one at Hassel Green, Sandbach (or vice versa), providing they were fully paid-up members and had obtained a clearance form. Hereinafter there is no mention of a saltworkers' union being in existence at Northwich until 1888, when a new union was formed.
34 Winsford Saltmakers' Association Minutes, 6 October 1900.
35 Northwich and District Amalgamated Society of Saltworkers, Rock-salt Miners, Chemical Workers, Mechanics and General Labourers Minutes, 4 February 1908.

V

1 P.P. 1888 XI, S.C. on Emigration and Immigration (Foreigners), Q. 3378.
2 P.P. 1876 (C 1443–1) XXX, Factory and Workshops Act, vol. II, Q. 9783.
3 P.P. 1876 (C 1443) XXIX, Factory and Workshops Act, vol. I, App. E, nos 10 and 11.
4 P.P. 1888 XI, S.C. on Emigration and Immigration (Foreigners), Q. 3317.
5 Ibid., Q. 3362.
6 J. J. Manley, *Salt and other Condiments*, London, 1884, pp. 74–5.
7 W. F. L. Dick, *A Hundred Years of Alkali*, Birmingham, 1973, p. 16.
8 *Birmingham and the Hardware District*, quoted in *Meliora*, vol. ix, 1866, pp. 371–2.
9 A. F. Calvert, *Salt in Cheshire*, London, 1915, pp. 684–92.
10 Ibid., pp. 631–2.
11 Ibid., p. 631.
12 P.P. 1888 XI, S.C. on Emigration and Immigration (Foreigners), Q. 3317.
13 In 1840, when these two families were working their flats on the Weaver, they each signed a successful petition requesting that churches for the use of watermen and their families be built out of the profits of the Weaver Navigation. House of Lords Record Office, Committee Office, vol. 29, 1840, Weaver Churches, p. 103. The Deakin brothers were particularly successful, and owned, in addition to salt works, extensive shipyards at Winsford, *Northwich Guardian*, 27 April 1889, p. 4.
14 See D. A. Iredale, 'John and Thomas Marshall and the Society for Improving the British Salt Trade: an example of trade regulation', *Economic History Review*, second series, vol. XX, no. I, 1967. The duties had been increased from 7*s*. 8*d*. to 10*s*. per bushel in 1798 and raised again to 15*s*. per bushel in 1805. Because, in normal circumstances, the duty had to be paid at the works, sometimes weeks or months before the manufacturers received

payment, it had the effect of excluding those with little capital from entering the industry, and small firms were driven into bankruptcy. The Marshalls considered this necessary to improve the trade, for, in their opinion, small firms only 'contributed to over-production, poor quality and low prices . . . evils the large proprietors wanted to eliminate'. Iredale, 'John and Thomas Marshall', p. 84; cf. also P.P. 1888 v, S.C. on Salt Duties, p. 519.

15 Ibid., p. 91.

16 Ibid., p. 92.

17 'A general meeting of the "White Salt Trade" . . . was held at the Angel Hotel, Northwich, on 4 March 1841, and began by taking steps to recover £1206 due to the late Association.' W. H. Chaloner, 'William Furnival and H. E. Falk and the Salt Chamber of Commerce 1815–1889', *Trans. Hist. Soc. of Lancashire and Cheshire,* vol. 112, 1960, p. 131.

18 A. F. Calvert, *Salt and the Salt Industry*, London, 1919, p. 55.

19 Chaloner, op. cit., p. 132.

20 Manley, op. cit., p. 73.

21 Chaloner, op. cit., p. 134.

22 P.P. 1888 XI, S.C. on Emigration and Immigration (Foreigners), Qq. 3232–35.

23 Manley, op. cit., p. 61.

24 Chaloner, op. cit., p. 135.

25 Ibid., p. 136.

26 Founded at the instigation of H. E. Falk the Salt Trade Committee became the official body for attempting 'to adapt the make to the demand and regulate prices. . . .' Although it was an offshoot of the Salt Chamber it was not officially recognized by that body. Ibid., pp. 137–8.

27 Manley, op. cit., p. 73.

28 Chaloner, op. cit., pp. 135–6. Prior to this the State monopolies in the manufacture and wholesaling of salt had been protected by high import duties. After 1863 the high import duty remained but with supply mainly in the hands of Cheshire salt manufacturers the position was virtually the same as that which had existed in England from 1805 to 1825 and the proprietors made the best of it. Even when prices fell they were able to maintain the price of 'Calcutta salt' at between 25 and 50 per cent higher than any other.

29 Manley, op. cit., p. 73.

30 Chaloner, op. cit., p. 139.

31 Manley, op. cit., p. 73.

32 Chaloner, op. cit., p. 139. The small masters were blamed by Falk for the defeat of the proprietors in the general strike of 1867–8. The strike was over a wage claim which the men put forward as a compensation for the recently legislated limitation of women's and children's hours under the Factory Acts

Extensions Act of 1867. The strike commenced either in late December 1867 or early January 1868 and it seems likely that before it began many of the 7,000 workers involved were already without work. Those on strike seem to have included all the workers from every trade engaged in the production of salt, and the entire salt industry of Winsford was paralysed. At first the salt-proprietors 'combined against the strikers very strongly' and for a time it must have seemed that the workers had miscalculated the timing of their strike (Calvert, *Salt in Cheshire*, p. 521; P.P. 1888 XI S.C. on Emigration and Immigration (Foreigners), Qq. 3326–8, 3237). However, the strike dragged on into a period of the year when trade was usually brisk, and after about two and a half months thirty-four of Winsford's thirty-five salt proprietors gave way. One reason given in later years was that they were 'largely men who had risen from the ranks themselves' and 'in great sympathy with the labouring population'. With the exception of H. E. Falk, the employers sought to bring the dispute to an end by whatever means they were able: the majority . . . certainly conceded the demands of the men in full.' Ibid., Qq, 3237, 3332. Only H. E. Falk gained a measure of success against the strikers. Determined to resist his workers' demands he imported a large number of Germans to take the place of his men, and kept them on when the strike was over. For the next twenty years Falk's works were black-listed by the wallers and labourers of the district (ibid., Qq. 3370, 3428–30).

33 Manley, op. cit., p. 73.

34 Calvert, *Salt and the Salt Industry*, p. 54.

35 Manley, op. cit., p. 73.

36 Ibid., p. 75.

37 Chaloner, op. cit., p. 140.

38 A. F. Calvert, *A History of the Salt Union*, London, 1913.

39 Calvert, *Salt in Cheshire*, p. 556.

40 Chaloner, 'William Furnival and H. E. Falk', p. 141.

41 Only *The Times* noted that the proprietors had not achieved a monopoly and observed:

'The firms that keep clear of the combination will thus be enabled to undersell the syndicate by a sufficient margin of price to enable them to get a leading place in the market.' Calvert, *A History of the Salt Union*, p. ix.

42 Ibid., p. xv.

43 Ibid., p. xvii.

44 Ibid., p. xviii.

45 Ibid., p. xxviii.

46 Ibid., p. xxviii.

47 Ibid., p. 263 and *passim*.

VI

1 To prepare the larger works to receive extra men many disused pans had to be repaired. At Messrs Verdin, Winsford, 'seventeen sets of smiths (were employed) renewing and repairing pans'. *Chester Chronicle*, 27 October, 1888, p. 5.
2 *Northwich Guardian*, 4 May 1889, p. 6.
3 Loading was largely governed by the state of the tide at Liverpool; for example, during neap tides no salt could enter the Mersey from the Weaver for four or five days. Therefore those wallers who had to load salt were occasionally required to arrive at 4 a.m. or stay after 4 p.m. in order that the vessels to be loaded could meet the tide.
4 This was a bonus formerly paid by the independent watermen and shared between the loaders. After the introduction of steam barges the practice was abandoned.
5 *Manchester Examiner*, 29 April 1889, p. 2.
6 *Chester Chronicle*, 2 May 1889, p. 6.
7 At this time Thomas Ward was Manager of the Cheshire district. He was elected to a directorship in 1891. A. F. Calvert, *A History of the Salt Union*, London, 1913.
8 *Chester Chronicle*, 7 May 1889, p. 6.
9 Ibid.
10 *Manchester Examiner*, 10 May 1889, p. 5, col. 8.
11 *Manchester Examiner*, 16 May 1889, p. 3, col. 4.
12 Loaders, it seems, soon took on something of the character of the wallers. Like the wallers they worked in small gangs, whose composition was apparently determined by themselves, and contracted to load salt according to an agreed list of prices. Their work consisted mainly of handling large tonnages of salt or coal, and they were paid on a piece-work basis. Thomas Ward, a leading salt proprietor, considered that as they were paid by the ton, 'It is immaterial to them what the hours are, whether from 6 o'clock in the morning till 2 or from 9 o'clock in the morning until later on . . . they are not working while they are waiting.' (P.P. 1892 (C 6795) XXXVI, Pt II, R.C. on Labour (Group C) Q. 21,065). However there was a lot of waiting time and they were not paid for it. One of their demands in 1900 was that 'all orders be out by 10 am. or the men waiting after that hour receive 1s per hour for waiting or the men will go home and be considered free for the day' (Winsford Saltmakers' Association Minutes, 1 December 1900). This same minute has a new list of prices issued by the loaders: '4d per ton for all salt loaded out of storehouses, 5½d. per ton for patent Butter Salt, 4s. 6d per 100 for all patent Butter salt in 1 cwt. sacks, 150 lbs sacks 4s. 6d per 100 and other sizes in proportion.'

13 The wallers were not completely cut off from loading. It became an 'extra' which they could undertake if they had time. Those who were able to have the help of wives and children in transporting the salt from the hurdles to the warehouse might finish early enough in the day for them to take on loading as a second job. It appears that, given early start and help from their families the wallers could finish their work by lunch-time; and afterwards, through custom and practice, they were permitted to spend the rest of the day working as labourers or loaders. Thus, to the agreed weekly sum received for the walling, could be added the pay for two or three hours' extra work, earned in the afternoons. This information was supplied by Mr Henry Thompson, salt proprietor, of Marston, Northwich, and was confirmed by Mr Harry Sutton, ex-General Secretary, Mid-Cheshire Salt and Chemical Industries Allied Workers' Union of Marston, Northwich, during an interview on 8 January 1974. (Mr Sutton began work in 1902, for the Salt Union, as an apprentice joiner. He continued in that occupation until about 1945, when he became a full-time union official; he also remembers that wallers, in particular, would leave off work at lunch-time and go to help local farmers with the harvest.)

14 The power of the foremen, or banksmen as they were called, increased over the years. In 1897 the foreman at the Over Works, Winsford, stopped the pans and put the men out of work after an argument with his workmen over the rate for the job. The Winsford Saltmakers' Association had to pay lock-out benefit to their members until the dispute was settled. Winsford Saltmakers' Association Minutes, 4 October to 13 November 1897.

15 P.P. 1892 (C 6795) XXXVI, Pt II, R.C. on Labour, Group B, Qq. 15,590–1.

16 Ibid., Q. 21,055.

17 Ibid., Q. 15,631.

18 Ibid., Q. 15,617.

19 Ibid., Q. 15,590.

20 *Northwich Guardian*, 20 August 1892, p. 4.

21 Ibid., 24 August 1892, p. 4.

22 Ibid., 27 August 1892, p. 4.

23 Ibid., 31 August 1892, p. 4.

24 Ibid.

25 Ibid., 7 September 1892, p. 4.

26 *Chester Chronicle*, 10 September 1892, p. 5.

27 Ernest Solvay, the son of a Belgian salt proprietor, between 1856 and 1861 discovered a method of producing Soda ash, a basic chemical with a multitude of uses, using brine, limestone and ammonia.

28 W. F. L. Dick, *A Hundred Years of Alkali*, Birmingham, 1973, p. 16.

29 Ibid., p. 83.

30 Stephen Koss, *Sir John Brunner, Radical Plutocrat, 1842–9*, Cambridge, 1970, p. 39.

31 Dick, op. cit., p. 103.

32 *Northwich Chronicle*, 16 February 1889, p. 5.
33 Tom Mann, *Memoirs*, London, 1923, p. 24.
34 *Northwich Chronicle*, 16 February 1889, p. 5.
35 Dona Torr, *Tom Mann and his Times*, vol. I (1856–1890), London, 1956, p. 274. '600 men toiled without meal reliefs, taking snacks of food as they worked twelve-hour shifts, seven days a week.'
36 A. F. Calvert, *A History of the Salt Union*, London, 1913, p. xi.
37 Dick, op. cit., p. 101.
38 *Labour Elector*, 2 February 1888.
39 Northwich Union Minutes, 15 August 1904.

Part 4

The Durham pitman

Dave Douglass

I

The miner ready for work squats on his 'hunkers', sits on his hat or leans squatting against a wall, enjoying his last cigarette. There is always somebody having a 'crack' or telling jokes. A lot of men will remain quiet while smoking, savouring the last drag, not only of the cigarette, but also of the fresh air before going down.

Groups of 'marras' or teams of men will be breaking away all the time and heading for the lamp cabin to collect the cap lamps and oil lamps. There will be a sudden break and the whole crowd will move in on the cabin. Walking towards the shaft in the darkness of the evening or the early morning, lamps dart to and fro and there is an orange glow from the oil lights, as the men make their way to the cage. Jackets fly out in the wind around the strange rhythmic stride. It is at this time that the bitterness is really felt. 'Bliddy life this bugger is', 'Whey th' hasta be sumit better than this though?', 'Aye, anyone knaa wheres there's a job gaaning from eight till nine wi' an hour for breakfast?' 'What? bloody great job this man, better than a teacher's shift anytime.' 'Ah knaa, they think wa bliddy owls am sure.' The banksman collecting the checks and smacking the pockets of the men in a semblance of a search shouts: 'Cumon me lucky lads! Any more for roond the pear'. The deputy in mock encouragement shouts: 'Lets get tiv it lads, wi'l meck a show today, wi'l kill it.'

The miner's only comfort underground is his baccy or his pinch. Some get that fond of the latter that they even put it on the end of the cigarette, to produce a poor man's menthol effect when it's lit.

The miner's task on the face is obviously improved or made harder by geological conditions. If the 'bottom' (or floor) is soft, it renders the work of the collier ten times more difficult; apart from his coal he will have to take a dint of floor up in order to get a hard base. The cutter will sink into the soft muck or stone shale and kick up terrible dust as it cuts. If the 'top' (or roof) is soft, and made up of loose, crushed stone, it is difficult and dangerous to support. A hard roof is safer to work in, although if it's very hard it won't budge when the supports are withdrawn; this is very dangerous as gas accumulates and when the waste (the 'goaf' or 'gob') falls it will come down in one great big bang. When the face was supported with props and bars, miners would listen for the sound of the weight coming on. As soon as the roof started to 'tinker' the men would retreat as fast as possible before the roof came crashing down, skittling the props from under it.

The Tyne pits had their own qualities, their own geological differences and dangers. Being close to the water of the Tyne, they were invariably wet pits. Certain of the water was of course indigenous to the rock strata itself, but in the whole range of pits along the banks of the Tyne – Friar's Goose, Felling, Wardley, Follonsby, Usworth, Hebburn and Jarrow – there was a perpetual battle with the never-ceasing water from the river. In certain of the collieries the water streamed from the strata higher up, and eventually came down into the lower seams. Others because of their closeness to the river, acted as involuntary, but none the less efficient, sumps, into which the water was constantly fed.

Friar's Goose colliery was sunk right on the bank of the river about midway between Gateshead and Felling shore. In the 1850s when it was the property of Messrs Carr and Partners, who at that time also owned Felling, it was pumping from out of its seams 1,000 gallons of water per minute or 6,000 tons; the weight of the coal excavated was 250–90 tons.[1] Wardley had to contend with water pouring in on it from all sides; the bulk of the water was coming from the Tyne via Friar's Goose, streaming down to join with Felling's and ultimately Heworth's waters; a virtual cascade, and as well as this there were Wardley's 'own' waters, which came in through the higher seams. Before closure, Wardley was pumping 500 gallons per minute. Now it is Usworth who carries the problem of water, but when she closes that should stop the line, unless Bolden ever starts to work back towards Wardley, in which case there would be a very grave danger of the water breaking away and flooding that colliery; however, such an eventuality as working in that direction is very unlikely to arise.

Wardley, to my memory, was always a cramped wet hole. Men lying on their sides were soaked right through, their boots sodden and flesh driven mad with the chemical-filled water and the abrasiveness of the sharp pieces of coal sticking to their arms and legs. Colliers kneeling in water faced an agony of soaked knee-pads rubbing into the bones and wet straps cutting into the skin. The tools were gummed up with the wet small coal which makes an extra toil of shovelling or picking. The slightest cut became filled with the salt-like water and the damp particles of coal 'curvings'.

The seams in north-east Durham are not uniform 'carpets' stretching under the earth. They will be top seams in some places, middle and even lower seams in other mines. Great geological convulsions have snatched them high up into the earth or sent them crushing down, twisted them round and broken them into short patches or a series of steps. The whole area round Wardley was a mass of faults and rolls. The seam changes quality according to where it is. A fault can in many cases spell the death of a colliery. If nothing else, a 'hitch' will mean hard graft, danger, and a real possibility of a reduction in piece rates. These hitches and faults were commonly called 'dykes' in the old days. At times a vein of stone will intersect the seam. It may break off sharp and continue at a different level either above or below. Two principal dykes intersect the northern coal field. Both run easterly. One of the dykes throws the seam (especially near to Newcastle and North Shields and Wallsend) perpendicularly down an additional 130 fathoms, so that two collieries within a quarter of a mile of each other could be working the same seam but at depths differing by ninety or 130 fathoms. The other was the Hemerth dyke, south of which the main seams rise twenty-five fathoms.[2]

The principal concern of the miner in all this would be the level, the danger, the dust. The following song describes what it can mean to the miner to meet with such a hitch.[3]

The Great White Wall

We have survived the great sweeping hand of Robens and his gang,
To each and every gaffer's trick we stood firm to a man –
And now the wage is just worth while, bad fate does us befall,
For the only face that has our hopes has hit a great white wall.

In many sour battles, and to gain a decent wage,
The bosses and the union bitter struggles did engage,
No quarter was expected, well no matter what the call,
But now its gone and licked us both that bloody great white wall.

Among the tall and shiny coal a message stark and white,
The dust as thick as fog, or steamer's white smoke in the night.
Just like a monstrous whale it sits so wide so very tall,
She'll burst our lungs or shut the pit that bloody great white wall.

When first we seen that little stone stuck in amongst the coal,
The gaffers said 'lads, have no fear, it's nothing but a roll.'
But daily now it's growing and we've had it one and all –
For there's very little coal to find amongst the great white wall.

In the poetic words of a miner in the radio ballad 'The Big Hewer': dust is a legacy from the past, it is a giant killer and it has destroyed an army of miners, so minute in its form, so destructive in its ravaging powers.

In Doncaster, it was my lot to be landed on one faulted face after another, the modern machines cutting and grinding their way through the solid stone. The picks throw up a hail of sparks filling the choking air with a foul burning smell of hot metal and burned stone. The dust comes in a thick impenetrable fog filling the nose, the eyes, and (of course) the lungs. One cannot see, quite literally, a raised hand put up before one's face. There is more dust about since the introduction of machinery and it is said that today's generation of young face workers will suffer even more than our forefathers from the dust, that our lives will be shorter and our end more painful. No figures are available that I have seen to prove or deny this . . . only the young men's lungs will make mockery of the official 'safety standards'.

As if the faulted nature of the coal in Durham were not enough, the seams at these collieries are generally very thin; even in 1965, when I worked there, the majority of faces were only 18 inches high, and while some were 2–2½ feet, still others were lower than 18 inches. In the Hazel seam in Doncaster coalfield, the average height would be about 4 feet, whilst the Barnsley seam was anything from 7 feet, and that leaving 2 feet of coal 'conneys' to act as a roof. When I was first transferred to Yorkshire, the training officer told me that he would take me for a walk along the coal face . . . a walk! I assured myself that this must be some part of an alien southern culture, and turned up ready with knee-pads and a couple of hand clogs (two wooden blocks of wood, grasped one in each hand), in preparation for the usual swimming 'belly flopper', or at best, ready for a long crawl. But sure enough they do make coal that high in Yorkshire and even higher, as I was to discover later. After working four years on my knees every day, I found it quite impossible to stand and shovel; once, when I was sent to the surface to fill bags up with sand for a fire burning down below, I dropped to my knees immediately I had the shovel in my hands, and this on the surface with the sky as a roof! Men who had been used to lying flat and shovelling all their lives found it agony getting used to kneeling up and working.

It is painful to watch a big man, who has been used all his life to a big seam, forced to kneel down and work. He will all the time try to stand up, and often ends up with his back wedged against the top and his legs spread-eagled for support, rather like a giraffe bending to drink water; anything rather than kneeling.

Apart from the seam height, there was truly a superabundance of fossils in our district. The men used to bring them home for their children and young miners would bear them like prizes to their girl friends. At Usworth was found one of the greatest animal fossils in the world – an amphibian named as an Anthracaurus, one of three found in the world; this beast would have walked the dense forests of Usworth and Wardley some 250,000,000 years ago.[4]

The thin seams of Durham are a nightmare, and many's the nightmare I've had about them since. Strange to say you don't really become aware of what they actually mean, how terrifying they are, until you've left them and look back on it. Then, in the form of a passing thought quickly pushed out of the mind, or in the middle of the night, when you wake up in a cold sweat and sigh for absolute relief, you know what filthy little cracks and warrens they are. When I first started work at the colliery (actual work, as opposed to training) I was early introduced to Wardley's thin seams in my job as line's lad.

At the time it was an adventure, and I can honestly count on one hand the number of times I was scared, but when I was scared I was near scared to death. My lack of fear did not come from any bravery, it came from absolute ignorance of my surroundings. Somewhere at the back of my mind I believed that the face, like a scaring fairground ride, was really quite safe; nothing could possibly happen that could hurt me. I was not then aware of the very real dangers that confronted me and the absolute insecurity of any coal face no matter how modern, of the temperament of the roof (particularly with the old-fashioned support method used at Wardley). The real situation didn't dawn on me until I was an actual coal face worker.

Crawling down the seam, only inches would separate the roof from your prostrate body, your head would be turned to the side, flat against the floor with maybe a two-inch space above before you made contact with the roof. Crawling was called 'belly flopper'; you would 'swim' forward, arms straight out in front, legs spread-eagled behind, pushing, riving, wriggling forward, and always the roof, bumping and creaking inches above your head. Working in Yorkshire, I found the seams at least high enough to crouch in, a thing which gave me great confidence when I was withdrawing the supports to allow the worked out area to collapse.

Occasionally the situation comes back to me when I am asleep, then I'm in the thin seam, my head flat to floor and the roof is slowly weighing down, down, down on my head and all efforts to escape are to no avail.

I've often wanted to return to Tyneside, to my North-east coast and countrymen, but when I think that the work might include work in a thin seam, I resolve that Donnie (Doncaster) will do me after all.

II

The Durham miner is described in the early nineteenth century as 'of only middling stature (few are tall or robust), with several large blue marks, occasioned by cuts, impregnated with coal-dust, on a pale and swarthy countenance'.[5] That description would go for his descendants today, but in the matter of dress there have been great changes. Most early nineteenth-century descriptions of the Durham miner make him out to be a colourful character. He might wear a coloured kerchief about his neck, a 'posied waistcoat', opened at the chest, displaying a striped shirt beneath, a short blue jacket not unlike that which seamen used to wear, only shorter, velvet breeches which would be left unbuttoned at the knee, blue worsted stockings with white 'clocks', and long, low-quartered shoes.[6] It was the custom of young miners at this time to wear their hair in curls, forming them at their temples by turning the hair round a thin piece of lead enclosed in paper.[7] The leads were taken out at the week-end, when the hair was given a thorough wash. The influence of seamen might be seen in the wearing of pigtails, which eventually died out.

The older miner of today, with his strict Methodist discipline, reckons little when some of his younger mates appear with beards, long hair, and coloured waistcoats. ('They divand look like pitmen these days, mair like pansies'); he would be surprised to know that long hair was for centuries a feature of the miner, as were his colourful clothes.

Aw now begin te curl maw hair
(For curls and tails were all the go)

Before the influence of Methodism, and the introduction of 'respectability', not only the miner, but his wife, too, struck a gay and wild pose before the eyes of the more conservative stranger. The pay Friday would be a grand day for merriment and extravagance on the pitman's fortnightly pay. The women would journey from their villages into the towns, with light hearts and full purses to get in the fortnight's provisions.[8]

> A long cart, lent by the owners of the colliery for the purpose, is sometimes filled with the women and their marketings, jogging homeward at a smart pace; and from these every wayfarer receives a shower of taunting coarse jokes, and the air is filled with loud, rude merriment. Pitmen do not consider it any deviation from propriety for their wives to accompany them to the alehouses of the market town and join their husbands in their glass and pint. I have been amused by peeping through the open window of a pothouse, to see parties of them, men and women, sitting round a large fir table, talking, laughing, smoking and drinking *con amore*; and yet these poor women are never addicted to excessive drinking.

Today miners dress for underground in the oldest clothes available – most of them descendants of 1930s suits or jackets which have been handed down to them. Old waistcoats are popular because they are light to wear but have numerous pockets. The older men often wear old suits, with great baggy pants patched roughly, many with a whole leg-piece missing, the scar of some tugging match with a strut which was sticking out, or a tub locker. Younger lads will wear old jeans cut from the turn-up to the knee, some with a chevron cut out. These make it easy to remove the jeans (or 'lang uns') when getting stripped for work. Two belts are worn, one attached to the trousers and removed with them, the other holding the battery pouch with pastry, the self-rescuer 'monoxide gas mask', knee-pads, dust mask and sometimes a small pair of pliers; generally, the bait tin will be attached to the belt as well. All of these will be needed through the day, so they will be put on again with the belt after the trousers are removed. The saying is 'owt dis th' pit'. It breaks a miner's heart to have to buy anything for the pit; whatever can be salvaged from a cupboard, a rubbish tip, a neighbour or a junk shop will be carried to the pit. The exception is the boots: safety boots are supposed to be compulsory; if the feet are injured and one is not wearing the regulation boots then no compensation is forthcoming. All the same, boots will be worn for years; one will often see miners with great holes in the uppers or with pieces hewn out of the boot; the soles will be wafer-thin and the slightest stone one stands on will sting as much as if one had bare feet. Rather than buy new boots the miner will spend hours sorting through the rubbish tip looking for old boots; these will be adapted to his size, either by cutting holes in the sides or splitting the backs and weaving them back with shot-firing cable. When the local Army and Navy stores was selling off clogs at 5*s*. per pair hundreds of miners bought out the stock. I myself wore a pair of these clogs the whole of the time I was on haulage work, finding them comfortable, but they were no good for face work, since they did not bend at the toes when I knelt down.

Old shirts are worn with arms torn out from being caught in some awkward object when the miner was walking inbye: shirts with no buttons and flying out wildly en route to the cage. Jackets are misshapen and present rather a camel-like appearance, since they cover the battery respirator, etc.

Pit stockings are like football stockings, nearly always navy blue; on top of these stockings are worn another pair with the feet cut out, the bottoms being a cover for the top of the boots, rather like a soldier's gaiters. These serve as invaluable aids in keeping small coal out of the boots when shovelling or filling wire bags (wire bags are filled with small coals and serve to build walls when stones of the right size are not forthcoming).

Underneath the 'lang uns' shorts are worn, rather like football shorts but shapeless, with baggy legs. These are called 'hoggers' in Durham and 'bannickers' in Yorkshire. When working at the face, the men will only wear 'hoggers', belt, boots and knee-pads; at times, in very hot places, men will work without 'hoggers', although this is nowadays discouraged, since it does not fit the NCB's image of respectability (but then neither does most of what happens at collieries).

The miner's helmet will be of a variety of shapes and colours, the NCB (for reasons best known to itself) changing them every so often. Regardless of shape or colour or material (plastic or fibreglass), they are hated by miners young and old, and have been ever since their general introduction in the early 1960s. They replaced the traditional helmet, which was black and made of pith or a type of compressed cardboard. The old helmets were wonderfully comfortable and after a few years they moulded themselves to the shape of your head, never wobbling or slipping when you were working. Although none of these helmets have been produced for many years, a good third of the men at most collieries will be seen wearing them. These helmets never find their way to the rubbish tip until they are entirely useless. A lot of them have been going the rounds since before the war and are passed on from father to son or from workers who are leaving to whoever can get his hands on them first; very much battered and bearing the scars of a thousand heads saved from a busting they are, nevertheless, as comfortable as a felt hat.

III

The Coal Board would like to introduce more and more supervision into the miner's work, and to reduce all miners, and all classes of their work, to a single status. They would like the mines to work as factories. The attempt is foolish and bound to fail. A coal mine has little in common with a factory. The entire 'plant' has to be moved forward three or four times a day; the working material is often wet, filthy, and crawling with beetles; the roof and walls are liable to cave in. All the time he is working the miner has his ears strained for ominous rumblings and his eyes on the alert for signs of collapse. He must be prepared to drop everything and secure his workplace before carrying on with the job. All equipment necessary for advancing the face, all roadways, all materials and all the men pass through two or three low and narrow tunnels. Most factory workers would regard the mine purely and simply as a black and filthy hole; funnily enough, the miner in turn regards the factory as a prison and its operatives as captives.

One of the few redeeming features of pit work, and one that miners will fight to maintain, is that of independent job control. The new face systems push more and more officials on to our backs. Where once a face would have a single deputy, and a district (comprising a group of faces) had one overman, we now have three deputies and an overman to a face, plus an under manager for each district. An absurd situation. Now there are three deputies playing hide-and-seek with the men and trying to keep out of their way, where before there was one. I remember only a few years ago, after the signing of the National Power Loading Agreement, which in theory transferred direction of the job to the under officials, workers walking out of the pit because they refused point-blank to be supervised. Carter Goodrich tells of a case arising from the Minimum Wages Act when an overman was called forward to tell the court whether a certain worker did his job properly. The overman answered that he didn't know: 'I never saw him work.' The magistrate insisted: 'But isn't it your duty under the Mines Act to visit each working place twice a day?' 'Yes', came the reply. 'Then why', said the magistrate, 'didn't you ever see him work?' To which the overman replied: 'They always stop work when they see an overman coming, and sit down till he's gone – even take out their pipes if it's a mine free of gas. They won't let anybody watch them.'[9]

Mr W. H. Sales, in a speech to the North East Deputies Conference in 1954, told them that their role in industrial relations was two-fold:[10]

(1) To get the workmen to act as collaborators . . . in a great enterprise with a common will and cause.
(2) To assist the workmen in developing sentiments of industrial pride, loyalty and a real sense of 'belonging'.

This is the deputy the Coal Board would like to see, but it isn't one that many miners would recognize (or many deputies either). In my own experience there have been three types of deputies: the sort of amusing (without knowing it) character two steps removed from the face worker; the hard but fair official for all the world like an NCO; and the character on the way up the promotion ladder who will stand no argument ('My decision is final', etc.). It is this latter sort who are often the cause of strikes.

Thomas Burt in his autobiography gives us a description of the first sort. I would be hard pressed to improve on what he says:[11]

> Thomas Mood, invariably called 'Tom Mudd', was deputy at a large flat where I was a putter . . . Tom was a man of varying moods, though as a rule he was kind-hearted and fair in his dealings. He was utterly illiterate, and knew nothing of the science and art of numbers. Tom's most trying ordeal was the placing of the work. He would rather have timbered the most rickety working-place, or drawn the most dangerous 'jud'. Knowing full well his weakness in arithmetic, he always tried to avoid this ordeal. 'Noo, hinneys', he would cry good-naturedly, 'ye've had a good bait-time; let's hev a start again. Ye'll get yer wark the syem as in the fore-shift.' 'Nay Tom', someone would . . . suggest, 'how can that be? We have four or five men more than we had in the fore-shift. We must have the work placed properly before we start. Here's a bit of chalk. Figure it out on the blackboard.' Tom was unwilling to acknowledge his incompetency to solve the problem. He would puzzle and perspire for a few moments. When he had made his final statement, he always had a few half men left that he did not quite know what to do with. All the time we . . . had mentally arranged matters among ourselves, and, after we had worried poor Tom sufficiently, and gained a few minutes extra bait-time, we cheerfully resumed the work of the day.

This puts me in mind of a deputy at the shaft bottom, standing beside some tubs with freshly arrived material for his 'gate'. He puzzled for a full half-hour over what name to give some sheets of corrugated iron, then picked up his chalk with sudden inspiration and wrote on the side: '4 tubs crinkly tin'.

The overman is directly below the under manager and regards himself as much more of an actual gaffer than the deputy. Some overmen are what we describe as 'bob-a-job' men; they are never content with the work you do, and the completion of one task sets them working overtime to find or invent another. They spend a lot of time running around after 'back bye' and 'datal' workers. I remember a set of material workers in Durham sitting down exhaustedly after unloading many trams of steel girders. The overman arrived just as they were sitting down. 'Right lads', he began, 'Ah want all them girders moving to the other side of the gate.' As he stalked off one of the workers said to his mate: 'Ah knaa what he wants'; the sharp-hearing overman whipped round: 'Aye, and what div ah want?' 'Oh', replied the worker, 'ye want all these girders moving over to the other side of the gate.' George Parkinson, in his book *True Stories of Durham Pit Life*, tells of his first meeting with an overman at the age of nine:[12]

> One man came through, wearing blue clothes, a cap with a peak behind, and carrying a stick in his hand. He looked very different from the other men, for his face and hands were clean, his jacket was buttoned, and his flannel shirt looked very white. I saw, too, that he carried a watch in his pocket, for the seals were hanging out, and altogether I was much impressed by his appearance. He was a big man, and seemed as 'one having authority'. He looked very sternly at me, as he held up his stick in a threatening way, and said, 'Now mind, ef thoo gans to sleep and dizzent keep that door shut, thou'll get it.' But as he went away I said to myself, 'Aw isn't gaun to sleep, and ef ye touch me ye'll get it when mi father comes out.'

Telephones for some reason seem to addle the brains of the simpler kind of overman. He will nearly take the handle off the phone winding it round and round as if it was an alarm clock. Once on the phone, he becomes aware of the great distance over which his call must travel, and invariably shouts at the top of his voice in the apparent belief that lung power and not electronics are responsible for his voice being heard all that way 'outbye'.

In the early days of mining, the miner was practically a self-governing agent. The degree of job control (though necessarily limited by private ownership) was almost complete. Under the 'bord and pillar' system, which was almost the only method of working the coal,[13] two men usually worked together in a 'stall'. The deputy would be expected to visit them twice a day; but his role was simply to guard against explosions and falls and to fire shots; what the hewers did with their time and how much work they did was a matter for themselves. Under the piece-rate system they could be safely left to it. If the carrot of more money would not extract more work, no official could ever hope to do it. When miners went in for restricted production, which they often did in County Durham, there was nothing the deputies could do.

The coal face, under the bord and pillar system, can be thought of as a checker-board of places, with broad pillars of coal isolating one stall from another. The stall was a room surrounded on three sides with walls of coal. Into these places the hewers would come. In some cases a hewer might work the face entirely alone; usually, though, he had a 'marra'. Each of these stalls was a totally separate unit, and except for the brief visit of the putter rushing in to hitch up the 'fullins' (the full tubs) and leave the 'chummins' (the empties), each would work alone. Maybe half a dozen places would be in operation on a single flat, two or three flats in a district.

Side by side with the bord and pillar system gradually crept in the 'longwall'. Instead of pairs of hewers working in separate drivages, there was now continuous face, with all the workers teamed up into a single assault squad. The face was perhaps 100–50 yards in length and would terminate at each end with tail gates or tunnels leading away; between these two ran the large 'mullergate' or main gate. The coal face would run horizontal to these three gates, and along its great length would toil the hewers and fillers, with the 'caunchmen' (stonemen) advancing the tunnels and setting the supports for it. The overall effect of the longwall system was to increase the number of marras; instead of two or three men, working together there were now perhaps a dozen. The longwall 'rationalized' the task of the deputy: all the places were now in one line rather than in separate stalls; but although under this new system he was always in or near the face, his powers of supervision were no greater in practice.

The average standard of job control remained very high, but now it was exercised by a team of hewers, advancing shoulder to shoulder against the face, rather than by lonely pairs of marras. The longwall team worked under its own elected leader, and according to its own methods. The team leader, or 'cavil leader' in Durham, was elected from the rank and file. He would sort out places, bargain over the price for a job, and pay out the team's wages. He was usually blessed with a little more intelligence than the rest; and needed the ability to calculate. He worked the same as the other men but was recognized as a sort of shop steward. The wages of the entire face were calculated on a 'Master Note', which recorded the total tonnage and the total wages of the team. In my time the cavil leader would collect this note on a Thursday and go into the pit canteen, where the money would be divided up. The first leader I remember at Wardley was a well-built little fellow. This man was an agitator. I remember him standing outside the canteen, with his marras round him, stirring them up about working conditions. Men from other teams would say to each other: 'He's at it again', but they had to admit that he was a good worker, even if he did have an 'open joint' (talked all the time).

It is doubtful whether another industry could see such a thing as the management's first line of command in the shape of the deputy being lambasted by a host of workmen because he hasn't managed to keep them supplied with tubs or material, and he profusely apologizing, offering explanations, and giving assurances that it will not happen again. If an attempt is made by an official to tell a worker how to do his job it will be couched in cringing terms; even then the response will be one of cursing and swearing. If the advice is taken it will only be because the worker knows it won't work; even if it might work he makes sure it doesn't, and after proving beyond a shadow of doubt that the deputy is 'wet behind the ears', he will carry on in his own way unbothered. Very few officials will take the responsibility in a tricky situation of telling a face man how to do a job. For if anyone was killed following a 'clever' bit of 'day release' advice, the official would be for the high jump. If an accident occurs the favourite defence of the deputy is: 'Well, they're experienced men – they know the job.' The deputy is in reality over a barrel. Management may have very clear lines of operation marked out for him, but the deputy himself is placed in the arena without quite knowing whether he is supposed to be a safety official, a shot firer, a general help, or a gaffer. Deputies don't take long to discover that what little authority they have is neutralized by the men's refusal to accept supervision. 'Little Hitlers' or 'Bob-a-Job Men' – the deputies who throw their weight about, or try to keep the worker employed with one job after another – are soon disposed of by the men, and are in any case usually hated by their own class of worker. The majority of deputies do the best thing possible, which is just about what they were doing a hundred years ago, namely inspect working areas, implement safety rules, and act as part of the 'supporting cast' for the face workers.

I remember a new manager arriving at the pit and coming down to inspect a team of caunchmen working. Gradually they all stopped work and looked at him; he looked at them, not knowing quite what was the matter. Then one of them said: 'Ey gaffer, does thee play chess?' 'Well, yes I do', said the gaffer. 'Well, thee gaan away an' play chess and we'll get on with woork.'

In the *Mineworker* there was a cartoon about the attempts to introduce supervision. It showed a worker being carried out on a stretcher. Another worker turns to his marra and says: 'What happened to him?' 'Oh' was the reply, 'sun stroke from the gaffer's lamp.'

If an under-official sits too close, the men will demand that he leave the working area. If he doesn't leave at once, a few props dropped close to his legs, or shovelfuls of coal 'accidentally' missing the chain and hitting him, and a hint or two about his safety will reinforce the message. If he still doesn't leave, the men will simply walk out of the pit: 'We divvind like gettin' spied on.'

The putter with his pony or hand-drawn tub used to be as independent as the hewer, though this was a job usually done by boys. He would be assigned to a 'flat' and delegated several 'places'. He made his money according to the number of tubs he led. Apart from this the hewers would bribe the lad with a bonus if he worked hard to get their tubs away, and sorely threaten his life if he wasn't fast enough. How could an official hope to do any better than that?

The putter's money depended on his speed. His pony made his wages and what he earned depended on his ability in handling it. Along the tracks would toil the putters, fetching the tubs from the flats to the hewers, charging out with full ones and then back again with the empties or 'chummins'. In certain places even ponies couldn't be used; the lads here would be hand putters, sweating and toiling in exactly the same way as young miners had for hundreds of years:[14]

> A'm on me way inbye, and working as a putter,
> Shifting tubs from flats and partings to the hewers.
> They're light on the flat, but coming back.
> Yi'd swear each tub weighed a thousand tons.

Up to the 1960s the 'gallowa' or pony was the chief means of conveyance in the pit. My own father was a pony driver for a long period of his life. Some of the ponies built up a strange relationship with their drivers, some a good one, others bad. What was impressive was the general intelligence of these beasts, second only perhaps to their downright stubbornness. George Hitchin, in his book *Pit Yacker*, writes of his pony 'Shot':[15]

> Shot was the most intelligent animal I ever knew, and if he did not always understand what humans said to him, he never failed to make his meaning perfectly clear to them. He could count, and what is more he had taught himself. The usual load of tubs was four and that meant three coupling-chains. If Shot heard a fourth chain click he would halt, and no power on earth or below it could make him move. And should anyone be foolish enough to hit him, all hell broke loose. He would kick out the hook with which he was attached to the tubs and, once free, his rear-end would swing round like lightning towards his attacker and hooves would fly in all directions. Then he would run off into the darkness.

During my early days in the pit I used to sit at bait-time in the darkness while the ponies ate their 'choppy'. One of them hated mice as much as any cat. On hearing them in his choppy box he would kick out at them, stomping about in the darkness and crashing his hooves, until he heard them running away. The pony could be just as good or as bloody awkward as the driver made him, although, to be honest, certain of these creatures could not be bribed, loved or bullied into work if they set their mind against it.

IV

The miner's wages (and his safety) depend not only on his strength, experience and guts, but also on getting good 'marras'.

The traditions of working in a team were all developed in the old days of bord and pillar, when the teams were often small. There might be only two marras to a stall, and even these might work on separate shifts, so that a man only saw his marra when he came on to relieve him late in the morning.[16] But even if they did not work together, they still shared their earnings. When Thomas Burt first went into the pit in 1855 the hewers grouped themselves in teams of four or so. Two would work each shift; 'Pay Friday' would come every two weeks, and the men would divide up the total of their earnings. Much depended on their luck in drawing a good place: Burt says that he had known as much as three shillings a day difference in wages simply because of differences in the stalls.

The 'Collier's Rant', one of the very earliest Durham mining songs, mentions the 'marra' relationship of two men working together.

As me an me marra was gannin to work
We met with the devil, it was in the dark;[17]
I up wi' me pick, it being in the neet,
An I howked off his horns, and his head an his feet.

Follow the horses, Johnny me laddie,
Follow them through, me canny lad, oh!
Follow the horses, Johnny me laddie,
How, lad, lie away, me canny lad, oh![18]

As me an me marra was puttin the tram,
The light went oot an me marra went wrang;
Ye would ha laughed to see the fine game,
Owld Nick took me marra, but I took the tram.

Nuw marra nuw marra well what does tha think?
Av' broken the bottle and spilt all the drink,
Av' put oot me light amang the big stanes,
Draw me to the shaft lad, its time a went hyme

Oh, marra, oh marra, nuw where hes thou been?
Drivin' the drift frae the low seam.
Drivin' the drift frae the low seam,
Haud up the louw[19] marra, de'il stop oot thy een!

Oh, marra, oh marra this is wor pay week,
We'll get peeny loaves and drink to wor beak;
Ay, we'll fill up the bumper an roond it'll go.
Follow the horses, Johnny lad, oh!

There is me horse and there is me tram,
Two horns full of grease'll meyek them gan;
An, there's me marra stretched out on the ground
You can tear up his shirt for his mining's all done.

There was a continuous relationship between the two marras even if they worked on opposite shifts. In John Wilson's early days the two shifts often overlapped, the back-shift man coming in to work 'fully two hours' before the fore-shift man finished, and the two of them working together. The man on the first shift would undercut the coal and fill it; when he had filled the coals and got the first jud finished he made a start on the second jud (a 'jud' was the measure or stint). When the back-shift man came in they would work together as best they could and when the jud was down the fore-shift man would go home. The back-shift man would complete the filling, and then start on a new jud.[20] Considering that the man would otherwise be working on his own all the time, this was a more humane way of working a stall. But even if the two never worked together they would still be closely marra'd and each would depend on the other's work.

With the introduction of longwall working the marraship increased in size. The numerous stall partnerships were assimilated into a few larger teams. Where previously two men had worked together, now many workers had to operate in a larger collective.

The members of this 'big team' were generally self-selecting, and the ability to recruit the right man was of great importance. A newcomer would have to suit the speed, character and style of his fellows. He would have graduated to the face through a series of jobs such as timber or girder leading, drawing materials to the face, road laying, etc. (Later on when he became too old to keep up with the face he would leave his marras and be employed on less demanding work 'outbye', such as restoring roadways or levelling off floor lift.) When a vacancy arose a youngster would be invited to join the existing group; or sometimes a group of youngsters would form a new team of marras on their own.

This principle of self-selection maintains group standards. If a workman does not meet the standard of the group he will be informally told by his mates to improve, get some more kip and plenty of grub, or at the next quarter he will have to find another set of marras, perhaps not as hard-working, and earning therefore less money. If the demoted worker is unable to find marras then he will be placed with other men of his same standard by the management; in this case he will be on an individual pay note and not the master note or group-shared wages. He may even find himself on datal work.[21] Young workers are usually earmarked for a particular team right from the time they are putters in the gates.

The custom of paying all the money to the team leader is probably as old as the cavil system itself. In the old poem, *The Pitman's Pay*, the author says that wages were paid by the viewer or overman to some one person who had sufficient 'lare' to enable him to divide it at the public house where the others met him.[22]

> I sing the pitmen's plagues and cares,
> Their labour hard and lowly lot,
> Their lowly joys and humble fares
> Their pay-night o'er a foaming pot.
>
> The dust wash'd down then comes the care
> To find that all is rightly bill'd
> And each to get his hard earned share
> From someone in division skill'd.

If there is any disagreement about the pay, then this man will go in to do battle with the manager, his clerk or the under manager. I have very vivid memories of the Thursday pays, with men surrounding their leader, discussing the master note, and often waving it about and voicing their discontent. Even if the note is right it never produces satisfaction: 'It's only just right, mind', is a common response when it is found they haven't been robbed this week. It was my privilege on a Thursday to help give these notes out and to see at first hand the working of this team democracy.

Within the team men work to each other's speed. Each man is expected to keep up, and there can be trouble if the tasks are not equally shared. Thus we read in the Minutes of Wardley (Follonsby Lodge) for 15 March 1964:[23]

> Mr. Robson asked if he could come off stonework, when asked the reason for his request, Robson stated his marras were not playing the game and some were not trying. A special meeting was to be called with the man's marras and a motion was posted to that effect. Three of the men were subsequently taken off caunch work by democratic decision of the men.

Each team after being together for a period begins to take on a distinct character; they become known for particular qualities or peculiarities. They may seem to be at each other's throats, or else have a reputation for being particularly 'red' and militant, or for being apathetic. Their standard of work also renders them distinctive. A first-rate caunch team, for example, may earn a good deal higher wages than others, owing to their skill and strength, and will be treated either with respect or as 'bliddy madmen'. A particularly poor team may be the object of derision, quite unfairly when one considers how hard they work compared to workers outside the coal industry.

Most people have heard of the almost legendary 'big hewer' in County Durham, the hewer who could work as hard as ten ordinary men and perform wondrous feats of strength, and who was blessed with a colossal degree of 'pit sense'. In effect there are many such 'big hewers', one at every colliery in fact. A team of hewers who are particularly efficient may be called after the name of some early big hewer, 'a real bunch of Bob Temple's men', for example. The same thing works in reverse, of course, when a team get a reputation as slackers.

A man can get a reputation as a bad worker as well as a good one. A whole legend can be built round a lazy or shirking miner, whose fame for being idle is complete. At Wardley there was such a man. Let us (to avoid embarrassment) call him Tony.

> Tony's crew is willing. Tony's crew is reet,
> Tony will be working all through the neet.
> He's worked for half an hour, already he's dead beat,
> He'll make a name for Wardley.
>
> Ah met some funny fellers when working doon below,
> But Tony is the slowest man yid ever want ta know.
> He can shovel with one hand, with the other have a blow,
> And shift more stone with hes hanky.

It was said of Tony that once he got cavilled with a one-armed man. Well, old Tony didn't like to hurt his feelings so he worked with him for a week. When he went to collect the master note at the end of the week he saw that their yardage wasn't too bad and the wage was decent. Up comes the one-armed marra, obviously a bit embarrassed. He says to Tony: 'Wey, ah reckon ah'll ask for the cavils to go in again and ah'll get a move.' 'Oh, that's all reet,' says Tony, 'the wage isn't too bad.' 'Ah know,' says his mate, 'but me one arm won't do for the two of wi.'

Sometimes a whole team of men will be the centre of a joke for years: 'Wi wadn't pay ye Mullergate men in washers or brass buttons.' A whole team of caunchmen might be met with a chorus of 'Crunch, tinkle, tinkle, tinkle,' (supposedly the noise of a few small lumps of stone coming off their shovels); or a lazy individual: 'Throw some breed aboot an' set the hens on scrattin' and thi'd shift mer stones than him.' Or: 'He's that thin he needs a sylvester (pull lift) ti lift a prop, and a help up wi thi hammer.' Sometimes the insults will be swapped: 'What, ye work? Ya that fat ye need a mirra to fasten ya buets.' 'Whey, whee's that talkin' . . . oh, ye can get a new set a false teeth now that Arkle's [famous racehorse] died.'

It was a general rule that if any member of the team had a dispute then the whole team came out. It is hard to imagine a situation where a worker found himself alone in dispute without it involving all of his marras; it is a very rare thing in a pit to have a class of worker on a face come out on strike without the other classes on the face joining them. You could for example have a group of caunchworkers on the tail gate. Perhaps water was flooding their caunch; or maybe the dust was excessive. If a new price couldn't be agreed to, or the management wouldn't do anything about stronger pumps or better ventilation, then all of the team would come out on strike, not just the stonemen.

If there was animosity in a team and it was obvious that they just couldn't work together, then the cavils might be put in again, but generally, despite any disagreements, the workers in a team have an unspoken affection for each other; a deep bond grows up between them. In County Durham there was never competition within a team; individual stints weren't marked off as they were in southern coalfields. On the contrary, it was one face, one team; the tonnage was shared as were the wages; if you had a bad section, or place, the others pitched in to help, when they had completed the work on their own.

In the Yorkshire pits the leader of the working group is called the 'puffler'. It is hard to trace the descent of this character and his role seems very uncertain. Most workers regard him as a man who looks out for their welfare. It is he who has to look at new support plans, represent the men at meetings, etc. One would think he was like the cavil leader in Durham, but he is not elected, he is not accountable to the workers, and the management pay him *2s. 6d.* per day extra. He also acts as a kind of very minor face supervisor. It is possible that he is a descendant of the old butty system, I don't know. The butty system was of course rife throughout the southern pits but never in County Durham. This was a system of sub-contracting whereby the owner/manager would let out the work to a butty man; it would be he who collected the money and paid it out (if you were lucky). Most men were robbed blind. I have heard older Yorkshire miners say that after working a full week, bloody hard on 'stints', they would have to fight the butty man in the pit yard to get their money. Others say that he might only work actually two days a week, extracting enough money from the men to live very comfortably. Still, this system is far removed from the puffler of today, since this man, despite the *2s. 6d.*, is a member of the NUM, and of the men, not the management.

Marras have a close relationship to those working the same face as themselves on the other shifts. Considering that all will be experiencing the same working conditions and suffering under the same geological deviations of the face, this is not surprising. Each depends on the other's work and how they leave the face. If a team of caunchmen found that their marras on the other shift had left particularly hard stone in the gate to be bored or blasted they might be angry. It would detract from their own working time and this would mean a loss of earnings (at least in the days of piece-rate). It would also be detrimental to their reputation for good work, which is the very soul of most caunch teams. When the shifts are changed the first question of the oncoming team to those coming off is of the nature of the 'lip' and 'Have ye done?', 'What's left?' With the information thus gathered the ingoing shift plots its course of action: who will go where, what shall be tackled first and what left back. If the job isn't as reported and the other team has left work over to complete, the next day will see violent exchanges between the men, since this would probably affect the third team as well; they would have extra work to do and would blame the second shift for it. The worst crime in the world is to be regarded as inefficient by your own class of workers.

Disputes affecting one team usually affect the corresponding team on the other shifts, and this is how strikes often spread. One team may decide that the water is too deep and complain that the pumps are too small to deal with it, or claim an extra allowance as compensation for the added difficulty of their work. Perhaps they will start the work, but then someone will start grumbling to himself and this gives the others a chance to join in. The chorus of complaint grows louder and more concerned until at a certain pitch someone will turn round and shout (to nobody in particular); 'Well, bugger me, are we gaana stand this lot?' 'Wi want shootin' for workin' in this, mind, wi dey'; somebody else will take it further; 'Whey anaar, but are we all of the same mind?' (this is always the catchphrase before a dispute of this nature). The strike will start with all the team walking out of the pit and storming the secretary's office. Then they will sit in the canteen and await their 'marras' off the following shift to explain the situation. After that they will go home, leaving the second shift with the responsibility for bringing out the third shift. The following morning a deputation led by a picked man would see the manager to negotiate a new price for the changed conditions. If the management refused, then the strike would almost certainly spread.

Marras become closely identified with a seam as they get to know its way. At work or socially the men from it will be 'Busty men' or 'Brockwell men'. If ever there is the necessity of a set of men being transferred to a different seam, all hell will be let loose, and it is usually many, many years before the men in the new seam cease to describe them as part of another crew. A man linked to a particular seam has the right to a place there before anyone from another, and the right to refuse a place in another seam while a place can be found for him in his own. If for some reason a man's place isn't on (i.e. available), he will be highly suspicious and annoyed, and will demand to know the reason. I have known workers going into their normal place, regardless of official orders, if they were not satisfied that there was good reason for a change. This explains Rule 4 of the Follonsby Cavilling Rules (1931), which were drawn up by George Harvey, the lodge secretary, and reads:[24]

> If any place is stopped from any cause whatever the men removed to return to the place before strangers. Should the place be reopened the place they fall to is to remain their cavil so long as their own place is stopped.

One usually finds that marras down the pit are also marras outside. They will be personal friends and go to the same places; their wives will probably become friends and arrange the same times to go shopping or to the laundry or to take the kids out. It is strange to see hundreds of workers in the early morning all stepping out for work around about the same time, and converging on the bus stops for the pit bus, or walking to work together; then all the shift reappearing at 'louse' (knocking-off time) and making their way back home together. At night they all go out together to the pubs and the clubs.

In the Doncaster coalfield, the identification with a particular seam is much greater than in Durham. Here the marraship (there is no equivalent term in Yorkshire) is generally secured to a seam for its life: once a team or group of teams is established on a face they remain there until the face closes down (anything between six months and four or five years). In Durham the men draw lots to change places within the seam (in some cases the whole mine) each quarter, even though the teams remain together. In Doncaster men will be associated with a particular face as well as with a particular seam. The High Hazel, for example, may contain four units or faces; each of the men in these will be known by the number of his unit, but at the same time all the units will be known collectively as High Hazel men. Being grouped together to a single lip or face section for a far longer period than on the Durham system, they can develop a much stronger seam tradition. In Doncaster you will find seams which are particularly noted for their militants, others which are comparatively slow to move and are downright conservative.

V

The cavilling system is the chief instrument of job control in the Durham coalfield. It goes back many years. Galloway writes of it as a mining custom 'peculiar to the north of England' and describes it as 'regulating the distribution of . . . working-places among the miners by lot, thus effectively preventing any partiality on the part of colliery officials'.[25]

Cavilling customs differed from colliery to colliery. The general method was the same; there was a ballot for places every quarter, but there seems to have been a distinction between choosing the members of the team one worked with or leaving it to luck. Lord Gainford (chief representative of the Durham coalowners in 1919) tells us: 'The men as a rule are allowed to select their own working mates, and they go into the place which has been selected by ballot.'[26] Galloway says that the names of workmen were written on slips of paper put into a cap and drawn out at random, 'each workman securing the work place, good, bad, or indifferent, which fell to his name'.[27] At Heworth Colliery in recent years we see that it was the first description that was applicable.[28]

> The Hewers shall at the commencement of each quarter marrow themselves into sets of four or such number as is required by the management.

Cavilling was the simplest, most democratic form of job distribution which the workers could possibly have developed for their purpose, given the conditions of private enterprise. It was a way of equalizing chances and sharing out opportunities. Because of geological conditions there are some good places in a pit, some bad and some more dangerous than others. A face of coal may be hard and close cleated; the roof may be soft or the stratum sodden with water. All of these things interfere with the miner's capacity to earn wages. Without the quarterly cavils one team of men might be stuck with foul conditions where the work was harder but the money less, because the stratum made production difficult. To even things out the cavils provided everyone with a swap. All work places were drawn afresh every quarter. There was no privilege for some and rough work for others. Of course it was the luck of the draw and a team might find itself more than once landed in a bad area of coal.

In Alexander Barrass's song 'The Putter' published in his *Pitman's Social Neet* (1897) we have the feelings of a putter who has been put on a rough cavil and bitterly resented it.[29]

There's a hawf-a-dozen gannins
At the flat that Aw'm at noo,
An if Aw'd me awn i' choosin,
Aw'd hev number one or two.
But dash me somehoo or other,
Hoo it comes Aw divn't knaw;
But as sure's Aw rub me kyevel,
It's the warst one o' the saw.

Thor's a hitch an' then a swally
Filled wi' watter like a ford,
An' a lot way aal twisted
I' the clarty gannin-bord;
Thor's law planks and raggy kanches,
Where Aw've sometimes got a snack,
An' it myeks ye twist yor gizzord
If ye chance te catch yor back.

Thor's a short plate an' a lang un
Near the double turn inbye,
They've been fettled wiv a closer, an' that closer winnet lye

O that short plate an' that closer!
Cud they speak, what wad they say?
For Aw've tell'd them lots o' secrets when Aw've tummel'd off the way.

Thor's a lang, law heavy pillor
Inbyeside the canvas door,
Where Aw horse the scrubbin full uns
Up for eighteen pence a score;

Hoo Aw bliss that lang, law pillor!
Hoo the awful hitch Aw dreed!
For it's fearful wark this stickin
An this shuvin wi' yor heed.

Cavilling was in stark contrast to the southern 'butty' system of sub-contracting, in which men literally fought each other for places, and the maxim was 'every man for himself'. A coal face would have 'stints' of coal; a stint was one man's place – it was marked off in whitewash; he was paid according to his strength; if he couldn't shift it all himself, he might sell his rights over it to others on the face. In Durham, with the cavil, the teams worked together; they pooled their earnings and all pitched in together, as we see by the following agreement at Heworth:[30]

> [On the Brockwell seam . . . the usual custom is to apply:] Fillers shall be allotted to a length of face and shall work to each other's hands and pool their earnings.

The cavilling system was a first line of defence against victimization. If the management had picked the places, union men and agitators would have stood little chance against the favourites. Without the cavilling system, the man with a tongue in his head would have been set to the most difficult and foulest tasks in the hope of breaking his spirit; he might not even get a face job at all. The reverse would also be true of course: the manager's narks would get all the good places and the chances to earn the best money with the least toil. Even the leadership of the Durham Miners' Association, extreme moderates and devoted to class collaboration, recognized the simple fact that without cavils the union would not last more than a brief period.[31]

> Lizzie, No. 22. Let the cavillings remain. Even with them, we have an amount of favouritism amongst overmen. Without them, our leading men might call for heaven's aid (Durham Miners' Association circular 4 Aug. 1882).

The idea of cavilling was not at all popular with the employers. They would much have preferred to have the cavil last for as long as the seam or the job, and in a few rare cases this may even have applied. We see this in a discussion on the question of a stonemen's agreement at Heworth Colliery:[32]

The Manager, Mr. Ridley, requested that once the men were cavilled they should stay in that gate until it was finished. Mr. Fleming the secretary replied that this was a question for the cavilling system and could not be part of the terms of any agreement.

The cavil might apply to one seam only and the cavil be drawn for different places in the seam. Or the whole pit might be cavilled, teams being changed each quarter to completely new seams and districts. This is what is known as 'cavilling through'. There could be a combination of the two: at Wardley, for instance, the Cavilling Rules of 1931 read:[33]

Pit cavilled through for all classes every 12 months. Seam cavils only for other three quarters, during which men keep in their seams.

Whilst it would be true to say that the employers hated the whole system, they were especially annoyed by the men changing its application. We may see from the minutes of the Joint Committee of the coalowners and the Durham Miners' Association just how much time this consumed. In the 1880s they were absolutely full of cavilling business, and the most frequent disputes which came before them concerned the question of 'cavilling through'.[34]

That the West Thornley Little Seam be cavilled by itself (Minute No. 32, Committee meeting, 19 April 1881).

That we seek at Joint Committee to have the Urpeth Low Main Busty and 5–4 seams cavilled through (Minute No. 50, Committee meeting, 3 May 1881).

Urpeth – Men desire to have the Busty Pit, Low Main and 5–4 seams cavilled through. Agreed, that the next two cavillings be through, and the third and future cavillings be separate (Joint Committee meeting, 11 June 1881).

That the Washington men be advised to let their cavilling through rest in abeyance (Minute No. 47, Committee meeting, 22 June, 1881).

That the Washington men be wrote to, about the cavilling through question (Minute No. 5, Committee meeting, 28 December 1881).

Boldon. (1) Owners ask that each pit be cavilled separately. Moved off the board. (2) Owners ask that the arrangement made with the workmen, 'that each seam be cavilled separately', may be carried out. Agreed, that the next two cavillings be through, and the third and all future quarterly cavillings of the Hutton and Bensham seams be separate; this to apply to both hewers and putters (Joint Committee meeting, 10 February 1882).

That we seek at Joint Committee for Wingate Drift to be cavilled for same as the pit (Minute No. 14, Committee meeting, 13 March 1882).

Wingate. Men ask that the Drift be cavilled for the same as the pit. Moved off the Board (Joint Committee meeting, 1 May 1882).

That an agent attend So. Medomsley and enquire into cavilling through question (Minute No. 96, Committee meeting, 22 August 1882).

That we seek at Joint Committee for No. Hutton New 5–4 Low Main and New Main Coal seams to be cavilled through (Minute No. 51, Committee meeting, 26 September 1882).

Margaret and Dorothea Pit. Men are at present cavilled all through, and the Owners wish each pit to be cavilled by itself. Decided, that the men work as they are now doing until the next cavilling; after which cavilling for both men and boys be kept separately for each pit (Joint Committee meeting, 4 March 1878).

That the men at Trimdon must resume work to-morrow, on the following condition:— The men who are now wanted to be taken by cavil throughout the entire colliery of men. Afterwards, we ask Mr. Watson to put in cavils amongst all the colliery of men whenever a batch is wanted. A deputation to attend today to see Mr. Watson and the men. And the question be adjourned until next meeting (Minute No. 2, Committee meeting, 12 August 1878).

Haswell. Owners ask that the men of each of the following seams, viz: – The Five-quarter, Main Coal, and Low Main, be cavilled separately. Decided that the men be cavilled separately (Joint Committee meeting, 22 March 1880).

Houghton. Men ask the seams to be cavilled separately (ibid.).

North Hetton. Men want seams in Hazard Pit cavilled through. Withdrawn (ibid.).

Washington. Men ask that all the seams be cavilled through. Adjourned for a report as to whether there is a difference in the nature of the two portions of the Maudlin seam (Joint Committee meeting, 30 September 1880).

Washington. Men ask that all the seams be cavilled through. That the East and West Maudlin seams be cavilled through; and the other seams remain as at present (Joint Committee meeting, 24 December 1880).

> That the Houghton men be written to, asking them why they desire their seams cavilled through (Minute No. 35, Committee meeting, 28 January 1880).
>
> Shildon Lodge, Furnace Pit – Men say this Pit has commenced in opposition to past arrangement, inasmuch as it has been cavilled by itself (Joint Committee meeting, 2 February 1880).
>
> That the Houghton question, on cavilling through, go to Joint Committee (Minute No. 27, Committee meeting, 18 February 1880).

The quarterly change of cavils was often a time of dispute. The workmen were in a state of flux and management sometimes tried to use the occasion to 'work a flanker'. This we can clearly see from the monthly report of the Durham Miners' Association of March 1880:

> It has transpired on several recent occasions, that when a very few men have been present to see the cavils drawn, the Overman and Manager have announced the most important changes, not only in the cavilling rules, but in the general arrangements of the Colliery. No Manager or Overman has any right to even suggest such changes at the time of drawing the cavils. Changes of the kind ought to be . . . considered and agreed to by the whole colliery of men, before the time that the cavils are drawn. If they seek to force changes upon you at the drawing of the cavils, without any previous consideration by the colliery of men, then leave the cavils undrawn until you have time to have the suggested change or changes considered by the whole of the hewers. I would further suggest that Lodges should procure and keep by them, a copy of the cavilling rules, signed either by the Manager or Overman or both.

An example of this kind of conflict can be seen in December 1913 when three of the collieries at Ashington, owned by the Ashington Coal Company, came out on strike. The trouble arose over 'bargain' work. The management wanted to fix the prices of all bargains and offer them out. However, the men would not accept a system which excluded them from the decision-making process, and refused to allow any of the cavils to go in. The strike involved 4,500 men and boys.[35] The same tactics were used at Ushaw Moor in 1879, when the manager tried to use the cavilling changeover to put in a reduction of *2d.* per ton in the low main seam. The men refused to put the cavils in.[36] Similarly, in the DMA minutes we read of a dispute at Burnhope colliery in January 1880 where the men refused to allow the cavils to be drawn 'because it was proposed . . . to let some winning places' (the Joint Committee sided with the owners against the men).[37]

Cavilling was also used as a means of coping with redundancy. In the case of a district becoming worked out, the management's simple logic would decree that wherever a number of stalls on a flat closed down, all the men displaced (if there was no other work for them) would be the ones to leave the pit. The men, however, tried to have redundancies decided not by the management, but by the cavilling system of drawing lots, 'so that they cannot turn men off that they call below the average'.[38] As the men of Silksworth moved in 1881:[39]

> That if any colliery wants to discharge any quantity of men at any time, they seek cavils to be put in for all the men, and the last cavils to leave.

The DMA for its part sometimes intervened in this way:[40]

> That Mr. Oliver be written to, asking him to cavil the Sacriston men, as to which of the other pits they must go to (DMA Monthly Committee meeting, 12 April 1880).
>
> Request – That if there be too many men at any colliery and some have to leave, cavils be put in.
>
> Reply – The owners cannot concede their acknowledged and undoubted right to employ such number and such men as they think proper (DMA circular to members giving 'requests sent to the Owners and their replies thereto', 17 May 1881).

We shall see from this, then, that the birth, life and death of the face worker's unit was regulated at all times by the cavilling system.

The refusal to allow cavils to be drawn was also used as an offensive weapon against the management. We find several occasions of its being used against owners who were wanting to open up a new flat. Once the flat was ready, the owners and management would be most concerned to get it under way as soon as possible; the workers, on the other hand, saw this as an opportunity to work in new agreements and conditions. Neither side could be absolutely sure that the new flat would have the same conditions as the others in the seam; and initial agreements were therefore usually open-ended to allow for change at a future date.

Conditions would only become clear in the course of operations. At the same time totally new teams of workers would be forming up to fill the new place, and all this added to the general flux of the situation, even more so if the workers had not particular preference for marras and were lumped together at random until they sorted themselves out.[41]

The 'bargain' or private tender worked usually side by side with the cavil but often they were awkward bedfellows and the management tried to set one tradition off against the other in the hope of smashing the cavil. The usual trick was as follows. The district would cavil as usual on the quarter; the men would accordingly be allocated to their flats and places. Midway through the following quarter the owners would open up a flat and ask for tenders. A group of men would accordingly be selected. They would move to the new place, leaving their old cavil empty. Then a new team of men (possibly made up of spare men) would be drafted to the empty place. If any dissent was raised the managers could say that there were six or seven weeks yet before another cavilling. There was no immediate reply to this, since it seemed to fall outside the bounds of normal cavilling rules, however lodges tried to bring it within them. The DMA sent out a circular on the subject in 1881:[42]

> A dispute has arisen at one of our collieries, relative to the bargains being let during the quarter, and other men being compelled to go to the cavils of the men, who have taken the bargains. In the case in question, some week or two after the cavils were in, a coal bargain was let, and taken by some coal hewers on the colliery. The manager then seeks to force other hewers, on the colliery, to go to these men's cavil. What is your practice? (DMA Circular 18 October 1881)

The men who normally acted under the tenders and group contracts were the caunchmen (stonemen, or, in Doncaster, rippers); it had always been their prerogative to tender for new jobs, and this was their opportunity to earn big money. Various tenders would be submitted but the workers were well conscious of the maxim, 'Cut the rate, cut your own throat', and kept a watchful eye to see that if anyone was going to benefit from the bargain it must be the workers. New faces being 'won' out might also be operated on a bargain system, although certain collieries included these in the cavils, as they might also new drifts and drivages.[43]

After the cavils were drawn and the places assigned, the last team on the last shift usually moved the gear either on a Saturday morning or else the first shift Monday. Tools of all descriptions would have to be moved to the new places; very often that was all the work done on the day. In certain places 'Cavilling Monday' developed into a holiday. Joseph Halliday remembers it 'an outstanding event' of the year:[44]

> Just as business firms close premises to take stock on certain days, the mine would close, not exactly for that purpose but that . . . all the cavils known as working places, would be lumped together and then drawn, raffle-like . . . On Cavilling Monday the schools closed down for the day and the sea-side was the Mecca. This day off was peculiar to Wingate. It was the only colliery in the county to avail itself of this facility and the custom was maintained until World War II.

Actually Joseph Halliday is not quite correct. 'Cavilling Monday', much to the annoyance of the owners, was observed in other districts too, as the Joint Committee minutes record:[45]

> Leasingthorne – Owners complain that the colliery was laid idle by the men on cavilling Monday. Moved off the board, but the Owners to be permitted to bring the question of the pit working on cavilling Mondays before the next meeting of the Joint Committee (Joint Committee meeting, 21 October 1881)
> Leasingthorne – Owners ask that the practice of laying idle on cavilling Mondays be discontinued. That the pit work on cavilling Mondays, but the men be allowed to go to work in the fore-shift (Joint Committee meeting, 9 January 1882)
> Westerton – Owners complain that the colliery was laid idle by the men on cavilling Monday. Withdrawn, but the Owners to be permitted to bring the question of the pit working on cavilling Mondays before the next meeting of the Joint Committee (Joint Committee meeting, 21 October 1881).
> Westerton – Owners ask that the practice of laying idle on cavilling Mondays be discontinued. That the pit work on cavilling Mondays but the men to be allowed to go to work in the fore-shift (Joint Committee meeting, 9 January 1882).

Another piece of evidence comes from 'The New Keviling Monday' in *Allan's Tyneside Songs*:[46]

The Kevil's draan at Parrington pit,
An' mat an' his marra hev myed a bad it
Doon the Sooth Crosscut we're shifting wor kit –
Oh, brother, the Kevilin Monday

Certain collieries might work the Monday but come to bank as soon as the gear was moved; others would shift the gear on the Saturday, as we see from the following DMA minutes.[47]

> That we seek at Joint Committee for South Pelaw men to be allowed to come to bank one hour sooner on cavilling Saturday (Minute No. 67, Committee meeting, 14 April 1882).
> That the Derwent men ride when they have got their places squared up on cavilling day (Minute No. 32, Committee meeting, 2 August 1882).

Almost the same as Cavilling Monday was 'Stoneman's Monday'. At Wardley the landlord of the White Mare Pool would set up a big bath tub, and every groceryman, bread man or potato man who came in would throw something into the tub. They would make a celebration of it. Occasionally the overman might pop his head round the door and jokingly implore: 'Come on, lads, gi' them a surprise, gaan ti work.' Apart from the stonemen, however, it was the practice at Wardley to ride (come out of the pit) as soon as the gear was moved on a Saturday rather than lay the pit idle on Monday.

Cavilling seems to be unique to Northumberland and Durham but the 'Priority System' in the Doncaster coalfield is similar in essence and intention, though no names are drawn out from boxes. A face worker when he first comes to the colliery will attach his name to the priority list drawn up for each unit. He will be classed as a 'spare' man until he is taken into a team. There will be a pool of these spare men 'on the market'. If a face worker of some description is off (as is nearly always the case), then a spare man will be given work for the day, and chosen according to the length of time he has been on the market. When the unit closes down, all of those workers who had a 'regular' job will become spare, too. When a new unit opens up those men who were spare last time become the regular men on the new unit. At the pit I was at in Doncaster, two priorities were in operation, one being taken for the High Hazel seam, the other for the Barnsley, these being different types of seam with different working conditions and of course different teams of men.

The cavilling system was the fundamental way in which the Durham miner managed to maintain an equitable system of work and managed to stave off the competitiveness, bullying and injustice of the hated butty system. In essence it was an embryo of workers' control, as can be seen from its ability to handle disputes between sets of workers without recourse to outsiders. It was a little Soviet which had grown up within the capitalist system. In this sense it was of necessity restricted in its development, but it is an example of the worker intervening in the productive process in a conscious way to say: this is how I run it, you adapt it accordingly.

VI

Cavilling was a periodical event, but control was by no means only exercised every quarter of a year or so. Because geological conditions varied so much from seam to seam, from face to face and (in the days of bord and pillar) from stall to stall, a separate price had to be bargained for each job. First there would be a seam price to be bargained for, before a 'winning' (a new working) was started; then there was on-the-spot bargaining to take account of geological peculiarities. The whole power of negotiation rested accordingly in the miners' branch and with the team itself. This became apparent in a study of the frequency of strikes in post-war Britain (the heyday of mining piece-rates and of unofficial strikes), where a researcher came up with the conclusion that the working out of wage rates was done not by any worked out code but by on-the-spot decisions 'of an arbitrary kind which seem almost invariably to involve a good deal of negotiation and bargaining with the workmen concerned'.[48]

A price struck for a particular yardage may be all right under the prevailing geological conditions, and the workers will toil hard to make the gate advance rapidly in order to make money. Then the coal starts to dip, the face takes an incline, the rock starts to twist or slip upwards, causing falls from the roof, the supporting of which takes time, not to mention all the extra stone that must be stowed and one is only paid for going forward. The propping materials may be inadequate, timber not thick enough or long enough, taking extra time to set, etc. All these things, changing from day to day, week to week, necessitate fresh bargaining, and perhaps conflict about the price. These are the vital causes of unrest. Bad physical conditions will be put up with by the miners if the price is right. The contradiction which generally gave rise to strikes was that, in order to encourage harder work, a piece-rate was established ('the more you do, the more you get'); then intervening factors (human or geological) would come in to render the work simultaneously harder and less remunerative, and a conflict would arise around the bargain.

Let us take, for example, the cutter men inching their way down the face with their jib machines. Suppose a fault or stone broke into the seam. The men could obviously not keep up the same rate of progress and would have to spend most of their time supporting the face. At some stage the men would refuse point-blank to operate any more and demand some immediate restitution for the extra toil – a new price to be struck there and then until something more substantial could be negotiated. If a higher rate were refused the men would start slacking until finally the overman caved in; after he had had yet another row with the under manager over the progress of the stall, he would charge into the face, where perhaps the men were eating their bait, and negotiate a new price for yardage on the spot, whereupon the men might throw away their baits and rush to do combat with the coal at a more joyous price. I have seen something of the kind actually happen.

It is the prerogative of the cavil leader to negotiate when there is a new bargain to be struck, though of course in any group there may also be an agitator, a man with fire in his belly and a tongue in his head; he would be spotted as a man who could do best in negotiation and lead a deputation if he hadn't already been recognized as leader.

The major source of dispute was without doubt piece-work, which set a premium on speed. Anything which slowed down the work immediately affected earnings. Let us take, for example, the Wardley putters' strike of the 1930s. The tubs in general operation at the colliery were wooden, the putters were paid by the 'score' or the number of tubs each lad 'put'. Steel tubs were introduced by the management; they were heavier, awkward and had two stupidly placed handles exactly in the place where one's back went when assisting the tub round a turn or bumping it back on the way after it had come off. At first the putter ignored the steel tubs and left them standing, using only the familiar wooden ones. However, the latter were gradually withdrawn (like money going out of circulation) and after various tactics such as smashing the handles off to make the steel tubs more serviceable had failed, the lads asked for an increase in the putting score since the new tubs interfered with their power to earn wages. When this was refused they struck in support of their claim.[49]

Another example from a far earlier time can be seen in the men's stubborn refusal to adopt the Davy lamp; such refusal continued until late in the nineteenth century. There were two main reasons for this. One, the Davy lamp severely cut down on their earning power, since its light was poor in comparison with candles; reduced visibility meant that more stone would be filled with the coal and more of the miners' tubs confiscated by the owners; also, dangerous places were foolhardily opened out and worked which would never have been attempted without the 'Davy'. (There was actually far greater loss of life from pit explosions after the introduction of the lamp than before.) Many miners believe that the Davy lamp was first and foremost a boon to the profits of the owners, and initially at least an agent in increasing the slaughter of the miners.

Piece-rate systems of payment were rather complicated in the mines compared with those in other industries. Yardage and tonnage prices were weighed against actual advance, with possible additions for extra work with timber and compensation for geological faults. An example is given in *Colliery Working and Management*:[50]

> Working the low main seam with a section of 3 feet 10 inches of clean coal, it was reckoned that in a 2 yd. 'place' an average hewer ought to 'take off' per shift a 2 feet 6 inch 'jud'. This calculated would be: 6 feet multiplied by 2·5 feet multiplied by 3·83 feet = 57·45 cubic feet = 2·136 cubic yards; 18·83 cwt. (the weight of a cubic yard of coal) multiplied by 2·13 = 40·107 cwt.; 40·107 cwt. 40·107 cwt. divided by 8 cwt. (the standard weight of a tub) = 5 tubs.

5 tubs at 14 per score	3s 4d
Yard work at 1·10d per yard, for 2 ft. 6 in.	1s 6d
per hewer per shift	4s. 10d.

As well as the payment by tonnage and yardage, there was also the issue of extras, of the 'overs and aboves', to be battled out between the coalowners and their deputies on the one side, and the colliers on the other. Extras arise from elements in the miner's work which cannot be calculated beforehand (because you don't know when they are going to arise) and which cannot be reduced to terms of tonnage or yardage price. Setting timber, taking up floor, setting chocks or hauling in material all detracted from the capacity to earn wages, and how much these subsidiary tasks were to be paid for often involved negotiation if, as was usually the case, the worker decided it was more remunerative to stay where he was filling coal than to run around doing unpaid extra tasks. An example of such extras can be seen from a master note put in my possession by John Oxberry, an old Wardley miner. The note is from Jarrow colliery. It is dated for July 1837 and records the fortnightly pay for a brakemen's team of four. Apart from the standard payments, it mentions the following special payments.[51]

> David Thompson and Guy Stephenson repairing the machine and waiting on 18 hours at 3d per hour, Bob Wilson and others pumping the West machine boiler 4s.

Some of these extras were fixed in amount. For instance, in the cavilling rules at Follonsby:[52]

> Beaumont Seam
> Ramble scale 1/4d per ton for each complete inch above 2 inches. Average of 3 measurements.
> Misfired shots – 1/- per shot.
> Wet working – All seams, 4d per shift within six feet of the face (top and bottom).
> Casting – if more than 6 ft. at commencement of fore-shift, 5d. per shift.

In the Follonsby hewers' cavilling rules an extra was paid under rule 14 when men were moved to a fresh working place: '1/- in case of a flat and 1/6d in case of a district'.[53]

Over and above these, there would also be unforeseen extras negotiated on the spot between the deputy and the cavil leader – 'waiting on' time would be paid if there was a shortage of tubs or a major belt breaking, or inability (for whatever reason) to have the face fired. Weight coming on to the face might involve the filler in far more work on extra supports than usual ('timbering'); the ventilation might become inadequate as the face advanced, and there would have to be a new price set for the face. When water came in at the seams extras were charged for 'bottom water' (water on the floor) and 'top water' (water coming in from the roof). The necessity for each and every one of these extras would have to be settled on the spot, with the overman or deputy possibly arguing that such extras were not necessary, and the cavilling leader insisting that they were. An example: 'What watter? The seam's moist not wet', 'Not wet ye begger, there's bloody alligators and pirana fish swimmin aboot on't, ye need a bliddy raft ti put a pack on.'

Under the day wage system, dreamed up in the 1960s by the National Union of Mineworkers and the NCB, all negotiations on wages were supposed to be national, and piece-rates and 'extras' were abolished. The branch was supposed to be left with little more than compensation cases for injured workmen. But local and pit face agreements have not been abolished in fact, even if they have on paper. Where extra effort is required on a job, bargaining over 'extras' is still a big factor. Extra money is still booked for carrying powder – this is a fixed price extra – and so is water money. Whether or not water money is payable is something which can only be settled on the spot. If a pipe on a hydraulic chock bursts, and a man is soaked, negotiation might take place as to whether or not this comes into its terms; and if it does, whether it is booked as 'top' or 'bottom' water or both. In the same way workers might agree to carry some piece of broken electrical or mechanical equipment to the repair shops at 'lowse' (knock-off time); for this favour the overman himself or the deputy might book in 'extras', but disguised as something else. All of these small prices as well as the bigger ones mentioned briefly before would require negotiation, argument or stand-up fights in the gateway of the mine. Another example: under the standard day wage system men are not required to carry tools into the job since they are not in theory paid for doing so. But without the tools the men can't work and have to wait for the material-haulage workers to fetch them in with the other supplies. Since this would not suit the overman one bit, the poor old deputy would have the task of carrying in the bulk of them himself; so what he does is to offer the workmen so many 'water' or 'powder' moneys for each tool carried, and perhaps a note to ride out of the pit early. I myself have had over 17*s.* payment booked in as 'water' on the pay note, for doing some task I would otherwise have done very slowly and badly.

Similarly, the NCB's attempts to render all face workers one class has effectively been defeated in most collieries – colliers, cutters, rippers, etc., maintaining their separate skills and job control. The National Power Loading Agreement, in essence nothing other than a productivity deal, was aimed at destroying job distinctions in the mines. All classes of workers, caunchmen, hewers, putters, were to be treated as performing the same kinds of work. This, however, is rubbish, since most of the separate tasks still exist in a power-loaded face and the workers remain on the selfsame tasks. Miners are very possessive about their work. Once they have a task it is *their* task and nobody else's; once they have a unit it is theirs; once they have a skill nothing will persuade them to perform somebody else's. They continue to keep their own class of work. At Doncaster I've known caunchmen arrive at work only to be told by the deputy at the kist: 'There's no work on for you, lads, today. The face hasn't travelled far enough and there is no caunch to work.' 'Wor marras just tel't us that their is a lip on', the men might reply, demanding to be allowed to do their own job. If the deputy still refuses, and tries to insist on the men engaging in other work, I've known them ignore him and simply get on the man-rider and go to the job, refusing to take any more notice of his pleas. The same kind of thing can happen if the deputy tries to send the caunchmen into a 'neuk' or stable. They will demand some assurance that no spare colliers are available for work. The caunchmen in Durham or, more particularly, the ripper in Yorkshire will always demand his own class of work and refuse to allow any other section to do that work.

In talking about day-to-day job control, one must mention the long-drawn-out battle about the supervision of the miner's tub, which led to (but was not ended by) the introduction of miners' checkweighmen in the 1860s. Many colliery disputes were centred round the master's crooked scales, and his zealous agents who confiscated tubs – or fined the hewers – on the excuse that stone was mixed with the coal, or that the tubs were underweight. The men's feelings about the 'keekers' (the masters' checkweighmen) can be seen in a song which is still sung in County Durham, 'Oakey's Keeker':

To do his duty is nothing but right,
But in hurting coal-hewers he takes a delight.
If he pleases the masters, that's all he cares for,
Suppose that he hungers poor men to the door.
At half-past six in the morning he starts
To fill up the box, which is only two quarts;
If he gets the first tub, how pleased he will go,
And say: 'That's a start for old Maiden Law Joe.'

This Maiden Law tyrant does nothing but shout:
Who belongs to this tub? Because it's laid out.

He smacks his old lips, his hands he will rub,
Because he has taken a good man's tub.

Amongst us coal-hewers how well he is known;
His hardness towards us he always has shown.
What makes him do it, I really don't know –
That cruel impostor, Old Maiden Law Joe.

I hope all the screeners, as well as old Joe,
Will think of the men that are working below.
Perhaps in a pit they may never have been.
That's where the hardship may daily be seen.
How would they like it, if they knew what we made,
When the Pay came, and the money not paid?

The question of checking and weighing is sometimes thought to have been resolved by the parliamentary Act of 1860, which decreed that a miner's checkweighman, elected by his fellows, should serve alongside the checkweighman employed by the owners. But this is not so, for although the miners' checkweighman was an important figure in colliery life, there were still many disputes over the weighing of the miner's tub, and the owners' fines for short weight. This was the cause of the Felling strike of 1887, according to a report in the *Durham Chronicle* of 23 September 1887.

> The grievances, the men state, are not creations of yesterday, but had existed for over two years, and dated from the appointment of the present viewer, Mr. Lee. The two causes of the complaint were the system of 'fines' in vogue and the disposal of the tokens. It is asserted that since November, 1884, when Mr. Lee went to Felling Colliery to be manager, and up to June of the present year there had never been a case heard from that colliery before the Joint Committee, everything having gone on with the best of good feeling. The system of fines was that for every 'box' (fourteen pounds of stone) found amongst a tub of coals the hewer was fined 3d. As many as six boxes were sometimes found in a tub, and as the top price paid to a man per tub was 8½d, he might often at the end of a day's work find himself 'owing' his employers instead of having anything to take. The men alleged, further, that when Sir George Elliot had this colliery the fine was only 3d per tub. There was an arrangement then by which all stone found in a tub over and above the sixteen pounds for which the 3d was charged

was to be weighed and kept off the total quantity of coal paid for, but the men state this rule was never enforced, and no more than 3d was ever kept off. On Messrs Bowes and Partners acquiring the colliery some four years ago, the old system was continued for fifteen months, and as long as Mr. Kayll was viewer. After that, however, it is stated that the 'keekers' were dismissed and others engaged in their place, and the present system put in force. Of this system the men complained strongly, inasmuch as the pit, they say, with the exception of the Hutton seam, contains a good deal of stone, and it is difficult working in a bad light to sort all the stone out of the tub, owing to its similarity to the coal, and to be fined so much pressed heavily upon them. As to the 'tokens' the men complained that any child working at bank is allowed to go and get these tokens whether he knows the correct number or not, and they state that there are many cases of men being fined for other men owing to getting the wrong numbers. The men asked that the tokens of the tubs that are 'laid out' should be brought back to the weigh cabin, so that both the checkweigher, the men's own representative, and the master's weigher might see that the numbers are correct. The men having preferred this request, and met with a refusal, 78 of them comprising the 'back shift', struck work, and their example was followed next day by the 'fore shift' men.

VII

The idea of work regulation or restriction of output is of course ancient, perhaps as old as cavilling. Welbourne, who incidentally makes his own disagreement with restriction clear, tells of a paper of 1708 which complains of the hewers and barrowmen being sometimes 'so roguish', as to set 'big coals hollow at the Corfe bottom, and cover them over with small'. Where supervision was lacking the hewers could defraud the owner of great sums of wages.[54] What the writer means is that hewers would fill up their tubs so that the big coal would block off the space underneath, and then cover the big coal with small, so that in effect the tub could be a third empty but give the appearance of being full. It seems slightly strange to me that the Newcastle Hostmen who held a monopoly of coal on the London market, who fought all attempts at competition for centuries, who implemented restriction or 'limitation of the vend' until 1846, were never thought of or described as 'roguish'.

An ancient form of restriction shows itself in so-called 'superstitions' which the miners harboured, and which indeed are still to some extent in evidence today. I am not able to explain where they originated, but I am certain that they developed through the miners' accumulated actual experience, even if to the outsider they appear weird and silly.

Innovations were always regarded by the miner with grave suspicion, not, as foolish writers of the time claimed, because of 'superstition' but from past experience. The innovation of metal props was resented because 'Steel dissent talk ti wi'. The men had come to understand the creaks and groans of wooden props just as if they were talking, while the metal supports, if they talked at all, spoke a different language. An earlier example is the 1842 strike at Wingate Colliery against the introduction of the wire rope. The *Durham Chronicle* commented that the strike was simply another instance of the foolish opposition of workmen to all new inventions.[55] Fair enough for them, since they in the offices of the *Chronicle* would never be perched on the end of such a rope above 1,000 feet of sheer drop. In effect the old hempen rope had proved itself to the miner over many years, and was to a large extent reliable. The miner never liked working with new equipment: for the fact that he was alive was proof to some measure that his old tools worked. Better the devil we know than the one we don't could well have been the maxim, then as now. Further, the wire rope of the 1840s must not be identified with the well-oiled, tested and inspected cable in use today: it was in service regardless of broken strands or worn-out covering. All it did was to serve the master's interests by winding the miner faster to his work and drawing coal up at a quicker pace.

Restriction of output had been part of the upsurge of trade unionism in the Durham coalfield in 1831–2, and it was adopted as a leading policy by Martin Jude's union in the 1840s.[56] It continued in many individual pits when the union declined. 'It cannot be said that the men have arrived at a fair maximum of work', complained a representative of the coalowners in 1854, and went on:[57]

> they are quite capable of doing more, although both extra work and higher prices are given to them than in 1844. But it is also a surprising fact, that the hewers generally prefer working for 4–6d or 5s. per six hours, when they could actually earn 7s. in eight hours, when the demand for coals for shipment is great.

The attitude of the DMA to restriction went through a number of changes. Crawford himself had some sympathy with the idea, but after the signing of the 1872 wage agreement, which marked the real arrival of the DMA to power in the coalfield, the union was committed to a virtual productivity deal. The owners conceded a large advance of wages, and the union, on its side, an abandonment of restrictive practices.[58] Everywhere the full-time agents preached against the doctrine of restriction, though this did not stop many individuals adopting it. In the depression of the later 1870s the demand from the rank-and-file for a policy of restriction became very strong, and at the September 1879 DMA Council it was moved 'that a general restriction of work should be adopted throughout this County'. Washington Colliery supported this and the lodge put a resolution to the Council 'that in future, the county average be our guide, and that no man or men be allowed to exceed it in any one day'.[59]

For a short period, in 1880, the DMA itself partially adopted the policy of restriction. In January of that year the Durham Miners' Council, a gathering of delegates from every lodge in the county, passed a resolution that no hewer should get more coal when his day's output was sufficient to assure him a wage of 4*s*. 2*d*.[60] This was to be the county average.

But this tentative move in the direction of restriction ran counter to the chief dogma and commitment of the leadership at the time – their adherence to the Sliding Scale. The resolution – and the county 'average' – were supposed to be secret, but at Seaham the lodge was so enthusiastic about the new turn that the pit-head buildings were plastered with a poster announcing it.[61] When the owners discovered what was happening they were able to force the leadership to go back entirely on the resolution. Crawford ordered an immediate abandonment of the restriction and the agents of the DMA were instructed not to act against it.[62]

The policy of restriction remained popular with the militant lodges and again and again we find them battling for its general acceptance. In June 1882, for instance:[63]

> We move that the time has now arrived for the National Association to seriously consider a general restriction of out-put, with a view of having it adopted throughout the whole of the Associations comprising the National Union. *Hutton Henry* (No. 32, programme for DMA Council meeting, 24 June 1882).

> That no colliery work more than 5 days per week, and 10 hours per day, as, in our opinion, nothing but restricting the out-put of coal

will improve the miners' wages; this to be submitted to the National Conference. *Elemore* (No. 34, ibid.).

Whatever the DMA leadership decided, individual lodges continued to pursue a policy of restriction on their own. A very good example comes to light during the Wardley Funeral strike of 1888. The strike originated among the putters, and spread to include the whole of the colliery, as well as the neighbouring collieries of Felling and Usworth. A putter named Wilson who was employed on 'bargain' work, had gone to a funeral, and asked permission for his brother, also a putter, to do the work in his place. The manager agreed, and the boy was put to the work. He earned extra money at it, and worked three shifts more than his fellow-putters. The putters came out on strike 'because the manager of the colliery would not lay the boy off work, to counter-balance the little extra money he had made'; in the subsequent dispute it appeared that there was a colliery rule among the men that no one should work more than twelve shifts in a fortnight, 'and this young man had worked a larger number'. The prosecuting counsel complained of a system of 'terrorism and intimidation' at Wardley, to keep the rule in force, but the entire colliery turned out in support of the putters, and when 181 miners from Wardley and Felling were taken before the magistrates at Gateshead, the lodges turned out with bands and banners to cheer them on and demonstrate their defiance of the law.[64]

The deliberate restriction of output was a major weapon of the miners all over the country. It was a means of fighting back without running the risks involved in coming out on strike, and when times were bad they used it to ward off the threat of unemployment, spreading a reduced quantity of output over the same number of man hours. The authors of *The Miners' Next Step* (1912) – the platform of revolutionary trade unionism in the coalfield – recommended it strongly, and wrote as follows:[65]

> Lodges should as far as possible discard the old method of coming out on strike for any little minor grievance. And adopt the more scientific weapon of the irritation strike by simply remaining at work, reducing their output, and so contrive by their general conduct to make the colliery unremunerative. . . . The Irritation Strike depends for its successful adoption, on the men holding clearly the point of view, that their interests and the employers' are necessarily hostile. Further that the employer is vulnerable only in one place, his profits! Therefore if

> the men wish to bring effective pressure to bear, they must use methods which tend to reduce profits. One way of doing this is to decrease production, while continuing at work.

A very good example of restriction is given in the same pamphlet. The management at a certain colliery wanted to introduce screens for checking small coal. The men who had previously been paid for large and small coal in gross took objection to this as a thinly disguised attempt to reduce their earnings. When the management persisted, the men reduced their output, and instead of sending four trams out of a place sent only two. The management immediately saw its output cut by half, whilst its running expenses were the same. It didn't take many days before the men won outright.[66]

At Wardley, during the unemployment of the 1920s and 1930s, George Harvey, the lodge secretary, introduced an elaborate system of work sharing and restriction to stave off the threat of the sack. At one time Bowes Lyon, the owners, were planning to sack fully one half of the men. Harvey drew up a rota system and doubled the number of miners working at each individual place. Instead of four there were now eight men working together: two of them spent a fortnight on the dole, while the other six were kept at work. Wages were divided amongst all eight so that those on the dole were brought up to an equal wage. Then another pair spent a fortnight on the dole. This was such an effective system that collieries all over the country wrote to Wardley for details of it.

The mechanization of the pits in recent years has made it more difficult to operate a systematic policy of 'gaan canny'. But it does survive in various forms. In the past it was done for the establishment of a better piece price. Today, under the day wage system, it takes the form of forced overtime. In my own experience, the rippers (a class faced with harder work and worse conditions than any other face worker) will 'gaan canny' to allow their job to fall behind; this will necessitate them staying a couple of hours over time, and give them three hours' extra money for the task they were supposed to do on the day wage; or else they will allow the job to accumulate in order to be brought on at a week-end, when they are paid double time. If there is a big pool of surplus rippers 'on the market' the men may 'gaan canny' for a month or so, leaving the task uncompleted week after week. When they are called before the under manager they may very well claim that they need another man with them. In the old days, of course, they might have asked for an increase in the yardage price.

Most workers will privately price a job in their heads; if they don't think the money is sufficient for the task required they will 'dodge' a quantity of the work. In the pits the main belts carrying the coal away from the face to the pit bottom will constantly spill over until quite a substantial amount of coal lies on both sides of the belt all the way outbye. A man is paid less than the face rate for clearing this up even though the amount of coal to shovel is much the same, and if he is given the job he will do as much work as he thinks the wage is worth and 'dodge' the rest. The problem is that on this task you are never far from an official; in the distance you can see his light, and that means he can see yours, too. One way of dealing with this is to remove the cap-lamp and swing it slowly from side to side, knowing that to a deputy in the distance this looks for all the world as though the wearer of the lamp is shovelling; in fact he is taking a rest, say, squatting against a wall.

In the days of bell-pushes we used to have 'ghost signals' as a method of slowing down the pace. In those days there was a bell-push on every face, which stopped the chain conveying the coal away. If a man sat down with his legs outstretched, he could place a small stone between the toe of his boot and the bell, and when he pushed it the bell rang and the chain automatically stopped. So long as he kept his boot there the chain remained stationary. The deputy would scream at the thought of all that money being wasted: 'Who's got chain, who's got chain?', and set off down the face trying to catch the culprit. When he got close, the offender withdrew his foot, the stone fell away, and the conveyor was off again. 'Well, bloody hell,' the deputy would say, 'bloody ghosts again.' This particular tactic was well known among miners, and easily employed, but it was only resorted to when serious discontent was brewing.

Sabotage is quite common in a pit where conditions are very bad and there are no militant workers at hand to heighten the level of struggle. Haulage, for instance, where there are no skips in use, is a job which often invites sabotage. The young worker is faced with an endless stream of tubs, minute after minute, hour after hour, day after day. The situation seems hopeless; the worker literally can't stop, even for a drink of water, because if he does so the streams of tubs bump into each other and come to a stop. They then have to be pushed from a dead start, which is very hard work, and for the rest of the shift he will be working to make up lost time. He may organize a smash-up to get a little rest. Creepers are mysteriously smashed; two lines of tubs collide; a piece of wire wedged into a bell will set it ringing and prevent it being used for signalling; or a piece of paper will stop it working entirely. While the deputy tries to find out what has happened the worker takes a little rest.

A more desperate form of restriction which I witnessed amongst young haulage workers was self-sabotage. The worker may be so desperate in face of the speed and intensity of his job that he hurts himself to escape from it and have a few weeks 'on the club'. He may wedge a piece of wood under an iron bar, put his hand under the bar, look the other way and with his free hand knock the wedge away, leaving the iron bar to fall on to his fingers. He may hang the iron bar on a cord, swing it away and then walk into it. These are, of course, very desperate measures and the job will have been sabotaged many and many a time before this would even be considered. It was usually done by younger workers who had not yet gathered the experience of older workers as to the best means of struggle. The great power of the haulage workers was that it only needed a five-minute strike by them to bring the whole pit to a standstill as tubs clogged up waiting to be removed. The lesson once learnt is never forgotten.

The most modern example of restrictive practice is probably that of absence, the 'gaffers' plague'. One will find in modern collieries great notice-boards displaying on one side the production and the potential of the district pits, and on the other the graph of absentees. The latter will always attract the most attention, and 'Woa lads, we're dropping behind, let's gaan yem' will be the miners' cry if their pit has a better attendance figure than the other collieries.

During the winter months a manager's notice was seen on the colliery notice-board saying: 'I am most concerned at the number of men staying off work'; the reply in pen underneath was: 'Well hang on, coz, the fine weather's starting.'

A mass meeting called on absenteeism by a Doncaster manager witnessed a gaffer on the stage imploring the men: 'Why do you work four shifts a week?'; back came the chorus: 'Coz we can't live on three!'

In years gone by the worker pleased himself when he went to the pit; he only worked for money; when the money was sufficient and no special needs were outstanding, he stayed at home and enjoyed the sun. There were times of the year when the young men and lads refused to work, and insisted on a 'gaudy day' which could be the first morning a cuckoo was heard, or peas had reached maturity. The day was christened accordingly and the lads would away to a public house where they would drink and enjoy themselves all day. In that early poetic description of the miners, *The Pitman's Pay*, we find a very clear reference to this:[67]

A cuckoo-mornin' give a lad,
He values out his plagues a cherry;
A back or knowe myeks hewers glad,
A gaudy-day myeks a' hands merry.

This, although unfortunately not named as such, is carried on to this day. In the fine summer months, on the afternoon shift, the sun high in the sky and the girls in summer gear, the young miner can be seen walking along, dirty old pit bag slung over the shoulder and full of disgust for the black dirty hole in the ground to which one must repair, away from the beauty and smells of the day; one might see in a field a dozen or so lads, bait bags used as pillows, snap tins open and the contents being swigged down to the accompaniment of bottles of beer from the nearby pub. 'Picknicking?' 'Aye, for a minute we thowt wi'd get there, ha'won lad, Danny, what's on thee snap?'

Billy Botchiby of Wardley recalls a marra of his called Geordie who would never work on Mondays. Never worked a Monday in his life. Eventually he was called before the manager. Geordie was informed that he'd lost fifty Mondays all in a row; this was a direct result of drinking. The manager informed him that beer drinking was not his business, for he had a mine to run. Geordie could have his cards or stop drinking so much beer. Geordie replied: 'That's terrible, gaffer, terrible,' but on seeing the manager sorting out his cards, agreed to the condition. Just as he was going out of the door the manager said: "Now remember, no beer'. 'That's reet,' says Geordie, 'but there's nowt wrang with a glass or two of stout, is there?'

The managers always claim that the best coal filling is done in the bars on a Monday; the men won't go into work but talk about nothing else in the pub:[68]

I'm a celebrated working-man from work I never shirk,
I can hew more coals than any man from Glasgow down to York;
And if you'd like to see my style then call a round on me,
When I've had several beers in the bar-room.

In the bar room, in the bar room,
That's where we congregate,
To drill the holes and fill the coals

And shovel back the slate;
An for to do a job of work O I am never late,
That's provided that we have it in the bar-room.

At putting, I'm a dandy, I hope you will agree,
And gannin' along the gannioboard I make the chummins flee;
Your canny sweeps and back-over turns they never bother me,
When I'm sitting on the limmers in the bar-room.

I can judge a shot of powder to a sixteenth of a grain,
I can fill my sixteen tubs though the water fall like rain;
And if you like to see me in the perpendicular vein,
It's when I'm setting timber in the bar-room.

Now my song is ended, perhaps we'll have anouther,
Now don't you fire any shots in here or we will surely smother;
The landlord here would rather sup beer than go to all the bother,
And to put up the ventilators in the bar-room.

All these job controls I have been writing about, all these freedoms, are under heavy attack from the National Coal Board, and for the last ten years and more, up to the great strike movement of 1972, the miners have been on the retreat. Productivity deals, and the rapid growth of mechanization in the pits, together with the multiplication of officials, have brought a whole series of creeping restrictions. The NCB have done everything they can to break the miners' control at the point of production. The National Power Loading Agreement, and, more recently, the Third Day Wages Structure, were intended to do away with the miners' on-the-spot bargaining power. They have rigged up a permanent spying system in the form of Tannoy systems, so that officials can constantly find out what is going on and what the men are doing. Instead of being able to take time off, the miner has been pushed by falling wages to work more and more overtime. The leaders of the NUM have been partners in the miners' humiliation, and for years were only too anxious to co-operate with the plans of Robens and the NCB.

VIII

When the Durham Miners' Association was founded in 1869, the county was divided into three districts, and an agent appointed to each. His wage was to be £1. 5s. 6d. per week, and the number of agents increased as the union prospered.

The full-time officials soon developed a particular character. Almost invariably they were drawn from the ranks of the moderate, self-educated, temperate miners. Once elected, they thought their role was to inflict upon the members their own moderation, and lead rather than serve. The members found that they were being policed by the men to whom they were paying wages. The officials became more and more preoccupied with arbitration and conciliation as the cure for all ills, and more and more impatient of local action which ran up against it. The leadership rejoiced in the formality of the conciliation machinery; they exalted the authority of the Joint Committee even when it was used against their own members, preferring any course of action, 'even simple submission', in preference to a strike.[69]

From this point on the struggle in the coalfield became three-sided. There were the men, the owners and, firmly between them, the full-time agents who negotiated on their behalf but came to totally unsatisfactory agreements and then spent the bulk of their time trying to ram them down the throats of the men. To make matters more complicated – and more infuriating to the lodge with a strong grievance – there was the 'Joint Committee' of the DMA and the coalowners, a high court which sat in judgement on them, and whose rules were supposed to be binding.

The lodges hated the idea of leaving negotiations in the hands of full-time officials. The constant appeals to abandon the Joint Committee, the suspicion of the Sliding Scale, the mistrust of the proceedings in the DMA executive itself, as well as the disputes between individual lodges and the miners' agents, were all based on a deep-felt resentment of delegated responsibility, whatever form it took.

From the earliest days the DMA leadership showed themselves deeply afraid of independent working-class action. In economics they accepted the total subordination of wages to the market, and were the upholders, defenders, and to some extent the devisers, of the 'Sliding Scale' system, which tied wages to the market price of coal. They were advocates of co-operation with the coalowners, however heavy the price at which it had to be bought, of conciliation, even at the expense of their own members. In June 1870, very shortly after the union had been formed, a new rule was introduced to say that any colliery which struck work in an 'unconstitutional' manner would be denied union aid![70] And as part of the 1872 wage agreement the DMA's full-time agents pledged themselves to do everything they could to 'prevent idleness', to 'reduce to a minimum the number of petty local strikes', and to advance productivity.[71] In fact they can be seen as pioneering some of the methods of modern industrial relations in which the trade union appears sometimes as a direct agency of the employers.

The DMA leadership were dedicated to the idea that arbitration must be substituted 'for the old and brutal judge of trade disputes', i.e. strikes. Time and time again we find that 'to avert a strike' the DMA submitted disputes to a court of arbitration. Their tone was conciliatory. We are told, for instance, that in an arbitration of 1874 'Crawford opened his case in a tone of apology for his presumption in joining battle with the owners. On behalf of the men he acknowledged that capital must be allowed a fair remuneration.'[72]

The DMA established jointly with the coalowners a powerful machinery of arbitration and conciliation, which, in intention at least, was to replace strikes and to account for every source of possible conflict. They clung to it in spite of strong rank-and-file opposition. In 1880, for instance, during the controversy on the 'hours' question, the Executive Council found themselves opposed by the majority of the membership, who voted against attendance at the Arbitration Court. But they were in no mood for taking notice of such trivialities as the wishes of the membership, and in a special circular they claimed that they couldn't look after the men's interests unless they had control of decisions:[73]

> The returns of this case have now come to hand. A majority of votes recorded are against the Committee attending the Arbitration Court wherever that may be held. This, we venture to say, is a most unusual decision, and carries with it . . . its own contradiction. At Council meeting . . . we were . . . appointed to prepare and conduct the case throughout. This, however, you have rendered impossible by your recent vote, inasmuch as we are deprived from attending the Court, and cannot, therefore, assist in the discussion or consideration of either the original cases or the rejoinders . . . by your recent vote, you deprive us of the power of seeing how the case is conducted, and prevent us from even seeing the most important portion of the owners' case . . .

In the early years of the DMA, the leadership came to their agreements with the coalowners through the machinery of arbitration. This is how they negotiated the wage advances of 1871–2 and the reductions of 1875. The procedure was abandoned, according to Welbourne, only because it 'excited too keen interest among the men, and provoked serious discontent'.[74] It was embarrassing to have the miners take too close and critical an interest in the proceedings, and the DMA turned instead to the 'Sliding Scale', a system which was to relate wages automatically to the market price of coal, and so remove them from the arena of dispute. The first Sliding Scale agreement was reached in March 1877, and the system was maintained, despite bitter opposition, until 1890. The Sliding Scale was introduced in the hope that it would prove less contentious than arbitration, but this is not what happened. The idea was that alterations in wage rates would be so frequent, and so small, that they would excite less turbulence than the steeper up and down movements of earlier years.[75] But this was not so. The Sliding Scale was bitterly opposed by many of the local lodges, who refused to accept that their wages should be determined by the laws of supply and demand rather than by their independent action.

Once the Sliding Scale principle had been adopted, the DMA had to devote its energies to seeing that the terms were respected, and that no individual lodge breached its terms. On 14 February 1880, for instance, we find the Durham Miners' Federation (the alliance of the miners' unions in the DMA and the unions of the colliery mechanics, enginemen and cokemen) bitterly complaining of the 'irregular' and 'unconstitutional' action of those who put in for wage increases 'in direct contravention to its provisions'. The following is the circular they sent out:[76]

> We deeply regret the action of a certain portion of men renders it necessary for us to address you at the present time. In the month of Oct. last, the Federation Board arranged a sliding scale for the regulation of the wages of all classes of workmen, and to continue for two years certain. This we did in keeping with rules, to which all our respective Associations had agreed. We said then, and repeat now, that it was for us, as workmen, the best and wisest step we could take, and feel quite sure, that in the end, it will be to our respective members, a very great good. We are sorry, however, that before it had been in operation the first four months, some men who subscribed to it are now violating it in opposition to all rules and arrangements. Today, the Federation Board has met the Owners Urgency Committee, for the purpose of hearing the extent of their complaints in this direction. We were first informed that the coke burners at most, if not all of Bolckow, Caughan, and Co.'s collieries, had given in their

notice for an advance of wages. That all classes of men at Browney Colliery had given in their notices for an advance in wages likewise. Another complaint was, that of laying the pits idle illegally and unconstitutionally. There was named as figuring in this category, Auckland Park, Urpeth, Gurney Pit, Leasingthorne, &c. As a matter of consequence, they complained of the very serious breach of agreement attached to these men's actions.

In the first place as to the action of men who are working under the Sliding Scale arrangement, but who have given in their notices for an advance of wages in direct contravention to its provision. It is impossible to do otherwise than regard with disapproval such irregular and unconstitutional action. We have been parties to an agreement, and we ought not to violate it with impunity. We have agreed that wages shall be regulated by the changed market value of coal; such value to be ascertained once every four months.
It must be clear to everyone, that having made this compact, we are bound to carry it out, and not violate it at the very first opportunity. It is this kind of imprudent action which during recent years, and even recent months, has so often brought working men's institutions into disrepute, and brought upon their members, ultimate suffering and loss. . . .

The second class are those who lay down pits illegally and unconstitutionally, thereby not only directly damaging special interests, but . . . laying themselves open to the law. We know of no law which is more equitable than the English law of master and servant, but it treats with the utmost severity those who lay down works in contravention to its provisions. It is not only setting at defiance, rules, which they have agreed, shall guide them, but it is heaping contempt and ridicule on the Executive bodies of their respective Associations . . . we, therefore, urge on all parties to act in future, legally and constitutionally.

A second arm of conciliation was the Joint Committee, set up by the DMA and the coalowners as a court of permanent appeal to which pits were expected to carry all their disputes for judgement. The Joint Committee met to fix an average county wage, and heard applications for reductions or advances in relation to it. It reviewed such questions as hours and cavilling disputes. Its purpose was to stop strikes through dialogue and 'common sense'. The following extracts from the Joint

Committee Minutes – bound with the Records and Minutes of the DMA – are fair examples of its interventions in local affairs.

> *South Pontop* – Men, as a body, having given in their notice without having brought their grievance before the Joint Committee. Resolved, that this Committee refer to the minute of January 5th, 1880, as applicable to this case. The minute is as follows. . . . It having been brought to the notice of the Joint Committee, that men are threatening to lay pits down for grievances affecting collieries they refuse to bring before that Committee, we consider such action to be most reprehensible, and desire that it should on no account be continued (DMA Joint Committee meeting, 16 January 1880).
> *Laying Pits Idle* – It was stated that the following collieries had been laid for various periods, viz: – Binchester, 2 days; Auckland Park, 3 days; Black Boy, 2½ days; Leasingthorne, 1 day; Urpeth, several days; Woolley, 1 day; that Trimdon Grange men had ceased filling. And that notice had been received by Owners at the following pits: – From the Hewers at Browney, and from the Cokemen at Newfield, Byers Green, Binchester, Auckland Park, West Auckland, and Browney, also that restriction was being practised at Nettlesworth. Binchester pit idle. It was decided, that inasmuch as the Joint Committee is the proper tribunal to take cognizance of all disputes, the action of the men in laying down pits is condemned (DMA Joint Committee, 13 February 1880).

The DMA set its face resolutely against local stoppages and refused to give strike pay to collieries taking action which flouted the consultative machineries. Its records, like those of the Joint Committee, are full of complaints against lodges who took hasty or 'illegal' action. Even when a matter of elementary trade union principle was involved, the DMA shied away from direct action. At New Herrington, for instance, in 1880, the men refused to ride either way with non-unionists. The response of the managers was to allow the scabs down and send the union men home. The Executive Committee issued telegrammed instructions to the lodge telling them that they were acting illegally and should leave the matter to the Executive Committee: 'Your action is illegal. . . . Go to work and leave the entire matter in our hands.' The men, however, weren't inclined to have their hands smacked and refused point blank to accept any such thing. The committee sent another telegram to the lodge, which read much the same as the first: 'You must resume work, and let the question be discussed by the Owners' Association and ours.' Once more the men refused and carried on with the struggle.[77] Another example was during the Wearmouth strike of 1875, when the workers struck against the employment of non-unionists (eighteen men among the 1,100 men employed at the colliery). The DMA declared that a quarrel among the men gave no excuse for a war with the owners, and refused both strike pay and official support. Sympathetic lodges levied themselves in support of the Wearmouth men, who stayed out on strike in defiance of DMA commands.[78]

Crawford, the first leader of the DMA, was a Methodist, highly respectable and convinced, like many of his kind, that the owners only needed to be reasoned with for an amicable settlement to be reached. From the start he protested his anxiety to work 'in harmony' with the owners;[79] he was strongly opposed to the strike of 1879, which he said would be 'suicidal';[80] he opposed any local action by the lodges whenever these might imperil recognized agreements or the constitutional procedures governing the settlement of disputes; on occasion he even apologized to the coalowners for 'unwarranted' action by local pits.[81] His statement in the monthly Circular for February 1880 is typical:

Stopping pits

> I wish to call special attention to this, as it ought to be avoided, unless the cause is one of existing and known danger. Both an Executive and Joint Committees exist where all grievances can be heard, and to which all matters of dispute ought to be referred. Stopping pits does good to no one, and is generally a loss to all, wherever it is done.
> 1. – Supersedes or supplants the work of both the Executive and Joint Committees.
> 2 – It throws either party open to be proceeded against in a court of law.
> 3. – In the case of the workman, when a fine is inflicted, and not paid, his household goods can be sold.
> Let this kind of thing be avoided by all hands, unless it arises, as I have just said, from a known source of danger.

Again, in August 1882, one finds the Executive issuing a special circular on the subject, boldly headed 'Pits illegally stopped'.[82]

> Fellow-workmen,
> Without discipline rigidly enforced in an Association like ours, we become a rope of sand. . . . If any portion of Society refuse to be governed by agreed rules, and established customs, no stability can be expected. Men entering an association, are both morally and legally bound to be governed by its rules and constitution. This however is not being done now, by a large portion of our members. Whatever turns up, pits are illegally stopped, and then seeking from the general fund, expenses incurred in consequence of being summoned.

The DMA policy of industrial pacifism did not go unchallenged. In the boom years of the early 1870s the Executive found it difficult to prevent local branches going over the top and pressing for advances beyond the district rate; and they were often at odds with local lodges over the question of restricted output. Opposition mounted in the later 1870s, as the trade depression deepened and the leadership accepted one reduction in wages after another. A good example comes from Bearpark, where the men struck work in August 1878 to resist a local reduction. They refused to refer the matter to arbitration, or to sanction an agreement which their own agents had framed. 'Even eviction did not change their attitude.' In September, at a mass meeting, the Bearpark men attacked the whole policy of the union. They said that 'arbitration had become a farce, that in every case the owners asked for twice as much as they expected to get, sure that the umpire would halve their demands. They complained that local lodges were too much under the domination of the central excutive of the union. . . . And they attacked . . . the Joint Committee'.[83]

The problem of 'illegal stoppages', which plagued Crawford in the 1870s, was still troubling his successors forty years later, as DMA records show:[84]

Collieries idle illegally in 1913

Feb 14	Littletown	July 8	Easington
March 10	Hetton	July 9	Washington Glebe
April 1 and 5	Greenside	Aug 7	Bearpark
April 10 and 30	Byers Green	Aug 11	Easington
May 5	Easington	Aug 13 and 14	Littletown
May 7	Heworth	Sept 12	Adelaide
May 28	Felling	Sept 12	Bowden Close

Sept 25 and 29	Oakenshaw	Oct 30	Auckland Park
Sept 26	Handen Hold	Nov 7	Heworth
Oct 21	Kibblesworth	Nov 24	Chopwell
Oct 23	North Biddick	Dec 20	Byers Green
Oct 27	Morrison	Dec 20	Hamsteels

Opposition to the DMA leadership was usually a matter of a lodge going it alone and ignoring the interventions of the full-time officers and agents. But sometimes the more militant lodges came together to give a more effective voice to their opposition. In November 1882, for instance, the DMA Monthly Report complains of certain collieries in the eastern portion of the county being called to unauthorized meetings with representatives from other collieries, 'for the unquestionable purpose of sending abroad organised discontent'.[85] During the big strike of 1892, according to Wilson, one of the problems of the Durham leadership was the unofficial meetings which were being held, where the Executive Committee was subjected to 'vile names' and 'slander'.[86] Crawford hated these meetings, and in 1879 had published a special circular headed 'Unconstitutional Meetings'. 'Having been placed as the custodians of the interests of our Association, we feel necessitated to call your attention to the irregular, and unconstitutional doings of some of the Lodges.'[87]

It is clear that during the strike of 1879 both Hebburn and Thornley had formed an alliance against the Executive and were holding regular dissatisfaction meetings which were being attended by other collieries. The first of them was in Durham City itself, the sacred precinct of the Executive. The Executive complained bitterly about these meetings, at which, we are told, 'a volume of black-balling, and malignant vilification was indulged in', and attacked them as 'pernicious . . . unconstitutional, and a direct contravention of the spirit of our organization'.[88] But the agitators were not to be deterred; they issued circulars in their turn, and adopted the style of the DMA in the conduct of their proceedings.

One of Thornley's leading agitators at this time was a man called John Lax, the lodge secretary. Little can be discovered about this man, and it is only when his activities became intolerable to the Executive that we have any record of his doings. What becomes apparent is that Lax published a series of circulars to the branches in which he gave militant criticism of the Executive and its policies. The Executive Committee summoned Lax to appear before them. He acknowledged the summons but refused to attend. He was summoned again but still refused to attend unless both expenses and wages were sent to him before he left his home. The Executive became so infuriated by all this that they expelled him and wrote to Thornley Lodge telling them not to accept his contributions. Thornley, true to spirit, replied that they could see no reason whatever why Lax should be expelled and refused to obey. The result was that the Executive Committee declared the lodge 'unfinancial' and accordingly expelled it, too.[89]

The biggest victory to the rank and file in the early years was the strike of 1879, when above 35,000 miners came out against the attempt by the masters to impose a 20 per cent reduction on underground wages and some 12 per cent on surface workers. Crawford had done everything to try to hold back the strike which he had described in advance as 'suicidal'. The Executive preached in two directions, one to the masters asking for arbitration, the other to the men asking them to settle for terms almost as bad as the owners were offering. The coalowners, who had made it obvious that only an unconditional surrender would satisfy them, must have been encouraged by the DMA's eagerness to compromise. But the workers voted against acceptance of the owners' terms, and in March 1879 a district strike was called.

The strike lasted nearly six weeks and involved the entire coalfield, and yet it was settled by less than half a dozen men. Arbitration was eventually agreed, and the owners were awarded a reduction of 8·75 per cent on the wages of underground labour and 6·75 per cent on those of surface workers. The anger of the rank and file exploded. In his book, *Memories of a Labour Leader*, John Wilson tells us that 'Immediately these offers were made known there arose a fierce agitation in the county, and on every hand mass meetings were held protesting against the terms.'[90] The opposition took the initiative and used every platform and gathering to condemn the sell-out. The Executive, for their part, were frightened men, and found public appearances dangerous. One of them was saved from the miners' anger only by an accident. The crowd were closing in on him at a mass meeting on the Durham sands, shouting 'lets put him in the river'; suddenly a man was knocked into a drum which was standing up on end. It went off with a resounding boom, at which the cry was raised 'They are firing guns' and everybody scattered in different directions.[91]

The lodges retaliated against the leaders with bitter attacks. They picked up everything they could lay their hands on. There were complaints from a variety of branches about Crawford going to America and agents buying drinks all round out of county funds. At the September 1879 Council meeting of the DMA, the Executive was faced with a barrage of criticism. Washington, in company with East Stanley and Auckland Park, attacked the Executive for having had the cheek to pay the clerks and agents of the union more money because they had worked extra during the strike. Washington, together with Usworth, Thornley and East Pontop, also put up the following resolution:[92]

> We move that the agents and clerks refund the money back again which was granted to them by the Committee for extra work during strike and arbitration case.

Thornley's motions included these:[93]

> We move the Federation Board be done away with, knowing it is no benefit to us as it is at present, but we seek to advise an amalgamation instead; and that all Branches pay to one Lodge, as we do think it a benefit both to us and themselves.
>
> We think that the E.C. should not have the power to give Mr. Wilkinson leave to draw £1000 and bank it in his own name; we think the county should do it, and then if anything should go wrong, they would have themselves to blame for it.

The wave of discontent gradually subsided, but the attacks on the Executive went on. A circular of January 1880 returns to the theme of 'Malcontents' ('enemies of all that is good . . . sowing dissension and dissatisfaction'[94] and in the following December, when the Executive faced a motion of censure from Silksworth colliery, they returned again to the complaint: 'There is a class of men in the World, who seem to take pleasure in fault-finding and grumbling, and suggesting votes of censure on people who simply do their duty.'[95]

The chief difficulty which the Executive faced was not mass, concerted opposition but independent lodge action. Perhaps a third of the special circulars issued by the DMA in the 1880s, and many of its meetings, were concerned with keeping them under control. The decisions of the Joint Committee counted for very little at the level of the colliery itself, even though it was supposed to be the supreme arbiter of local disputes. In effect, what happened was that wherever the lodge did not ignore the Joint Committee right from the outset, they would ignore it if the results were not to their liking. Sliding Scale agreements were also at risk, though the Durham leadership threatened to expel lodges which ignored them.[96]

The DMA leadership clung to the Sliding Scale even after the system had been swept aside by the Northumberland miners in 1887, but a revolt was brewing, and it broke out when the Sliding Scale came up for renewal in 1889. When a ballot was forced, the Executive tried to use one of its reserve powers (which incidentally was to be used again) to stave off the revolt; namely, the wording of the ballot papers. Lodges were asked to choose whether the Sliding Scale was to be continued in its existing form, or to be amended: no place was left for members to signal an outright rejection. There were bitter protests throughout the coalfield. 'Every colliery called a meeting of protest'; and the agents were forced to issue an amended ballot form which produced a huge majority for outright abolition: 'In place of a petition for the renewal of the agreement the agents were ordered to make a demand for a general wage increase of 20 per cent.'[97]

The upsurge of discontent in 1889–90 was greatly helped by the short-lived boom in the coal market, which gave miners a favourable bargaining position, but it survived into the depression of the early 1890s. The miners' spirit was too high to follow the market curve of coal prices when it turned downwards, and in 1892, when the owners decreed a reduction in wages, the miners refused to bargain. The coalowners had demanded a reduction of 10 per cent; faced with outright refusal they reduced their demands. The DMA put the question to the ballot, under the following questions:[98]

(1) For acceptance of the 7·5% reduction.
(2) For acceptance of the two reductions of 5%.
(3) Strike.
(4) Federation board to take full power to settle.

The reply in terms of voting was

(1) . . . 926.
(2) . . . 1,153.
(3) . . . 40,468.
(4) . . . 12,956.

The overwhelming vote for a strike was reported in the local press, very correctly, as showing a lack of confidence in the leadership; the leadership had advised the men to leave the question for them to negotiate, but the ballot showed a preference for an all-out fight.

The defeat of the strike, after ten weeks, and the intervention of the Bishop of Durham, which ended it, led to the re-establishment of the paraphernalia of conciliation. By December 1892 the Joint Committee was again in session, 'under new rules which departed little from those of the past'.[99] In 1895 a Conciliation Board was set up to decide on wages very much on the principle of the Sliding Scale (it decreed two wage-reductions in the course of its first year of operation).[100] Very soon the old battle lines were drawn, with a leadership even more committed than its predecessor to a policy of industrial peace, and militant lodges, on the other hand, pressing for a forward policy.

One of the most interesting rank-and-file revolts came in 1910, with the strike against the Eight-Hour Day, which was as much a strike against the leadership as against the coalowners. To avoid confusing readers who have read the history of other coalfields, let me say that the question in Durham was inseparably associated with the introduction of night-shift working, and of a three-shift system in place of a two. The implementation of the Eight-Hour Day in other coalfields may have been a progressive step. In Durham it was not. It had been bitterly resisted by the men (also by the leadership) for twenty years, on account of the peculiarities of the Durham system of shifts, and when it was finally introduced, after an agreement between the DMA and the coalowners, there was an immediate crisis.

The miners' opposition to night-shift working was deeply felt; most miners in Durham had been used to a seven-hour day, though the putters often worked as many as ten. Ten hours a day in a colliery is no picnic, but a shift system which drives a man down the mine at dead of night is a damn sight worse. Day-shift working had meant many things to the Northern miners. First, there was the question of safety and working conditions. Day-shift working gave the ventilators a chance to clear the mine of gases, to cool the air, and to clear it of dust, therefore rendering the mine less liable to explosion. It allowed time for repairs and strengthening of hewing places, stalls, etc. The stoneworker had time to take up 'bottom' (floor) and to make sufficient height in the thin seams to get the tubs as near the face as possible. (The nearer the tubs are to the face the easier it is for the hewers to fill.) Second, there was the question of wages (notoriously affected by the selling price of coal). Day-shift working restricted the amount of coal on the market and kept up wages; three-shift working would produce a glut.[101]

The introduction of three-shift working threatened the miner's home. This fact was of great importance. Any man who has worked down that miserable hole knows the value of a home and a chance to be quiet with his wife; the Northern miner, perhaps above all others, felt this dear to his heart. The miner's wife valued greatly the company of 'her man'; the mine thrust its way into their partnership in so many ways that the thought that now they could not even share a bed at night must have been unbearable. Under a three-shift system the miner is forced to go to work at all hours of the day and night. If there were three or four miners of a family on opposing shifts the poor wife would hardly get to bed, as if the terrible fear that put knots in her belly was not sufficient, every time 'her man' was late:

> Manys the time av sat by the fire
> And thought how the coal was won,
> Waiting to hear his knock on the door
> When another day's work was done,
> Manys the time av listened and trembled
> To hear that warning bell,
> Dreading to hear that knock on the door
> And fearing the tale it might tell.

There were big families at that time in County Durham, with many pit lad sons. There would be clothes to wash every day, both pit and surface gear; meals to prepare for both the husband and the sons; but there was only one wife and she needs must adapt to them all, to rise with them whenever they went to work and to be ready with a meal whenever they came home.

These things were of supreme importance to the working miner; but the leaders of the DMA had been in a huddle with the owners for six months, debating and discussing the Eight-Hour Day question in secret without any reference to the membership. Nothing whatever passed outside the doors of the council chambers until two or three weeks before the agreement was to come into operation; nor was there any mandate for the agreement that they reached.

On 24 December 1909 the terms of the agreement were released in the local newspapers under the heading: 'Eight Hours Act: Agreement between Owners and Miners'. The agreement preserved the county's seven-hour shift, but gave everything else away. The number of shifts, the number of men in each shift, and the number of coal-drawing shifts, were to be 'such as may be deemed necessary by the management at each colliery'. Coal-drawing time in the pits was no longer to be subject to limitation – 'the shafts or other outlets of the mines may be used by the owners at any time of the day or night for the purpose of drawing coals or for any other purpose without limitation, except that no person shall be required to remain at his work in any one shift longer than the hours recognized by the two Associations'. The times at which the various shifts should descend the mine 'shall be such as may be fixed by the management to suit the circumstances of each individual colliery'. The meetings between the DMA and the coalowners had been conducted 'in a friendly spirit' on both sides 'under the tactful chairmanship of Sir Lindsay Wood' and it was confidently predicted that the terms of the agreement would meet with the 'ready acceptance of the men'.[102] John Wilson sent a circular to members, outlining the terms. After dealing with possible objections he finished up by giving himself and his fellow negotiators a round of applause.[103]

> Let me, in conclusion, say that if ever a body of men deserved universal praise from their constituents, your Executive Committee are fully deserving of yours. There is no self-praise in that; it is simply what is due to a body of men who have passed through times of great anxiety, and who have brought a very complex undertaking to a successful and beneficial result. Your good has been their aim, and they can with satisfaction assure themselves such has been attained.

As soon as the terms were announced there was a storm of protest; all over the coalfield angry meetings were held, and when the agreement came into operation, on 1 January, many pits refused to work under it. The Executive was condemned for negotiating an agreement on behalf of the men without ever consulting them, and the agreement itself was denounced as a signing away of the miners' liberties. Mass meetings of the individual lodges were called to condemn the agreement, and on 28 December, a conference of pit delegates at Chester-le-Street, called by Pelton Fell, and attended by representatives from twenty-three lodges, unanimously condemned the actions of the Executive. A parallel dissatisfaction meeting at Anfield Plain, attended by representatives of a further twenty-six lodges, unanimously adopted a resolution calling for a special county meeting on or before New Year's Day, when the new agreement was supposed to come into operation. In the West Stanley district, too, a hastily summoned meeting of pit delegates carried a resolution against the Executive 'almost unanimously'.[104]

At first the opposition carried all before it. A further round of mass meetings reaffirmed the rank and file hostility, and carried the opposition a stage further with an outright attack on the 'autocracy' of the Executive, and a mounting demand for strikes. On New Year's Day a delegate meeting called by the ever-militant Thornley Lodge, and attended by delegates from twenty-six lodges, unanimously called upon the Executive Committee to resign 'en bloc'. When the new agreement came into force on 1 January a number of pits refused to go to work under it, whilst others staged one-day strikes, and tried to negotiate satisfactory terms with the individual colliery managements. Alderman House, President of the DMA, and one of the new generation of Labour Party leaders, was hopeful that the unrest would die away after the first week of the new agreement.[105] But in fact it extended, and reached its high point in the second week of January, when some pits, which had started work under the new agreement (hoping to be able to negotiate better terms) decided to join those who had already come out on strike. On Saturday, 8 January, a series of mass meetings at different places in the coalfield – and a delegate conference at Chester-le-Street attended by representatives from seventy different lodges – resolved to come out on strike, and at the start of the second week of the agreement the strike movement was general.[106] At the beginning of the third week in January some of the pits returned to work, but feeling amongst those who stayed on strike was running high. On Monday, 17 January, an army of miners 10,000 strong invaded Gateshead to demonstrate against John Johnson, a DMA official who was standing in the parliamentary election. The men arrived at midday headed by three bands. They carried banners, one of which read: 'We are the South Moor miners. Down with Johnson, the three-shift candidate, the miner's ruination.' On their way home at night, the miners attacked the pit at Birtley, where work was going on under the new agreement. The authorities had been forewarned, and the miners found the gates closed against them:

> The railings were speedily demolished, and a rush was made for the pit, but there was a surprise in store for the miners, as they were met by a force of police and men in the employment of the Birtley Iron Company, numbering about 100. The defenders were all armed with heavy walking sticks, and many of the miners carried pieces of the broken railings and stones. Several scrimmages took place, and then

> the police and their supporters made a rush, and laid about them vigorously. There was a strong reply, and two of the policemen and several of the miners were somewhat injured about the head by blows from sticks and stones. Another force of police appeared in the rear, and they eventually succeeded in driving off the invaders and preventing them reaching the pit.

If this was the conclusion of the march it had started in similar spirit. Marley Hill Colliery, owned by Bowes and Partners, was raided by young miners. Between 4,000 and 5,000 miners left the Stanley district to take part in the Gateshead demonstration. As it reached the Marley Hill Pit about 400 'young-un's' broke away and charged the colliery. They turned over tubs and smashed windows in the weigh cabin. Registers were captured from offices and hurled down the shaft, pony fodder joined the registers, the crowd advanced and smashed windows in the tub shop and pick shop whilst arming themselves with picks and props. Oil lamps were swept from the tables, and the lamp cabin devastated; tools and other equipment were carried away. The entire colliery was held under siege for above an hour before the raiding party returned to the main body of the march making for Gateshead.[107]

On the following Thursday, 20 January, a much more serious disturbance broke out at Murton, one of the largest collieries in the county, employing 3,000 hands. Here the miners had been on strike from the beginning of January, and a party of police had been drafted into the village to protect the coal company's property. The disturbances began at noon with a mass raid on the coal heap by many hundreds of men, women and children, who were able hopelessly to outnumber the police, and to carry away coal for their homes 'in every conceivable kind of receptacle'. The police immediately called for help from surrounding districts, and by 3 p.m. a big force had arrived:[108]

> The crowd who had ceased to take any more coals in the face of such a show of police resistance, stood surveying the officers with curiosity, and swarmed upon the ballast heap and up to within a dozen yards of the police. The crowd on the ballast heap must have numbered between 2,000 and 3,000, and another crowd of nearly equal proportions stood within the gates of the colliery premises. Something occasioned the crowd to make a rush further up the ballast heap, and at this moment Supt. Waller gave the order for the remaining officers to drive them back. The police got behind the crowd, and for a minute or

> two the officers patiently endeavoured to force back the people. Then several stones were thrown, and the order was given to the police to draw their batons, and an indescribable scene followed. People rushed madly from the ballast heap, whilst the officers, who had been reinforced by their colleagues from other parts of the colliery, belaboured the crowd right and left. In less than a minute the crowd, helter-skelter, had fled from the ballast heap, and then it was seen that five or six men were lying helpless or rolling on the ground from the effects of the blows from the batons. They were taken home by their relatives, blood streaming from their faces in some instances. A minute later the crowd commenced to throw bricks and stones at the officers again, and the policemen had to seek shelter behind some waggons. Some of the more determined of the crowd then got right above the waggons and threw more stones until the policemen were forced to charge in the face of a fusillade of missiles.

This was followed by an attack on the colliery offices; on Cornwall House, the private residence of Mr E. Seymour Wood, the manager; and finally on the coal depot office, which soon became a wreck:

> Stones were hurled through the windows, and doors and gates broken down. Books, papers, etc., were thrown about.

The final visit was to the residence of Mr John Bell, assistant manager:

> Here the damage was tremendous. Stone after stone was hurled through the large windows and the out-houses. The railings enclosing the gardens were pulled down, and altogether the damage was serious.

On Saturday, 22 January, a delegate meeting was held at the head offices of the DMA, attended by delegates from all over the coalfield (and strongly guarded by a patrol of Durham City police). The meeting was continued on Monday, and at the end of it a motion of censure was narrowly defeated by 344 votes to 338. The agents, clearly shaken by the vote, decided that it was 'too near to make it pleasant for men in the position of your agents to retain their places', and put the issue of their resignation – and that of the Executive Committee – to a lodge ballot.

This was the high point of the opposition. The ballot, when the vote was counted by the agents, showed a majority of 173 against the resignation of the Executive (426 votes against, 253 votes for); the strike movement lost momentum, and by the beginning of February was confined to a hard core of oppositional pits, who carried it on for a further four or five weeks, when they were driven back by hunger. The striking collieries made the best terms they could with individual managements, but the fight against night-working, and the three-shift system, was lost.[109]

In the rank-and-file movements of the 1870s, a village which stands out for its militancy is Thornley. It was an old Chartist centre and the only pit village in Durham to take part in the Chartist 'Sacred Month' – the attempted general strike of 1839. In 1843 the Thornley miners were up in arms against the slavery of the yearly 'bond', crying their defiance of the owners when they were taken to court; they were defended by W. P. Roberts, the Chartist leader and pitmen's attorney.[110] It was the men of Thornley who first went on strike in the movement of 1869 which finally put an end to the system of the bond.[111] Thornley was well named so far as the DMA was concerned; it was a real thorn in their sides, and took a leading part in rank-and-file movements. Welbourne calls it 'always a storm centre'.[112] Thornley, together with Hebburn, played a forward part in the rank-and-file opposition during the strike of 1879, and it is singled out for complaint in the DMA circular on 'Unconstitutional Meetings' issued in August of that year. The Thornley men were determined that the Executive Committee should not be allowed to sell the strike short; they circulated the branches with an attack on the Executive's handling of the dispute. In their reply the Executive complained that the Thornley circular had much to say in the way of grumbling, but nothing to suggest as a remedy for their imaginary wrongs. The Thornley circular argued that the Executive: 'Should have more regard for the men's interests and less for the interests of the masters', adding that their lodge 'has entered on a great struggle to regain their liberty'. For their part the DMA replied that they would find no liberty by promoting disunion, discontentment and imposition.[113]

Thornley was still troubling the Executive in the following year. The lodge had been receiving an allowance from the DMA for men out of work through damage or breakage to a shaft. The DMA claimed that coal was being drawn again and the allowance, which had been given for two months, was stopped. Thornley issued a circular denouncing the Executive, who replied by complaining that 'a few unquiet and mischievous spirits are again at work at this place'. They alleged that Thornley had been trying to spread disaffection throughout the coalfield by their circulars. Crawford's remarks on the subject are interesting: 'I need not say, that a few men at Thornley have always stood pre-eminently high as malcontents', he wrote in the Monthly Report for April 1880. 'This fact is so well known, that I think even they themselves will not deny it.'[114]

Chopwell was another militant lodge. In the 1920s it was known in the Durham coalfield as 'Little Moscow' (my own village of Wardley was also given this name) and famed for its communist following. Chopwell's tradition goes back a long way. Like Thornley it was a pest to the Executive in the 1870s and 1880s, putting forward motion after motion against the policy of conciliation. On 12 June 1886 the Executive replied to one of Chopwell's motions of censure in the following way:[115]

> Chopwell, No. 9. Votes of censure are quite the prerogative of this Lodge. To judge by their constant action, they alone are perfect. It would be well if some men would pluck the beam out of their own eye before they begin to find fault about the mote in the eye of somebody else. . . . We have no hesitation in characterising this motion as the outcome of sheer vindictiveness.

Seaham stands well as a militant lodge. In 1854 when the Durham miners were collecting signatures to petition for better inspection of the mines, the Seaham men went straight ahead and struck. Martin Jude, the area leader of the time, offered to arbitrate for the men. They didn't object to him doing this but ignored the award he reached. When the blacklegs arrived the strikers threw themselves upon them and smashed down the doors behind which they cringed. The strike was subsequently to close with a long list of arrests for riot.[116]

The Seaham men, because of their vigorous unionism, necessarily ran up against the staid leadership of Crawford. Crawford had already announced his intention to work in harmony with the owners when the Seaham strike of 1872 broke. The manager of the colliery was insisting on a third shift; the men got tired of the argument and walked out, 1,500 strong. Crawford remonstrated with the disobedient miners, but reaped only a vote of censure from the lodges for doing so.[117] In 1880, after a terrible explosion at Seaham, in which 164 miners lost their lives, the Seaham Lodge and the Executive were to clash repeatedly. After the accident the men laid the pit idle the following day, as is the customary mark of respect. The Joint Committee – the committee of the DMA and the owners which sat in judgement upon local disputes – reached the following amazing decision:[118]

> It is the opinion of the Joint Committee, that the action of the men in laying the pits idle on the following day the accident occurred, be condemned; and should not be continued.

The Seaham disaster disturbances continued for a number of months. After the explosion there was a fire raging in the pit, and the owners immediately suggested their age-old cure of sealing off the seam that was on fire. This was violently opposed by the miners and their wives, burned up with anguish at the thought that the owners were walling up a seam in which there might be miners still alive, and mass meetings were held. Tom Burt, MP, the Northumbrian miners' leader and a great industrial pacifist, had a cousin at the colliery. A stark contrast was Tom Burt of Seaham, for he was a real spokesman for his fellow miners, an active militant and subsequently blacklisted by the owners. He questioned the need for stoppings and spoke with great bitterness about the haste with which the workings had been sealed after a colliery explosion in 1871, in spite of the belief that there were a number of men in the colliery still alive. Strong feeling was building up about the latest stoppings and it broke into open riot when they continued. Many men were arrested. In February 1881, when a special court was convened to deal with these arrested men, all the village turned out in sympathy; the court and streets were packed with people. One of those charged was a woman who had rushed out and blinded one of the blacklegs with pepper. Crowds turned up to see the prisoners driven off to gaol.[119]

The trouble did not end there. When the pit commenced work again there were still thirty bodies unrecovered in the Maudlin seam, and the men refused to work in it until all of the bodies were brought out, and went on strike when the managers tried to order them into the seam. Instead of backing the men's action, the DMA attacked their attempts to win support. The lodge balloted and overwhelmingly decided to stay out on strike.[120] Support for Seaham was beginning to spread and soon the Executive Committee was facing motions from other lodges:[121]

> That a vote of censure be paid at Council meeting on our Agents and Ex. Committee, for the very unfavourable result to which they have brought the voting in the county, and upon the Londonderry collieries, in relation to the Seaham difficulty, by keeping the county

ignorant of the vital interests involved and not adhering to rule 56. *Silksworth*.

The Tyneside collieries between Gateshead and South Shields, had a strong fighting tradition, amongst them my own village of Wardley. They were some of the oldest collieries in Durham, and were very much to the fore during the miners' movement of 1831–2: a magistrate was done to death at Jarrow Slake in 1831; and at Friar's Goose Colliery, Felling, the miners armed themselves to face the detachments of troops and police sent to evict them from their homes. In the big strike of 1879, Hebburn, alongside Thornley, was a centre of rank-and-file disaffection. The Tyneside pits were amongst the largest in Durham, and the cosmopolitan nature of the area, and close contact with the river mouth ports, perhaps helped to promote advanced opinions.

The Tyneside collieries were particularly combative at the time of the great strike of 1892. In the vote on whether or not to take strike action, local feeling was very strong:[122]

	Strike	*Settlement*
Hebburn	830	49
Heworth	516	86
Wardley	525	59

The water supply to the miners' cottages at Washington was cut off at the start of the strike. This also happened at Heworth colliery.[123] At Boldon colliery the wives and children charged about rattling their tins and pans, striking the fear of God into any unfortunate blackleg.[124] As the strike progressed (Welbourne tells us) the men of the Tyneside pits were advocating 'revolutionary action', seizing the mines, restarting the machinery and working the coal for themselves.[125] Perhaps it was this as much as the hunger of the miners' families which prompted Bishop Foss Westcott to undertake his urgent mediations in favour of peace.

Across the fields from my own village was Usworth, where I worked for a period while undergoing various training schemes. Right from the start of the Sliding Scale agreement Usworth made its presence felt against it, and in a motion to the DMA Council meeting in November 1879 moved:[126]

That we protest against the agreement of the Sliding Scale, and that we don't stand by it.

Some ten years later 'the unruly lodge', as Welbourne calls Usworth, was to be instrumental in bringing about the downfall of the Sliding Scale. The Executive, being quite pleased with the Scale, were very annoyed when Usworth made a demand for a local advance, backed this up with a strike and at the same time issued a circular to its brethren explaining the case.[127] Usworth's action precipitated a general assault on the Sliding Scale and everything it meant.[128] Again in 1892 Usworth comes into prominence: two days before the county strike was due to start a man was killed at Usworth so the pit was immediately laid empty by the miners.

In the strike of 1879 there was a split at Usworth. A party of hewers decided almost immediately to accept the owners' terms and go to work. However, the majority of the Usworth men had different ideas. On the Monday evening one of the hewers, George Ramshaw of Waterloo Square, was on his way to work when he was encountered by about 100 union men who paid him off with a volley of stones and shouts. Another hewer, Simon Crane, arriving a little earlier, had been treated in the same way. This went on all day till one after another blackleg hewers were gently dissuaded from carrying out their intentions. A little later on, the cottages of those who had got to work were visited by the strikers, who smashed the doors and windows. The following day a concerted effort was made by the blacklegs to get to work; a riot ensued and the police were attacked with stones. A mass meeting of the Usworth strikers was held, at which it was obvious that they would resist any attempt to reduce their wages. After this meeting a party of young miners went to the house of one of the blacklegs with the intention of persuading him to come out with them. The man came at the deputation with a gun, but this was wrested from him and smashed to pieces, as were the door and windows of his house. Other fights ensued throughout the village, in which pokers, amongst other things, were used. The strikers had virtual control of the village, although a batch of police was being used in the affray. Several scabs leaving the pit were met with the customary hail of stones and slag, and a shotfirer was badly beaten. During the whole time the men from the neighbouring collieries were with the Usworth men, rendering assistance. The women of course harassed the blacklegs in the usual way, with rough music, shouting, and carrying clothes props with flags attached. It appears a man called Wilson was the leader of the Usworth pickets.[129]

Hebburn, a mile and a half to the east of Wardley, was a centre of disaffection during the strike of 1879. A little earlier Hebburn miners had been out on strike against a reduction in wages. Crawford had stated that no funds could be given them, since they had been carrying on negotiations without having consulted the Executive Committee; the men, it was said, had been operating a work restriction.[130] Hebburn is singled out (together with Thornley) in the DMA circular against 'unconstitutional meetings' in August 1879, and condemned for 'envy and ignorance', and for seeking 'mischief and destruction', in particular because of a motion which proposed that 'when Mr. Crawford is out of the county on arbitrations etc. his wages be stopped'.[131] Later in the year Hebburn was circularizing the coalfield with anti-executive demands: on 15 December 1879 the Executive felt bound to issue a counter-circular attempting to reply to Hebburn's charges and accusing them of misrepresentation and gross impositions.[132]

The most southern of Tyneside's rebel collieries was Washington, which at times the Executive must have wished was in Yorkshire. In a strike of 1875, when the union miners went to jail rather than pay fines, the DMA told them that after all 'they should obey the law and not absent themselves from work'.[133] In 1897 a strike at Washington became the focal point in a long-standing dispute between the Executive Committee and the lodges over strike pay. There had been numerous stoppages, which had taken place against the Committee's advice. The lodges ignored the advice, but still demanded strike pay out of the general fund. Wilson asked 'Should the rule be the guide, and the Executive Committee have the management, or should the Lodges be allowed to stop their colliery in opposition to the constitution, and suffer none of the consquences?'[134] The 1897 strike at Washington brought the issue to a head. Wilson tells us that the defeat of the strike was 'of great importance' to the DMA in the maintenance of rules; he gloats on Washington's defeat at the court of appeal as if he personally derived satisfaction from it. He went on to tell the lodge in his worldly-wise way: 'My advice to you is to consider carefully every amendment which may come before you. Trades organisations will prosper most when they are founded upon, and guided by, business principles.'[135]

It was not long before the Washington men found themselves before the courts again for withdrawing their commodity from 'the market'. On Saturday, 17 July 1909, 181 men from the Glebe Pit were brought before the Gateshead county police court for refusing to pay damages imposed on them for taking part in a strike. All of the men refused, saying that they would not pay, on principle. Mr Lampert, a solicitor, represented the owners and pointed out to the court that the pit was always being laid idle and there was a terrible loss to the owners. The magistrate sent the miners to jail.[136] There were so many that it was decided to send them in batches. Each was given a rousing send-off, the carnival flavour of the occasion serving as a kick in the teeth to both owners and law courts. The men were escorted by the lodge band to the railway station, accompanied by many of their fellow-villagers. Houses were decorated, flags fluttered from windows and thousands clustered around the Glebe colliery, which was laid idle on the days the men were leaving. The police tried to prevent the people getting on to the station platform to give their comrades, husbands and brothers a send-off; they failed and the platform was swamped. All manner of singing and dancing was in progress. As the train pulled out of the station the band struck up a lively tune to cheer them on their way.[137] Considering that there were only thirty in each batch; that each batch was given a big send-off and a welcome on its return, and that for each of these occasions there was a general stoppage of the pit, the owners could hardly have been pleased with what had happened. When the first twenty-nine prisoners returned, a procession was formed. Speeches were made by Mr T. Craggs, branch president, and other members when a halt was called at the premises of the Washington Station Social Club. 'The men of Washington', said Craggs, 'have proved that the miners of Durham are not cowards as they have been called by some of the judges. They have gone to prison for a principle in claiming something like equal rights from the management of the colliery.' Further speeches were made by the men on their prison experiences but at the same time they assured their listeners that they would go right back again for the cause they had in their hearts. Then the procession formed up again and set off round the village, led by a prison loaf which was held high for all to see.[138]

Washington led another movement in 1915, when the owners were forcing one reduction after another on miners' wages. The men could take no more, and on 13 April 1915 the first revolt came. On that day the Joint Committee was to consider no less than nineteen applications from East Durham alone for reductions in tonnage and piece-rate prices. On the initiative of Washington's Glebe colliery the delegates of the affected collieries gathered at the Coal Trade Hall in Newcastle on the same day as the meeting of the Joint Committee. George Harvey of Wardley moved, and Washington seconded, that they allow no business to be done that day seeing that the union representatives on the committee had refused to carry out the wishes of lodge delegates. The Joint Committee meeting was successfully disrupted, and after heated argument and recrimination the proceedings were abandoned. The Executive Committee managed to pass a motion of censure upon Washington but the reductions were suspended.[139]

My own village of Wardley belonged to the select company of 'red' villages; the lodge banner proudly proclaimed it in the 1920s, when it carried portraits of Lenin, Connolly and A. J. Cook, together with the hammer and sickle: at that time it was known in the coalfield as 'Little Moscow', and it elected George Harvey, a self-proclaimed Bolshevik, as lodge secretary. In 1888 a local magistrate called Wardley 'the worst behaved colliery in the country'. At that time it was leagued in a triple alliance with the two neighbouring collieries of Usworth and Felling. 'The three collieries were federated, and it was understood that when a grievance occurred at any one of them the three collieries should take joint action'.[140] An example of this solidarity occurred during the Felling strike of 1887. The strike was originally over the weighing of miners' tubs and at first affected only the men of Felling. But the owners proceeded against the strikers in court. The men were fined for striking without notice, and when a number of them refused to pay the fine they were sentenced to fourteen days in prison. All three villages now struck work and resolved not to resume until their fellow-miners had been liberated. Between 600 and 700 men came out at Wardley, 180 at Felling, and 700 at Usworth:[141]

> The men affected held a meeting in a field opposite the Railway Hotel, at Wardley, on Wednesday afternoon. Mr. W. Bottoms (Usworth), presided. . . . Mr. W. Richardson (Felling), moved:– 'That we do not resume work until the six men are released from prison.' – Mr. R. Bolam (Usworth) seconded the resolution. The speaker said it was no use disguising the fact that the law was against them. They had entered on the warpath, and they must continue their course. They knew that they had done wrong, but they had suffered, and they were determined to make those who had caused them to suffer to suffer in turn (applause). . . . Mr. John Errington (Wardley) supported the resolution. Individually speaking, he regretted the position they were placed in. The law of the country was against them, and so was the

constitution of the Durham Miners' Association (hear, hear). So their position was by no means an enviable one. . . . As it was they stood by themselves. – The resolution was put to the meeting, and carried *nem. con.* – Mr. R. Johnson (Felling) then moved:– 'That we deeply sympathise with the wives and families of the men now in prison.' – Mr. Coulson (Wardley) seconded, and Mr. S. Ferguson supported the resolution, which was also carried unanimously. – The meeting then ended.

There was also a meeting on Wednesday afternoon at the White Mare Pool. The attendance of miners from Usworth, Wardley, and Felling was very large. Despite the unfavourable state of the weather, the men kept well together, and, after speeches were given, motions were moved and carried that the pits still remain idle as long as the six men are kept imprisoned. The pits therefore, were to continue to lie in.

Only a year later, during the Wardley Funeral Strike, the three lodges were once again involved in sympathetic action. When the men of Wardley came out on strike they were immediately joined by those of Felling, and when the men of both were summoned before the Gateshead magistrates, all three lodges marched to court, with their bands and their banners, to defy the prosecutions. *The Times* commented:[142]

> A combination has existed for some time among the men of Usworth, Felling, and Wardley, the object of which is that when any dispute arises between the men and the masters at one pit, the men at the other shall assist by enforcing the terms demanded by their fellow workmen.

The owners brought prosecutions against 181 miners. The men were charged with illegally absenting themselves from work, and the owners demanded 5*s.* a day damages. The owners asked the magistrates to try three test cases as a start, but the miners refused to let this happen. When the name of the first of them, John Quin, was called, there was no answer 'and a message was brought to the magistrates that he was in a public-house, and declined to come out, and that the others would not let him come'. The cases proceeded and orders were made against all but ten of the 181 men, but many declined to appear in court to answer to their names, and of those who did so most of them declared they would not pay the court's fines. On the following day the strike was resumed, and when the police attempted to collect the court's fines they were met with a firm refusal. Wardley was the leader in these proceedings, and John Errington, the Wardley checkweighman who represented the men in court, was told by prosecuting counsel, Mr Cooper:[143]

> There can be no possible justification or excuse for laying collieries idle. There is a regular tribunal . . . the joint committee, before which you have to bring all these questions. . . . You are the worst behaved colliery in the country . . . and you cannot be allowed to lay a pit idle like this.

The miners' unions were among the first trade unions in the country to develop a class of full-time officials. But at the same time as relying upon full-time representatives, more than other working men at the time, miners also distrusted them deeply. Their very distance at area office was a sufficient cause for mistrust, and their actions plagued the miner with annoyance. In all the disputes within the DMA there is a basic pattern, a recurring fear of being misrepresented which can be seen in the type of motion calling for mandates, card votes, records of who voted, which way, etc., and also in the general suspicion of the conciliation machinery which the DMA worked so hard to set up.

The miner's mistrust of 'delegated representation' is a carry-on from his traditional self-reliance and independence at work. He dislikes the idea of anybody deciding what he should do, apart from his 'marras' and himself. It is at best something that has to be put up with. His wages and the toil to get them are a big thing to trust to another man's hands. He prefers the open direct representation of 'we are all leaders'. Even in the case of the working miner who is a branch official, yes – they can see he is a worker, but they would still only trust him if he was seen to be acting on their behalf and the results were coming in. When the lodge secretary goes to the gaffer, a deputation of marras will often go with him to see that he really does represent them. At branch level the union committee is, of course, elected by the rank and file, subject to dismissal, and up for re-election every two years, but even these men have to be watched. The branch committee man, if he is a good one, even the agitator, can be got at in many ways by a gaffer; he may be offered a pleasanter place, away from the centres of trouble, if he will only show himself to be 'reasonable'.

The attitude of the rank and file towards the lodge committee has always been – if the branch officers are not negotiating the right terms, kick them out, or build an unofficial committee to take their place. It is the rank and file itself that takes the initiative in all events. On many occasions strikes start in a wild-cat way long before a branch meeting can even be held. Men walk out of the pit as soon as there is something they won't put up with. A group of marras, or a shift, will go to see the manager on their own, and have often refused to take an official of the branch with them. If the branch official is brought in it is sometimes only as a formality. They prefer to speak for themselves. Rarely, if ever (and never in my own experience), do the workers carry out the rule of working under protest while the union official carries out negotiations on their behalf; they would have to be there themselves (there is also, of course, the simple fact that the union official has no bullet up the spout or power in his elbow until the men come out on strike).

If all of this is true of the branch officer, a working miner, how much more true is it of the area official, a man who has removed himself to another sphere. If the branch officer can be regarded as distant, even after a year of office, the area official inhabits another world. The worker can trust the man who works next to him, even if he is wary of giving him too much power. But the man who removes himself from the pit to an office becomes an alien. You can see this time and again when the area official comes down and tries to give a 'lead'. The men may listen to what he has to say but what they are thinking is '*we* are the union, not you'. The gulf becomes much greater when the men take action on their own. In the words of George Harvey: 'the religion of the area official is compromise'. These people detest the branch strike. In the first place this is because they cannot themselves control the activity, and in the second place it undermines the machinery of conciliation which they are paid to keep well oiled. By and large miners' leaders have been renowned for their respectability and almost all of them have ended up moderates, even those who were firebrands in their youth.

IX

Histories of the miners usually centre round doings of the area executives. The activity of full-time officials is presented as that of the 'miners' as a whole. Rank-and-file movements are treated as nothing more than a 'flash in the pan', as though they were somehow 'incidental' to the whole study. Yet had it not been for the activities of the rank and file the historian would be writing about starkly different organizations.

Men such as Burt, Fynes and Wilson write of their fellow miners from the lofty height of leadership (their own). The rank and file, when it makes an appearance in their pages, is portrayed as somewhat immature and in need of firm paternalistic guidance and control. The leaders in the localities, the 'agitators', who get in the hair of the area leaders, appear only as a pestilent phenomenon getting in the way of 'responsible' leadership. George Harvey, a great figure in my own village of Wardley and a leading agitator in the coalfield for thirty years (he is still remembered as a 'hot heed' at the area offices of the DMA), gets only a two-line mention in Garside's new book on the Durham miners – perhaps because he was a great thorn in the flesh of the DMA, or perhaps because he was not the kind of man who left his mark on bureaucratic records, or perhaps because Mr Garside was more interested in the doings of Sam Watson, the secretary of the DMA, which occupy page after page of his book.

I first heard of George Harvey when I went down the pit at the age of seventeen. At this time, having already been through the Newcastle Young Communist League, I was (or so I then thought) firmly embedded in politics as a 'wobbly' IWW supporter – a Tyneside Syndicalist. My politics were already well developed, and no amount of bribes or threats from work-mates, hard lads, gaffers and lodge men could encourage me to stop pumping them out at every available and inopportune moment I could. On one of my many bait time 'chunterings' I was surprised to find a lot of qualified support from the older miners. They informed me: 'Ye talk like Georgie Harvey, ye knaa.' Slowly, in tender but stirring terms, I was told of the 'Wardley Lenin', who had led the miners of my village in great struggles before I was even a twinkle in my father's eye. His activities were well remembered, and it seems everyone had something to add about them: 'Ney Communist, ye knaa, a Bolshevik, that's what he was', one old miner told me. I was soon filled with the desire to find out as much about him as possible.

George Harvey arrived at Follonsby (Wardley) Colliery in about 1912–13, where he was very soon made the men's checkweighman (at the age of 22–3, this made him the youngest checkweighman to be appointed in County Durham). During this time he kept up with his political work and published a stream of first-rate booklets of an agitational and political nature. Amongst these were *Industrial Unionism and the Mining Industry*; *Capitalism in the N.E. Coalfield*; *Capitalism in the S. Wales Coalfield*; *The Mighty Coal Kings of N.E. England and How are the workers to meet them?* The owners, Bowes and Partners, really knew they had a union when George was elected lodge secretary from 1919 to 1920; almost as a matter of course at the end of every lodge meeting the windows were thrown open and the 'Red Flag' or the 'Internationale' sung with great gusto.

In the village he conducted mass open-air meetings on the sports field, and different speakers were invited every week. A. J. Cook, the miners' leader during the 1926 strike, was a regular speaker. In Wardley's field he stood pointing at the headgear and said his famous: 'Not a minute on the day, not a penny off the pay', and 'The grass will grow on those wheels before we submit to tyranny.'

During the 1921 and 1926 strikes Harvey organized teams of men to griddle and shovel the pit-heaps for coals, which were sent away for sale after the fair coal had been separated from the slag. It was worked in the style of bord and pillar in two-hour shifts. Lorries were hired and the coal was sold to hospitals and factories. Thousands of tons were shifted in this way: the money went to Harvey's soup kitchen to feed the men and their families. Old Wardley men say that because of Harvey's fine organizational capacity they fared not much worse during these strikes than while they were actually working. The men working on the heap were getting £2 per week, as much and sometimes better than they earned at the colliery.[144]

Harvey's leadership of the lodge scared the rice out of Bowes-Lyon, the owners, and in 1935 they made an all-out effort to rid themselves of Harvey and the Follonsby Lodge committee. Fourteen hundred men were issued with their notices. Almost immediately the owners posted a further notice saying that they would re-employ 400 of them. But they were to be men of their own choosing, and they were to sign on without knowing what their conditions of employment were to be. The 400 whose names had been posted refused. The DMA suggested 're-employment' to the others, which they refused, on the grounds that the grading system ('A', 'B', and 'C', 'Good', 'Moderate', and 'Very Bad') was contrary to all county custom. The lodge had proposed that preference for work should be given wherever possible to long-service miners The lodge further objected to the shift system that was being introduced, and to the fact that out of a normal quota of thirty only two of the lodge's deputies had been included amongst the 400 men.[145] The representatives of the DMA met the owners at the Urgency Committee, but Wardley went out on strike.[146] Three days later, when two scabbing deputies went in to work, crowds demonstrated against them and sang hymns outside their doors.[147] About 200 police were in action during the strike: the streets were lined with above 5,000 miners, the men of Usworth, Felling, Boldon and Heworth, coming down to help the Wardley lads.

The DMA and the owners announced that they had reached an 'agreement': it was only the lodge which didn't agree, they said. *Only* the lodge: it was they who were out on strike! The 400 men now had their unemployment benefit stopped, since they had, according to the State, refused employment.[148] The lodge had absolutely no confidence in the agreements the DMA was trying to draw up. The Lord Mayor of Newcastle offered his services as mediator, but the DMA declined them. In a letter to the Lord Mayor, Harvey said:[149]

> The real reason for closing the door to you as mediator appears to us to be that both the D.M.A. officials and those of the coal trade office are anxious that the miners should quietly acknowledge the authority of the 'Urgency Committee' and other machinery created to dampen the rebellious instincts of the workers. These officials are greatly concerned about their prestige.

The DMA at this stage refused to sanction the strike or give strike pay.[150] They complained that Harvey's circulars about the strike were 'in bad taste'. Harvey replied saying that the Executive were leaving the men without protection and encouraging non-unionism in the pits. 'Members want protection after they pay their officials as much in a week as they themselves earn in a month.'[151] Local lodges rallied to Wardley's support, and seventy of them sent regular aid while the strike lasted.[152] Wagons were organized from the village to go collecting for supplies, and they returned loaded down with bread and vegetables. One night, after touring the Easington area, they returned with enough groceries, provisions and loaves of bread to feed 400 men.[153] The village men still remember the food from the other villages: 'bloody loads of it, they kept us like fighting cocks for six weeks'. Various tactics were used against the village's few blacklegs; women in football jerseys and other comic costumes demonstrated against the blacklegging deputies.[154] The women organized football matches to raise money for the strike; the games, although lacking order, made up for it in spirit, and large crowds attended.[155] The men in the village who were receiving unemployment pay each gave a third of it to those who were receiving none.[156] On

2 October, after the best part of seven weeks' strike, the owners gave way completely.

But Harvey wasn't finished yet; he took upon himself the role of fighting the 'Unemployment' for dole money, which they had refused to pay to the 400 men. He fought the case all the way up to London, where he defeated a QC and won £3,300 back money. Then he succeeded in getting the DMA to pay the £700 strike pay which it had refused to pay out while the strike was on. He returned to the village, commented a local newspaper, 'somewhat a hero'.[157] In 1936, the year the money was won, yet another strike broke out at the colliery, but there is not the room to tell its story here.

Many workers have told me that they would have followed George to the end of the earth and fought every gaffer in England if he had called for it. I was hard-pressed to get any information from the older Wardley miners about George's political life and views in any detail, and it was only from investigating other sources that I discovered the great political maturity of the man. But from the personal reminiscences I found the name of Harvey taking on shape and colour, a mental picture forming itself from the deep memories that his comrades had stored away in a safe department of their brains. And from one of George's surviving brothers I discovered much about his early life.

He had started work at the colliery of Pelton Fell, and was soon an active trade unionist. At the age of fifteen he had his first encounter with the law, when he was brought before the courts for the popular but illegal activity of riding his pony outbye at 'lowse'. The usual fine for such things was no more than £2. His mother gave him the money to pay the fine. Long hours later he hadn't returned home; on inquiry it was discovered that George was in the cells refusing point blank to pay a penny. He had asserted that the courts played on the ignorance and fear of the working man and facilitated robbery by the owners – he would take jail, before parting with his mother's £2.

The Wardley miners' memories of comrade Harvey are a curious mixture of awed admiration of him as a revolutionist and a fascination with the details of his ways as an ordinary man. Billy Botcheby, one of George's friends and a fellow trade unionist, told me in Wardley dialect: 'it was a funny way he had, son. He wadn't look reet in your fyce when he was talking. . . . He was a-dam-ant with the gaffers, bet (beat) the managers, bet the owners and betted all ther tricks'.

My father's memories of him are as a quiet man, never a full-blooded speaker – very often he couldn't be heard on the platform. But his logic was as clear as a bell and as easy to understand as you could wish. Others tell that: 'He niver went a'l roond Waalsend ti get ti Jarra.' (He came straight to the point in argument.) An old miner who died recently, and who was one of the Durham agents of the DMA, used to say: 'when Geordie Harvey walks through the door of the council chamber there is dead silence. He knew too much for them to dare say anything'.

In good times as well as bad he kept his army clad and in good stead. He had a shed at the back of his garden which was full of uniforms: busmen's, postmen's, railwaymen's, etc., which were all sold cheaply to the men, along with boots and caps, etc. They resembled for all the world a workers' militia marching off to work (or to picket), dressed in uniform and boots.

One of Wardley band's favourite tunes was 'The Great Little Army', and this they would play as they marched into the streets at Durham for the big meetings, proudly bearing their banner with the portrait of their captain, the hammer and sickle and the red star emblazoned under the photographs of Lenin and Connolly.

The Wardley men were determined to remember George, and at Wardley today we find that between Morris Gardens and Thorne Avenue runs Harvey Crescent.

George, we are told, had a quick temper, and the domestic battles between the great man and his wife (an intellectual) were a great source of amusement to his comrades. It was one of those little things which fascinated them, to think that a man like George, a Christ-like figure to some, could do such ordinary things as engage in domestic squabbles! 'Poo; pots a ink and byueks wad gaan fleing a'l owa the hoose.'

His wife was from the Outer Hebrides and the daughter of a crofter. She was sent to Edinburgh University, where she became a socialist activist and subsequently a great comrade of the pitman from Wardley. She and George were both members of the Socialist Labour Party when they married, and she carried on her political work with her husband; together, man-wife, companions and comrade, they would tour the whole area of the Durham and Northumberland coalfield, selling revolutionary socialist literature. In later years, for some unknown reason, she seemed to drop out of the political struggle and became 'like a hermit' to many in the village. She developed a wide range of interests, including that of spiritualism, and addressed meetings at Newcastle City Hall on a variety of subjects.

Harvey kept goats, which fed along the banks of the railway lines and were a great novelty and centre of attention to the village folk. Like the other little points about his domestic life, the goats helped people to understand a man in many ways distant from themselves. Songs and jokes made by Wardley's 'poetys' were constructed around 'Geordies Goats' and it is not yet unusual in the bar room of the White Mare Pool on a dinner time to hear Joss Bainbridge, the village songster, give a rendering to the delight of the older village folk (and the younger ones).

George, I was told, had gone to Ruskin College, a college held then in high esteem by the workers. His example was one of the things that set me thinking about going to the college. When I got there I resolved to use as much of the college time as possible to re-discover my village's leader and what seemed to be an impossible political animal (in my ignorance at least); namely some kind of a Syndicalist-Bolshevik-Wobbly.

A strange sequence of farcical events were to take place upon my entry into that masquerading establishment. I discovered that under the auspices of an institution designed to serve working-class politics and trade unionism, Ruskin's major role is to take working-class militants away from their jobs and communities to 're-educate' them. It attempts to cut off their class roots and fill them instead with notions of competitive achievement, and generally renders them useless for anything except full-time service in the ranks of union or management bureaucracy. I further discovered that time was not just for the using of, and the type of research I wanted to do was frowned upon. 'The proper course' had to be adhered to unless one simply decided to 'sod them' and get on with real historical inquiry and discovery. How the hell, I asked myself, did George survive this place and still come out a Red?

Of course my realization was not unique. Harvey and his fellow students had come to it sixty years ago, and it led them to take part in the 1909 strike at Ruskin, and the breakaway of the left-wing students to found the Central Labour College. The students had been attempting to resist the introduction of examinations to the college, and to put a stop to the college's growing links with Oxford University. They wanted the college to stand for independent working-class education. The split began when the students demanded that Marx's economics should be admitted to the curriculum as an optional alternative to the orthodox Marshall. And when the college principal, Dennis Hird (a parson turned rationalist), complied with the demand, virtually all the students opted for Marx and there was trouble in large lumps. The principal was instructed by the governors to exclude Marxian economics from the curriculum, and, when he declined on principle, his resignation was demanded. This also being refused, he was sacked out of hand; and the students almost to a man refused to attend any classes conducted by his appointed successor, and came out on strike. The college was reconstituted, with closer links to the university, and the imposition of a university examination as the be-all and end-all of the Ruskin student's course. The left-wing students, backed by the South Wales Miners' Federation and the National Union of Railwaymen, set up a rival Labour College, which soon moved from Oxford to London – a real nursery of revolutionary socialists and trade unionists in its early years.

Imagine my utmost excitement at finding Comrade Harvey in the very forefront of this struggle, and at reading the books and articles he wrote in these years. Sitting reading them word by word, every full stop and comma, was to be the nearest I would ever get in communication with this fellow mineworker who squatted down the same hole as me in Wardley, talking politics not a kick in the rump different, yet sixty years previous.

After leaving Ruskin, Harvey returned to the coalfield and also joined the newly-founded Socialist Labour Party. He became Durham Secretary of the Party, and was for a time editor of *The Socialist*, the SLP paper.

The SLP was Marxist, but much more revolutionary than Hyndman's Social Democratic Federation, of which it was a breakaway. One of the London organizers was Con Lehane, who had been at one time the secretary of the Cork Branch of James Connolly's Irish Socialist Republican Party. James Connolly, who sympathized with the party and later became its first national organizer, agreed to print *The Socialist* on his printing press in Dublin.[158] The SLP took its inspiration from Daniel De Leon in America. Tommy Jackson, in his great book *Solo Trumpet*, tells us of his impressions of De Leon, whom he met in London in 1904:[159]

> De Leon was physically quite a different type (from Hyndman). His ancestry was Jewish, but his immediate forbears were Latin American. He looked the keen, intellectual, professor he originally was and specialised in lucid, non-rhetorical logical exposition. But it was easy to see from the flash in his eyes that he was a bad man to cross.

The SLP was one of the bodies which in later years came together with other small groups to form the Communist Party of Great Britain. But long before then the party had anticipated the development of soviets and called for socialism on the basis of industrial unions, two ways of saying the same thing. Walter Kendall in a brief account of the SLP in his book *The Revolutionary Movement in Britain* says that:[160]

> Alone amongst the political organisations of the British working class it declared against the existing system of parliamentary government. Alone, also, amongst the socialist parties, the S.L.P. directed its appeal to the industrial working class, expressly rejecting the 'political state of capitalism' and demanding in its place the 'Industrial Republic of Labour'.

De Leon reached conclusions very similar to the Bolsheviks independently and a decade before the October revolution. He rejected the idea that political action was sufficient to emancipate the working class. 'No bunch of office holders will emancipate the proletariat. The emancipation of the proletariat can only be the mass action of the proletariat itself, "moving in" and taking possession of the productive powers of the land'.[161]

As a leading activist Harvey became the editor of the SLP paper far in advance of most of the lefty so-called 'revolutionary' socialist papers of today. Earlier he was the paper's mining correspondent. Had I been a blood relation I could not have felt more pride than in discovering *The Socialist* in the British Museum newspaper library in Colindale and turning the pages to see week after week George's hard-hitting articles on the miners, reviews of agreements, perspectives for the future, appeals, demands, and sometimes international articles on the miners' fight in other lands. Every word is directly written and right to the nerve.

In one such article, 'Does John Wilson, M.P., serve the interests of the working class?', he struck the point so hard that he was dragged up before the courts on a libel charge by the DMA 'leader'. Harvey, in the article and in the court case which he conducted himself, maintained that Wilson was an enemy of the working class and a servant of capitalism. This he illustrated with various examples, including Wilson's agreement to a 5 per cent reduction which even an umpire deemed unwarranted. But the members of the jury sided, not surprisingly, with Wilson, and he was accordingly awarded damages of £200 with £100 costs.

With the disappearance of the Socialist Labour Party the fire to some extent went out of George's life. No more analysis and perspectives on the world political struggle came from his pen. It was not until the owners started to put the screws on him at his lodge that all the old fire and determination reappeared.

This has been only a brief glimpse of one of the real unsung leaders of our class. Of the little I have discovered, only a fraction is reproduced. Not all the rank-and-file leaders were as political as Harvey, and not all of them were as successful as Harvey in the strike of 1935. But until we know something about them we shall know little about the real history of the miners, even though the shelves of the libraries are lined with heavy books about them.

Notes

1 *Morning Chronicle*, 29 January 1850, Supplement, Labour and the Poor.
2 Ibid.
3 This song was written in the closing weeks of Wardley's life.
4 There is a drawing of it in the Usworth Surveyor's office, kindly shown me by J. Robinson.
5 *History of the County Palatine of Durham*, vol. 1, p. cxiv.
6 Ibid, pp. cxiv–cxv.
7 Thomas Wilson, *The Pitman's Pay*, Gateshead, 1843, p. 55.
8 *History of the County Palatine of Durham*, vol. 1, p. cxv.
9 Carter L. Goodrich, *The Frontier of Control*, London, 1920, p. 137.
10 National Coal Board Archives, W. H. Sales, 'The role of the deputy in labour relations', typescript dated 18 June 1954.
11 *Thomas Burt . . . an Autobiography*, London, 1924, pp. 106–7.
12 George Parkinson, *True Stories of Durham Pit Life*, London, 1912, p. 20.
13 In 1935 bord-and-pillar methods still accounted for nearly half of the total coal output in the country. J. H. Goldthorpe, 'Status and conflict in industry', *British Journal of Sociology*, X, 1959.
14 BBC Radio Ballad, 'The Big Hewer', by Ewan MacColl, LP Rg. 538 Mono.
15 George Hitchin, *Pit Yacker*, London, 1962, pp. 92–3.
16 Burt, op. cit., p. 111.
17 In the early days men half believed that the deeper they mined the more they were trespassing on territory belonging to the devil.
18 The miner knew that if his lamp went out all he had to do was to follow the horse for it instinctively knew the road.
19 Louw = light.
20 John Wilson, *Memories of a Labour Leader*, London, 1910, p. 82.
21 E. L. Trist *et al.*, *Organizational Choice*, London, 1963, p. 34.
22 Thomas Wilson, *The Pitman's Pay*, p. 4.

23 Wardley Colliery, Follonsby Lodge, Minute Book, kindly lent me.
24 Follonsby Colliery, Hewer's Cavilling Rules, 1931. Document in the writer's possession.
25 Robert Galloway, *Annals of Coal Mining and the Coal Trade*, London, 1904, II, p. 358.
26 Goodrich, op. cit., p. 155.
27 Galloway, loc. cit.
28 Durham County Record Office, N.C.B., 3/178, Heworth Colliery Agreements 1953. Stone Men's Agreement, Brockwell Seam, Prices and Conditions.
29 Alexander Barras, 'The Putter', in *The Pitman's Social Neet*, Consett, 1897. There is a recording on Topic Records, 12T 189.
30 Durham County Record Office, Heworth Colliery Agreements.
31 Archives of the Durham Miners' Association, Red Hill, Durham City (referred to in later notes under the abbreviation 'DMA').
32 Durham R.O., Heworth Colliery Agreements.
33 Follonsby Colliery, Prices and Conditions, February 1931.
34 Extracts from the DMA archives.
35 *Newcastle Daily Chronicle*, 26 December 1913.
36 *Durham Chronicle*, 1 March 1879.
37 DMA, Joint Committee meeting, 16 January 1880.
38 DMA, Minute No. 87, Programme of Annual Council meeting, 3 December 1880.
39 DMA, Programme of Council meeting, 23 July 1881, p. 20.
40 DMA archives.
41 An example of the formation of a new team can be found in the Follonsby Lodge Minute Book for 28 July 1962, which records the following rule: 'Maudling seam – the first 10 out to be the first placed. The second 10 out to be end up end. The first five to go along with the first 10 men out and the second five to go with the third 10 out.' The sets of fifteen would then go back into the hat for claim on the seam and would be numbered 1–15 for the fore-shift, 16–30 back-shift. This system was applied to all seams and districts in the pit.
42 DMA archives.
43 Trist *et al.* op. cit., p. 50.
44 Joseph Halliday, *Just Ordinary, But . . .*, Waltham Abbey, 1969, pp. 31–2.
45 DMA archives.
46 *Allan's Tyneside Songs*, Newcastle upon Tyne, 1891, p. 570.
47 DMA archives.
48 Goldthorpe, op. cit.
49 Oral information from old Wardley miners.
50 H. F. Bulman and R. A. S. Redmayne, *Colliery Working and Management*, London, 1925, pp. 96–7.
51 Document in the writer's possession.

52 Follonsby Colliery, Prices and Conditions, February 1931, pp. 2–3 (document in the writer's possession).
53 Follonsby Colliery, Hewers Cavilling Rules, December 1925, p. 12 (document in the writer's possession).
54 E. Welbourne, *The Miners' Unions of Northumberland and Durham*, Cambridge, 1923, p. 8.
55 Quoted in ibid., p. 65.
56 Ibid.
57 T. Y. Hall, *A Treatise on the . . . Northern Coalfield*, Newcastle upon Tyne, 1854, pp. 167–8.
58 Welbourne, op. cit., p. 152.
59 DMA, Programme of Council meeting, 29 September 1879, nos 19, 29.
60 Welbourne, op. cit., pp. 206–7.
61 DMA, Special Circular on Restriction, 25 March 1880.
62 Welbourne, op. cit.
63 DMA Archives.
64 *Newcastle Daily Chronicle*, 1, 2, 5 March 1888.
65 *The Miners' Next Step*, Tonypandy, 1912, pp. 26–7.
66 Ibid.
67 Thomas Wilson, *The Pitman's Pay*, p. 67.
68 'The celebrated working man', the Elliots of Birtley, Folkways record album, FG. 3565.
69 Welbourne, op. cit., p. 242.
70 Ibid., p. 148.
71 Ibid., p. 152.
72 Ibid., p. 164.
73 DMA Special Circular, 22 July 1880.
74 Welbourne, op. cit., p. 177.
75 Ibid.
76 DMA, DCFA, Special Circular, 14 February 1880.
77 DMA Special Circular, 11 October 1880.
78 Welbourne, op. cit., p. 170.
79 Ibid., p. 153.
80 Ibid., p. 188.
81 Ibid., pp. 222, 252.
82 DMA Special Circular, 12 August 1882.
83 Welbourne, op. cit., p. 185.
84 G. H. Metcalfe, 'A history of the Durham Miners' Association, 1869–1915', p. 358 (typescript in the DMA offices, Red Hill).
85 DMA Monthly Report, November 1882.
86 John Wilson, *Memories of a Labour Leader*, p. 234.
87 DMA Special Circular, 30 August 1879.
88 Ibid.
89 DMA, Special Circular, 27 February 1880.

90 John Wilson, *Memories of a Labour Leader*, p. 272.
91 Ibid., p. 273.
92 DMA Programme of Council meeting, 27 September 1879.
93 DMA Minutes of Committee meeting, 29 September 1879.
94 DMA Monthly Report, 10 January 1880.
95 DMA Special Circular, 3 December 1880, N. 43.
96 DMA Monthly Report, 5 February 1880.
97 Welbourne, op. cit., pp. 250–1.
98 *Durham Chronicle*, 18 March 1892.
99 Welbourne, op. cit., p. 289.
100 Ibid., p. 295.
101 George Harvey in *The Socialist*, Edinburgh, February 1910.
102 *Durham Chronicle*, 17, 24 December 1909.
103 Ibid., 31 December 1909.
104 Ibid.
105 Ibid., 7 January 1910.
106 Ibid., 14 January 1910.
107 Ibid., 21 January 1910.
108 Ibid., 28 January 1910.
109 Ibid., 18 February, 4, 11 March 1910. For a general account of the movement see George Harvey's articles in *The Socialist*, February to June 1910.
110 R. Fynes, *The Miners of Northumberland and Durham*, Blyth, 1873, pp. 37–49.
111 Welbourne, op. cit., pp. 144–5.
112 Ibid., p. 174.
113 DMA Special Circular, 4 July 1879.
114 DMA Monthly Report, 29 March 1880.
115 DMA Appeals, 12 June 1886.
116 Welbourne, op. cit., pp. 109–10.
117 Ibid., p. 153.
118 DMA Joint Committee meeting, 10 August 1880.
119 John E. McCutcheon, *Troubled Seams*, Seaham, 1955, pp. 81–127.
120 DMA Special Circular, 10 March 1881.
121 DMA Council meeting, 23 July 1881; for a later dispute between Seaham and the Executive Committee, see DMA Programme of Council meeting, 30 July 1887.
122 *Durham Chronicle*, 11 March 1892.
123 Ibid., 18 March 1892.
124 Ibid., 25 March 1892.
125 Welbourne, op. cit., p. 283.
126 DMA Programme of Council meeting, 22 November 1879, No. 29.
127 Welbourne, op. cit., p. 249.
128 *Durham Chronicle*, 18 March 1892.
129 Ibid., 11 April 1879.

130 Hebburn Report, Coal Trade Hall, Newcastle upon Tyne, 30 August 1879.
131 DMA Special Circular on Unconstitutional Meetings, 30 August 1879.
132 DMA Special Circular to Lodges, 15 December 1879.
133 Article by G. Harvey, *The Socialist*, Edinburgh, August 1910.
134 John Wilson, *A History of the Durham Miners' Association*, Durham, 1907, p. 282.
135 Ibid., p. 288.
136 *The Times*, 12 July 1909.
137 *Durham Chronicle*, 23 July 1909.
138 Ibid., 13 August 1909.
139 G. Harvey, *Industrial Unionism and the Mining Industry*, Pelaw-on-Tyne, 1917, p. 106.
140 *Durham Chronicle*, 23 September 1887.
141 Ibid., 16 September 1887.
142 *The Times*, March 1888.
143 *Newcastle Daily Chronicle*, 1 March 1888.
144 Information from old Wardley people.
145 *North Mail/Newcastle Chronicle*, 17 August 1935.
146 Ibid., 20 August 1935.
147 Ibid., 21 August 1935.
148 Ibid., 30 August 1935.
149 Ibid., 30 September 1935.
150 Ibid.
151 Ibid.
152 Ibid., 13 September 1935.
153 Ibid., 28 September 1935.
154 Ibid., 7 September 1935.
155 Ibid., 13 September 1935.
156 Ibid., 14 September 1935.
157 *Newcastle Evening Chronicle*, 1 February 1936.
158 T. A. Jackson, *Solo Trumpet*, London, 1953, p. 65.
159 Ibid., p. 69.
160 Walter Kendall, *The Revolutionary Movement in Britain, 1900–21*, London, 1969, p. 75.
161 Ibid., p. 76.

Part 5

Pit talk in county Durham

Dave Douglass

Dialects

Wor 'twang', there's mony was say, is deed an clay cad; people will say that our language, the language of the Geordie and of his traditional industries, the mines and the shipyards, is very much dead; they will tell us that a language so inseparable a part of a way of life now fast waning, is a lost heritage good only for the literary museum – or the philologist's handbook. That is true in part at least, and is painfully obvious to we who delight in our culture and our people. And yet, picking up an example of a new feature of north-eastern culture in the shape of the underground paper *Muther Grumble*, 'The North East's Other Paper', I find reprinted in full an old song which is the very essence of wor ilk. What J. Harbottle, the author of the 'sang' would have thought in 1891, about finding himself in print in a modern alternative paper I do not know. Certainly his prayer that 'wor speech may ne'er depart fra aad CANNY TOON!' has been answered.

> Oh! cum! ma canny lads, lets sing another Tyneside song,
> The langwidge ov each Tyneside heart, wor aad Newcassel twang,
> Ne doot its strange te stuck-up folk, and soonds byeth rough and queer,
> But nivvor mind, its music sweet untiv a Tyneside ear.

The whole nature of the Geordie miners' speech has been one of adaptation and change. Why even the term we north-eastern Durham miners describe ourselves as, 'Geordies', is a gradual adaptation of a term which applied originally to the supporters of King George against the Scots in the English garrison of Newcastle; we Gateshead folk would never in the past have readily associated ourselves with that royalist title. Here we see, then, that the word has been retained while its meaning has changed; the same is also true of a good many of the pit terms I shall be looking at.

In compiling this glossary, I was of course drawn to the earlier glossaries of Tyneside talk, and particularly to Greenwell's *Glossary of Terms used in the Coal Trade of Northumberland and Durham* first published in 1849.[1] I had thought that the Durham pitman of the 1970s would have used much the same terminology as the Durham pitman of the 1840s; after all they were still down the same dark holes in the ground doing in essence the same filthy work. In effect many of the terms are the same, but their meanings have altered with changes underground, and the new mining methods. What has in fact changed in many cases is not the terms, but the objects which they describe. They say a Durham pitman is a man of few words. If this is really so, it is not surprising that he was not prepared to let a perfectly good word go out of existence when there was a new object which needed a name. If the reader compares this glossary with Greenwell's, he will find sometimes different meanings given for the same terms. Both are correct. The language, like the people of Tyneside, has simply moved over and adapted to new circumstances. More usually however the term was widened to include meanings other than the original. 'Cracket' for example was defined by Greenwell as a low wooden stool on which the hewer worked; later on the same term was also used to describe the wooden 'crutch' which the miner fashioned to support his neck or arm while lying in a place too low to sit in. Later the word drifted into wider Tyneside usage for any wooden stool used in the house. Again Greenwell describes 'dadd' as referring specifically to the dashing out of a small underground fire with a jacket; later the term was extended to mean hitting anything, generally with a flat object such as a shovel or a plank of wood. This word is in fairly common Tyneside usage for hitting somebody, 'A'll dadd yor nose for ye. . . .' 'Hand-fill' used to mean filling by hand, i.e. picking up large coals to throw into a tub to separate them from small coals; later the term was applied to mean filling the coal with a shovel on the coal face as opposed to machine cutting. A 'hogger' in Greenwell's time was a 'wide leather pipe used to deliver water into a cistern'. In my time it was applied to any rubber pipe which looked like a hose, whether or not it carried water. 'Mullergate' (mothergate) today applies to the main gate on a face unit, the one in which all of the belts or chains carrying the total produce of the face are housed. In Greenwell's day it was just 'the continuation of the rolleyway beyond the flat into the workings'.

Pit talk in Durham is not a uniform language; it has accents and dialects of its own. Moving from one village to another can involve the pitman in real difficulties of communication, and coupled with a strange dialect the words he hears can be almost unintelligible.

When the Geordie pitman is transferred into another coalfield it is like another country. The most confusing thing is when the 'foreigners' use a familiar word to mean something totally different. Being asked for a plank the Geordie will pass up a short piece of wood about 4ft long, when what the Yorkshireman is asking for is a long baulk about

20ft. Other words which we use have extra meanings; a 'sprag' in Yorkshire is any piece of wood used as a temporary wedge or support; in Durham it is nearly always applied to haulage only, either a piece of wood under the front wheel of a tub to prevent it moving or else as another word for 'drag'. In Durham 'snap' is a term for the instrument formerly used on the screens to chip off brass or shale from the coal; in Yorkshire it takes the name of 'bait', the Durham miners' term for lunch.

The differences within Durham itself were mainly those of accents, but the differences from villages only a few miles apart could be quite surprising. Basically the main variations I encountered were Gateshead eastward to Shields; Gateshead westward and northward towards Newcastle; Wearside; and the coastal pits. All of these tendencies would have a multitude of village variations.

Some of these differences may have resulted from the influence of people and settlers from other parts of the world and the degree to which the villages absorbed more of one tongue than another. The Northumberland and north-westernly Durham men's 'How's thee fyeld?, sairly fyeld' or 'fairly fyeld' seems to me at least to have a strong Norwegian ring to it. We are told in the modern introduction of Greenwell's book that 'chumuns' (empty tubs) comes from the Norwegian 'tom' having the same meaning as empty. Other words in general use like 'gaan' meaning to go, and 'hoose' (house) it seems are also Norwegian. In my brief visits to the mining areas in Holland and Belgium I found some words which we use in Durham to be the same as theirs. Our 'keeker', the man in charge of 'viewing' the tubs to see if they were full to capacity, has an equivalent in the Dutch 'keek', to see, and 'kikt', looked. The older more exact usage of the word in its Dutch form was in common use in Gateshead and Newcastle in the 1820s. Typical is the song 'The Colliers Keek At The Nation' written in Newcastle in 1829 and found in *Allan's Tyneside Songs*, p. 177. The Dutch 'lopen' and the Geordie term 'lowp aboot' both mean to run about, and have almost the same pronunciation. Brockett, in his extensive although highly technical *Glossary of North Country Words*,[2] tells us that our word 'alane' comes from the Dutch 'alleen' both pronounced the same and meaning alone. Our 'ald' for old is Saxon. As is our 'amang' meaning among. Backbye, inbye, outbye, used extensively in the Durham and Northumbrian pits, and adopted as an official mining term, is of German origin. Our diversified heritage of Geordie tongue it would seem, like the spoken 'English' itself, is formed of many influences.

It seems a pity that in the light of all our village variations many well intentioned writers on the Geordie tongue have lumped them all together as one speech. The efforts are of course to make the Geordie language seem more extensive and full. Similarly we have folk singers and writers who for example sing Gateshead songs in Northumbrian style, with a burring 'r' i.e. 'woverg' instead of 'rover'. It is true that certain of the north-west Gateshead pits round Dunstan and Blaydon areas do use this, but the attempt of the folk singers is to present many of our songs as a hotchpotch of dialects and sounds in order to make them less intelligible to the southern listener and therefore more appealing. A broad distinction must be drawn out between the surface language of the towns and villages, and the pit talk underground. It is doubtless true that owing to the gross concentration of miners in the north-east of England more than half the working population would be involved in mining. Owing to this certain words that the pitman used would have drifted into common usage among town and village folks who weren't miners themselves. It is in that situation, that we would find a pit term being absorbed into the Geordie dialect as such. The pitman's 'cracket' eventually became everybody's word for a wooden stool. But some of these words remained peculiar to the miner underground and it is a great pity that one or two of the modern glossaries of Geordie words include pit terms as if these were in general usage by the whole community.

The main strength and weakness of Greenwell's glossary is that it gives classified, technical definitions of the terms in use in the coalfield at that time. It does this with great precision and technical accuracy, but because of its technical language and brief definitions no one except a Geordie miner could understand it or use it to get a picture of life and surroundings underground. If my own glossary suffers from time to time from technical imprecision that is because I have attempted throughout to build up a mental picture in the reader's mind by using similes and examples, that he might see the work and surroundings of the miner. I am not a philologist and make no claims in that direction whatever. I have not tried even to discover the origins of our words and terms: what I have tried to do is to show what they mean and how they are used. A glossary of work and the working environment is what I have attempted to create. Many, many words, which because of their technical nature, or my inability to break them down to give a simple understanding, have unfortunately been left out.

This glossary is offered as a small peek at the miner's language, but as it is made up of separate words and phrases one can never get the effect of total speech. The miner's politics and culture are one phenomenon, not separate things. The isolation of their communities, the inheritance of their speech and songs, their independent job control and relative freedom from supervision, the hatred of full-time union officials and of the gaffers, all of these things must be seen in relation to each other, formed in the dialectic of struggle.

Glossary

AFTER DAMP The gas which results after an explosion of methane. It was, and indeed is, deadly poisonous.
AIR The miner's way of referring to circulation; 'the air's foul', means that the circulation is bad. In most cases the air travels down the main gate, and along the face and out the tail gate; by the time it gets there it is hot and dusty, and it will have picked up a quantity of gas on its travels. At this point 'air blarers' are needed to freshen it up and make it move along.
AIR BLARER (air blower) This is an air mover which works like a vacuum cleaner taking in air at one end and redirecting it out at the other. In Yorkshire it is called a 'ventura'. There is a larger version of this, a giant tube, about 2ft in diameter, which is hung on the roof by metal clips and looks for all the world like a caterpillar, like a drawn-out concertina, and has a great fan pulling air in at one end and pumping it out at the other into the working. The air blarer may be used to break up pockets of gas and disperse it into the general body of air.
AMAIN (away) Tubs are said to be 'amain' if they break away from the set and are freewheeling down the way. Quite a lot of deaths in mines occur in haulage work, often through tubs getting amain. All classes of haulage workers face this hazard, whether pony putters in the old days, or loco drivers today. The last fatal accident I can remember at Wardley was when a maincar 'got amain' and killed a miner when it got to the bottom of the incline.
BACKBYE The galleries and districts away from the face. Backbye workers would include road-layers (the men who lay the tracks), certain haulage workers, like the lads who work on haulage ropes, clipping on material for the face, men levelling off floor-lifts. Many backbye workers are older men who have lost the strength needed to be face

workers (caunchmen, hewers, cutters, etc.), although certain backbye workers still do very heavy work compared to other industries. Many of them are also known as 'datal' workers, which under the old grade system was the lowest and worst paid class of work, except for surface work.

BACK SHIFT Originally the second shift, that is the afternoon or near afternoon shift. With the advent of three shift working the back shift became the afternoon shift. Now in Doncaster, with the introduction of the hated four shift system, we have two afternoon shifts so perhaps the later one can be called the 'back shift', although I have heard immigrant Geordie miners announcing 'I'm on the back shift' to which his marra replies 'which one, late or early?' So we see a new application of the word taking place.

BACK SHIFT OVERMAN The head man on the back shift; he usually sticks to the same shift although there seems to be a tendency today (when there are so many more overmen) to assign an overman to each team: he moves around with them as they change from shift to shift every week.

BAIT (in Yorkshire SNAP). I suppose the only equivalent in standard English is 'lunch', but workers don't eat lunch (a petty bourgeois concept), and bait is hardly the same as dinner. Bait time is the time the miner takes off from the shift (twenty minutes today) to eat a couple of sandwiches, and get a drink of water from the bottle. In the days of piecework – as recently as the 1950s – there was no special time for bait, it was eaten when the belt broke, or the miners were waiting on for tubs or mine cars; if the shift went well it might not be eaten at all. Nowadays it is eaten in or around the middle of the shift. Men usually fetch two sandwiches, often with jam, sometimes with dripping, very often just butter, not in this case out of poverty but out of experience, knowing that a good few hours slog at the shovel has still to take place after 'bait' and it can't be done on a bloated belly. There is also the factor of taste, men will eat things underground they would never have the wish to on the surface, like a raw onion or a lemon for example. Men generally won't go to work if their bait isn't made for them. They very, very rarely prepare it themselves – this being one of the jobs left for the women.

BAIT POKE (bait bag) The small generally ex-army bag which some miners carry their sandwiches in. Workers outbye, haulage workers, and older workers usually carry their bait in a poke. Their jobs being more alienating and monotonous than that of the face workers, they

carry more to help them through the day than the face workers – flasks of tea, a bag of sweets, more sandwiches, etc.
BAIT TIME At bait time miners want comfort, a bit of height and some cool; men may spend a third of their bait time crawling to a cooler place.

> . . . The secretary of wor lodge
> Says its time that ye had yor bait,
> So wi tak wasells tiv a quiet spot,
> Wiv a plank and a chock forra seat,
> And the crack at last flies thick an fast
> Of the die'ins in the club last neet.
> (*The Collier Lad is a Canny Lad*, Johnny Handle, Newcastle upon Tyne)

In the gates at bait time, the men will eat their bread and still have time for a laugh, a political discussion, a 'bullet' (sweet) or two and a pinch before the recommencement of the day's work. Some men like to sit on their own at bait time, they never speak, and wouldn't thank you for speaking to them; they like to put their lamps on dim and sit in absolute silence away from their marras; this kind of privacy is respected down the mine and never questioned.
BAIT TIN Most face workers (caunchmen and hewers) carry their bait to work in a tin. The tin itself is rounded at one end – perhaps 8in long and about 4in deep. It has on the back of it (which is flat) a wire with which it can be connected to the belt and carried to work. It was always a strange sight before showers were introduced to see miners walking along the streets with their coats bulging out at the back like a hump over these tins. The tin has a special hook (a kind of wire handle) and can be carried on the belt leaving both hands free for walking inbye, if you have to be travelling about all day. Caunchmen and hewers hang the bait poke with their clothes (they work semi-naked) in the gate (tunnel). If they can, they come off the face at bait time, out of the heat and cramped working places, and sit in a quiet corner to eat their baits. When the miner comes home the tin is washed by the wife and left out in order to remind her to put the man's bait out. On occasion bait tins are used to carry home bits of wood for the fire. It has been known that a miner has carried the tin home with a bit of wood in it, and left it on the table; his wife forgot to clean it or put the food up, the man came down

in the morning, felt the tin was heavy and brought it to work only to find he'd brought a lump of wood back to work with him.
BAND This is what is known as inter-stratification; it is a soft vein of coal often with a furry appearance and feel. It is very breakable. Some 'band' is a grey colour and feels like clay when you handle it. It comes between the seams of coal. My father says I am wrong on this point: to him band is a 'dark, heavy stone'.
BANKSMAN The banksman is a worker who is employed on the pit top. He is in charge of the cage when the men are getting in or getting out; of course he's also in charge of the cage while they're lowering it for coal and when the coal is drawn to the surface. He communicates with the bottom of the shaft – and with the winder – by a series of raps. For example three raps would be a sign for the 'riders' (or men) to get on. Only the onsetter at the bottom of the shaft or the banksman at the top would use these raps. The banksman also has a telephone for emergencies to phone the winder.

> Me father use'd to ca'l th turn,
> When the lang shift wes ower,
> A'l the way ootbye, y'id hear him cry
> Does tha, kna its efter fower,
>
> He'd cry, rap'per ti bank
> Me canny lad,
> Wind her away keep turnin
> The backshift men are gaanin hyem
> Thi'l be back in the morning.
> (*Rap'per ti Bank*. Gateshead song)

Both banksman and onsetters (although relatively lower-paid workers) are usually, because they are in highly responsible and trusted positions, sour-faced blokes who cannot stand a lot of fun – at times when men have been larking about in the cage, they have been known to stop the cage and bring them back up to the surface, because they thought somebody had been hurt in the shaft. After one incident like this at Hatfield (Geordies were the culprits) the banksman had a notice hung on the cage saying 'No rowdyism in the shafts.'
BARRIER This was the term given to the boundaries established between various owners' royalties. A barrier of coal was also left between collieries to protect one colliery against excessive weight or

flooding from another. Barriers have still to be left today between collieries to prevent flooding especially in the case of a disused colliery. In the early days poaching of coal went on, and an owner might steal large chunks of the barrier and go over into another owner's royalty.

BAT This is an expression used in a Doncaster coalfield, and refers to haulage. The 'bat' is like a cricket bat, except that it has a shorter handle (or shaft) and a longer blade, and there is a piece of belting hammered on the end of it. When the set of tubs is running in – maybe twelve or twenty-four at a time – they get up quite some speed on a self-acting incline. In order to slow them down when they get to the pit bottom a bat is pushed or slid under the wheel of the first tub. A haulage lad does this, walking with the tubs until they grind to a halt.

BEATLE This was a very small engine, fixed on a prop, and was used to pull the tub out of a low side, when the 'gallowa' (pony) couldn't manage it. The putters had to operate it.

BELL PIT These were the original pits which once honeycombed the whole of the Tyneside area. They were simply a shallow shaft sunk into the seams near the surface with a small amount of working at the bottom. There was no special ventilation – the only air which reached the miner was that which nature allowed down the hole. When the coal got too far away from the shaft the pit was allowed to cave in and another one nearby was worked instead. Coal was normally extracted from these pits by means of a winch similar to that on a water well. It was hoisted up in a bucket or 'corf' the rope being often turned by a woman or child. Later on gins were invented to take a greater load; the gin was a circular pulley much like a capstan except that it had only one shaft; a horse was harnessed to it and walked round and round to hoist or lower the corf.

'BET' HAND, KNEE or ELBOW 'Sciagma', I think is it's medical name, the plague of the collier working in the cramped quarters of the thin seam. It comes from the constant chafing of the stone floor against the knees (which are covered with pads) and elbows, or the wear of the shovel against the hands. This results in the festering of the joint or limb and a huge inflammation area develops.

BEVIN This is a term used to describe a day off which you get paid for. You might say you had a 'Bevin' if you were gas-bound, that is if there was too much gas on the face to work it. You would also have a 'Bevin' if you were waiting for the colliery bus and it didn't turn up; if you'd stood for a long time in the cold you'd say you weren't going to work or

to wait any longer. Bevins, nowadays, are very few and far between but quite enjoyable when they occur.
BEVIN BOY A term applied to anybody who is inexperienced, or who does a job badly. Another example of adaptation – Bevin boys were originally drafted into the pits during the last war.
BLOWER The name given to a sudden and extensive discharge of gas from the roof, seam or floor. Quite recently in Wales a group of unfortunate Welsh miners were all gassed together as a result of just such a 'blower'. Highly dangerous phenomenon.
BOB-A-JOB MAN This is a term used for a deputy or overman who won't let the men rest from one job to another. When you complete one task you are immediately found another. This is a new phenomenon in the coalfields, since the worker previously had one job to do, and when he finished that job that was it – that was him done. He would sit somewhere once he filled off his coal; his task was done – he couldn't be sent anywhere else. Now, with the growing amount of supervision in the pits, this is brought into question. Bob-a-job men spend their time chasing young workers or low-paid workers around the pit trying to find them jobs to do – one job after another. These characters cannot stand to see a man not doing anything, even if he's just finished a job: they try to keep the men working non-stop. The normal patter is: 'you can do this and when you've done that you can do the other – and when you've finished that you can just pop over there and see if there's anything else wants doing.'
BONNY LEETS This is a Durham term which has rather the same meaning as bull's-eyes in Yorkshire. At one time the officials had spotlights on their cap lamps which gave out a long beam. The workers could thus see them coming at a great distance and would shout out. 'Here's a bonny leet coming' or 'here's a bull's-eye'. Now most miners carry a spotlight, so this way of identifying officials is more or less dead.
BORD AND PILLAR The system of mining in operation before the advent of longwall working. This system was that of driving main roadways in through the coal and working it in vertical tunnels, in an opposite direction to the mainroads. The working area looked like a chess board, with pillars of coal left in to support the roof and gates, and separating each stall from its neighbour.
BORING Drilling holes in order to fire shots. Boring is also carried out to establish the nature of geological peculiarities – the extent of a seam, or fault, to judge the reserves of coal in an existing colliery or prior to sinking a new one.

BRAKEMEN In the old days, the men were employed to work the steam engine, or other machinery used in raising the coal from the mine, nowadays we call them enginemen and winders.
BRATTICE A kind of a cheap canvas used to make temporary doors or 'stoppings'. The purpose being to redirect the air, sometimes to prevent it going round the pit too quickly, sometimes to prevent it going into a particular area and at other times to direct it into the roof where gas may be accumulating. Being usually waterproof it has a variety of uses for the men who take it away home with them. It can be used for waterproofing chicken houses, pigeon sheds, and coal houses. It makes a first rate gun case or fishing rod case, it can be used as a ground-sheet. If one looked very carefully at some of the paintings done by Newcastle art students about 1966 one would see behind the frames 'NCB Wardley Colliery' stamped all over the back of the paintings. That's what comes of worker-student alliances.
BULL WEEK The term is of Yorkshire origin, and refers to the week before a holiday when the miner will work every shift so as to have as big a wage packet as possible. That week is known as bull week.
BUTTERFLY A very important safety device used in the shaft. The butterfly is connected to the cable which hauls the cage. If for some reason the cage is over-winding when coming up the shaft the butterfly will stop it. If it was left unchecked it would carry the cage right over the top of the pulley wheels and destroy it. This butterfly as soon as it gets past a particular height will disengage the rope and hold the cage in the air until it can be recoupled.
BUTTY SYSTEM The hated system of sub-contracting in Yorkshire and some of the southern coalfields where contracts were hired out to a buttyman – to a sub-contractor who was paid a price by management and then decided how much the workers who actually did the job were worth paying. He managed to keep his power by building up a collection of blue-eyed boys, his favourites, paying them a little extra, or finding them 'better' jobs. In many cases the men had to literally fight in the pit to get the money they were entitled to off the buttyman.
CABINS These are occupied by the overmen and fitters. They are generally places hewn out of the stone and whitewashed. Long tables and benches will be in these along with desks for the overmen's plans, books and deployment tables. Men usually meet at overman's cabin to be 'deployed' as the NCB term it. Men with complaints of wrongly booked prices will crowd into the overman's cabin to ask the reason for their short pay. One overman, it was said, was fond of feeding the 'gallawas'

(ponies) with 'bullets' (sweets) before the men came down the pit. A pair of hewers who had come down early to see why their pay wasn't right, looked round the door of the cabin and saw a group of ponies with their front legs on the overman's table, leaning over towards him. One hewer turned to the other and said: 'We've got some chance here, marra, even the gallawas are complaining.' The fitters keep their tools and spare parts in their cabins (major jobs are done on the surface). Haulage workers at the pit bottom usually have a 'bait cabin' where they can go for a sandwich and a flask of tea and take shelter from the sweeping cold air that is drawn down the shaft.

CAGE The means of conveying the miners up and down the shaft; it is also used for coal. Cages vary in size; Wardley's had two decks but only the top one was used, Follonsby's had two decks in use, which took one mine car in each deck. There was a powerful steam winder to ram the cars on to the cage at the surface and underground. In other collieries two cages are in operation, one coming up when the other goes down. Some have anything from two to five decks, some are very, very narrow and cramped. In Yorkshire they call them 'chairs'. The cages are always packed tight coming up with hardly room to breathe. I've known myself to be carried on to the cage by the crowd and my feet not touch the floor all the way to the surface, we've been that tight. Miners can't get out of the pit fast enough once their shift is over.

CANNEL COAL In Greenwell's *Glossary of Terms used in the Coal Trade of Northumberland and Durham*, we learn that cannel coal is: 'A fine compact description of coal with a conchoidial fracture: It burns with a bright flame, like a candle whence possibly its name.' This was very rare coal and to the best of my knowledge is exhausted now.

CANNY One of those words which seem to have innumerable meanings, yet on every occasion when it is used the circumstance and tone of voice clearly indicate its meaning.

A Canny good . . . Not too bad.
B Canny wick . . . 'Neygood' (No good at all).
C 'She's canny, mind' (Nice looking girl).
D 'She's a canny body' (usually applied to old folk meaning a sweet old lady).
E 'Will het ti tek it canny' (We will have to take it slowly).
F 'Will be canny aboot this job' (We will be smart in doing this job; careful).
G 'How duw?' 'Canny!' (How are you keeping? Well!)

Many other examples could I am sure be found. 'Goan canny'; I should add, is the Durham term for going slow and working to rule – limited industrial sabotage (exemplified in 'E').

CANTEEN (A) The place where the men can get a cup of tea and a pie either before or after going down the pit. At Wardley a beautiful dinner with soup and pudding was served. On a Friday, pay day, men used to meet their wives with the children in the canteen to save the wife the cooking of a meal. She would get dressed up, have a dinner 'with her man' in the canteen, and then get an early start to the coast for a day out. Union subs were paid over a table set up in the canteen, master notes were divided up here by the 'cavil' leaders. As a little lad I would march down to the canteen to meet my Dad going out of the pit as would many other kids.

CANTEEN (B) Name given to the 'Waater Bottle' a vessel for carrying your drinking water, it varies in content from 3 pints to 5, although somehow or other some men manage on less than a pint a day. While working on a particularly hot place at Hatfield main I used to drag a gallon of water in every day. The men would joke that if a stone fell on it, it would flood the district. Water bottles are known as 'Dudleys' in Doncaster, after the name of the firm that makes them. In keeping with the pitman's hatred of buying anything for the pit, the old miner used to fetch his water in a pop bottle. The Coal Board banned glass bottles as dangerous, but the new innovation of plastic bottles for everything from vinegar to disinfectant saved the thrifty miner expense. My gallon container was in fact an empty bottle which had contained photographic fluid. When walking across the yard with my bottle one day, the name on its side – 'Development Fluid' – attracted the comment 'didn't dey ye much gud, son, did it?' (I'm not the size of two pennyworth of chips, by the way.)

CAP Refers to the gas reading on a lamp. 'To have a cap' means to have a reading of gas usually when it is above 3 per cent. The gas is ignited above the flame in the lamp. At about 5 per cent it looks like a perfect pyramid; at less than 5 per cent it is more like a volcano, and after that the flame starts to 'tail' up. Men who carry lamps must take a test to show their capability in reading caps. One in every five men in a working area is entitled to carry a lamp.

CAP PIECE (SOLE PIECE) Two names given to the same object according to whether one uses it on the top of the prop or at the bottom. It is a square piece of wood about 6in by 6in and 2in or 3in deep. Wooden props would nearly always have one of these on the bottom and perhaps one on the top before they were banged in tight against the roof. When setting iron props under iron bars and girders a cap piece must always be used because steel flies in all directions when the weight comes on to it.

CARTRIDGES The early name given to 'pills' of powder. In some areas it was used to describe the detonator. Nowadays 'cartridges' nearly always refer to the metal disc which is fired into the coal by compressed air, and has the same effect as powder. (I have not seen this instrument myself – it is only used by the shot firers.)

CAUNCH Pronounced 'canch'. The term is applied to three working areas. The face caunch is the edge of the coal face and is much higher than the coalface itself to allow materials and conveyors to be brought up to the face. The caunch is the end of the tunnel; it is secured by arch girders and lagging (wooden planks or metal sheets). The caunch is known in Yorkshire as the 'lip'. The men who work on it are known as 'caunchmen' in Durham, 'rippers in Yorkshire, 'brushers' in Scotland and 'stonemen' in Northumberland. There is the 'back caunch', this is usually where the tunnels have become broken in and sunk or else are too low; they are then heightened. These would be worked by 'backbye caunchmen' who would rip down all the old gate yards and replace them with either new or bigger arches. Third there is the 'bottom caunch'. Both the backbye caunch and the face caunch could be bottom caunches. When he is faced with them the caunchman, instead of ripping down rock from the roof, takes it from the floor to provide more height.

CAVILS The system of job control which operated for hundreds of years in Northumberland and Durham and took the form of a kind of lottery to allocate working places. The system was evolved of drawing places out of a hat along with names of men to work them. This lottery gave everyone the same chance of good and bad places, and prevented union men and agitators being victimized with bad and dangerous work places and 'crawlers' or 'gaffers' men' from getting the good places as a reward for their collaboration. (A fuller description of this system is given in my chapter on 'The Durham pitman'.)

CHOCK (A) A single oblong lump of wood about 2½ft long, 10in square and 10in deep. (B) A square structure of interlocking chocks which holds up the roof. The chock is constructed upon a pair of 'trips'. These are a release mechanism which provide an easier means of knocking the chock out, than if it is set directly between the roof and the floor. When the face moves forward after the strip of coal has been taken off, the chock is knocked out and reset further forward.

So the belts are set and rolling in the new tracks,
And wi'l draw off the chocks while lieing on we're backs,
Then wi'l set them again in their new place, fot'ta
Stop the weight'in on the face,
Doon the Brockwell seam in the north of number five west.
(*Stoneman's Lament*, Johnny Handle)

(C) A hydraulic system of roof supports which runs the entire length of the face and is operated by handles. The hydraulics operate on a mixture of oil and water.
CHOPPY The corn-like food which the pony eats at bait time from his choppy box and also at lowse when he comes out.
CHUMMINS Empty tubs. When tubs are empty they are said to be 'chum'. The putter runs his chummins into the hewers for filling.
CHUM NEUKS This was the putters' last run in, when the work places would be left empty for the week-end or at holiday times. He would go in with just the limmers, and no empty tub. He would shackle up the last full tub and take it out. No more empties would be brought in until the start of the new week. When going in like this the putter was said to be 'gaanin in chum neuks'.
CLAGGY Sticky, adhesive and usually wet. A claggy roof for example. Claggy in wider Geordie parlance means 'sticky'.
CLARTY Muddy, wet, semi-liquid coal dust resulting from very wet seams and small coals teaming up against the miner.
CLAY CARRIER This was the worker who cleared away a road down the face for the shot firers and generally assisted the shotfirer in his operations. They were called clay carriers because the 'stemming', that is the stuff that is rammed into holes to make the charge go the right way, was usually clay. It is still the best form of stemming although in recent times we have had the advent of water-filled plastic stemming.
CLEAT The 'grain' of the coal. Coal crystallizes in lines and cleat is the name given by miners to the direction of the lines. If a hewer caught these cleats running horizontal to him he would have one hell of a job breaking into it, but if the cleats run towards him he can tear the coal down in huge lumps. In Durham coal hewn with the cleat was known as 'Boardress Coal', coal hewn the hard way was 'Headress Coal'.
COALRUNNERS This is a term used in Yorkshire for the lads who go down with the tubs. A set of tubs will be filled up by the loader and pushed through on to gradients and round corners by teams of haulage lads. The coal runner passes the coal from one lad to another. As one hops

off the tub to go back for a new set another young lad jumps on the back of the tubs to ride out with them. To warn the men a new set was coming you would bang on the pipes and this would echo through the areas relaying the message. The coal runner rides down along the back of the tub (he isn't supposed to ride on the back of the tub but he does this rather than try to keep up with them). At various places, at turns and junctions etc. he would 'locker the tubs up' to slow them down. He would also have the back-breaking job (if the wheels were blocked by loose coals which had spilled on to the track) of pushing sets of tubs to get them on the move again.

COUP Ancient term in Durham for an exchange or swap of cavils. This might be done for a variety of reasons not least if a pair of men had drawn the same bad cavil twice or thrice. Another pair of marras might swap to give them a chance.

COWS/MONKEYS The young miner is never too surprised to find ponies or even the odd horse down the pit; cats and mice and rats are quite common, but on being sent for a cow one begins to wonder what sort of a place a mine really is. One discovers later that a cow is a safety device used on haulage; as is the 'monkey'. The 'monkey' is a braking device which prevents the tubs running backwards; the 'cow' is a long iron bar which is trailed behind the last tub of a 'set'; if the tubs break from the rope when going uphill the cow throws the last tub off the way and stops the set from 'getting amain' and running back down the incline ('cow' is pronounced in Geordie 'cuw').

CRACK A light hearted conversation. Men on their long journey inbye will stop to hear 'the crack' before ganning on a bit further. Crack can include the news of the day underground, the newest batch of jokes, and most other things.

CRACKET A cracket was a piece of wood which the collier used, in the days of hand hewing, to support his neck or elbow while he was working. It was wedged in the ground like a shooting stick, but instead of being wedged between a racecourse backside and the turf, it was set between the hewer's neck, elbow or arm when he was working lying down. A cracket was also a stool, on which the hewer squatted in 3 or 4ft seams. I've seen old miners today 'luck for a bit cracket' at bait time and come up with a small piece of wood on which they squat. This again is a modern application of an otherwise redundant word.

The bonny pit laddie, the canny pit laddie,
The bonny pit laddie for me, O!

He sits on his cracket and hews in his jacket,
And brings the white siller to me, O!
(*Allan's Tyneside Songs*, Newcastle upon Tyne, 1891, p. 3)

CREEPER A long chain with teeth on it usually on a gradient but sometimes on a level. These are to give the tubs a start. The teeth hook underneath the tubs and pull them forward. The creepers might be wound by hand, to slow the tubs down when they were on a gradient, or they might be powered by compressed air to push the tubs round a difficult place. If the pressure of work is too hard, haulage lads have been known to break the creepers by wedging wood beneath the teeth and the wall and letting the weight of the tub smash the creeper so they could get a rest.
CROSSCUT An excavation or 'drift' driven in any direction between headings, air course, and broadway coursings. In the bord and pillar operation, the major excavations would be driven in a square fashion, numbers of drifts 'heading' forward and others at right-angles to them. The cross-cut intersected them in a diagram fashion.
DADD To hit something (or somebody) with a flat object. 'You dadd' a mouse with a shovel – or someone who is provoking you. In Greenwell's glossary a more specific meaning is given: 'Dadd, to dash out a small fire of gas, or a small accumulation of gas with a jacket.' A regular scene before the advent of pithead baths, was to see the women folk 'dadding' their men's pit clothes against the outside wall of the house to knock the coal dust from them. It is not uncommon today, at holiday times when the man brings his pit gear home while the baths are being cleaned out.
DATAL WORKERS Non-face workers, backbye workers. They have a general day rate for the job. They do general work such as loading props into tubs and maintenance, or they may be working on the haulage. Most surface men are datal workers. Generally you will find that these people are older workers, too old for the face, who have finished with the face, or workers who have been partially injured, or young workers who haven't yet come on to the face. You'll always get a class of worker, young lads, 22 or so, quite satisfied not to go on the face, doing these jobs with the old men. These workers, because they're usually older men and because usually they're hanging on for a pension and don't want to be hurt, are very difficult men to move in a strike. Datal workers are probably living from hand to mouth, and with their wages cannot afford a sustained strike of two or three weeks as a face

worker perhaps can. Consequently they usually fight against any strikes of the face workers, even though the latter, in most recent strikes, have been striking on their behalf. We have had the strange phenomenon of lower paid men saying they didn't want any more if it meant a strike.

DAY HOLE A small exit or vent to be found in the shallow drift mines, particularly of Northumberland. It makes the mine cooler and provides another means of exit from the pit. It did other things too, because of the face worker's well established freedom from supervision and control of job speed, etc. Some of them would work 'like the clappers' (very fast) and finish their 'stint' in double quick time; they could then creep away up the day hole and have a lie down in the sun. In a certain Gateshead colliery (which I will not name in case it is still open and I might give the game away) on the afternoon shift the men would finish early and creep up the day hole which cropped out in the side of a hill: the hill faced a nurses' hostel and many a quiet pinch of snuff (so it was said) could be enjoyed while watching the girls undressing. Needless to say the pressure to finish early must have sent many a strong young man to an early grave. In America the Molly Maguires used to emerge from forgotten day holes to carry out acts of sabotage on the surface.

DE'IL The old Durham pitmen's name for 'the devil'. He half believed that the deeper he went the more of the de'il's territory he was encroaching on. A miner who went missing, would have been 'te'an be the de'il'.

DEPUTY The deputy has over the countless years of mining changed his role repeatedly. In the 1850s he was employed in a supervisory capacity but his main job was that of setting timber and supports for the face workers. He also had the job of extracting them. In later years he was also a safety official in charge of seeing that the mines acts were obeyed. In these modern days of increasing supervision there are more and more deputies and supervisors foisted upon us, and we have a deputy and a couple of Grade 2 deputies to a coal face and another in the gate looking after material – a ridiculous state of affairs. These deputies of course are not provided to help us in our tasks but rather to spy on us and act in the same capacity as a foreman in the hated factory system, which the Coal Board holds so much as its ideal.

DILLY LINE A railway line used on haulage systems both on the surface and underground. The dilly ways conveyed the coal directly from the pits to the staiths whereupon they were shot in to the keels or colliers which had come up the river. The dilly ways and the wagons have a distinct place in my own memory, the simple reason being that they ran 15 yards from our back door. Our house was one of an unbroken line from the junction of the dilly lines to the last house of the 'raw' some 30 yards from the pit tip: the street was called Wagon Way Street . . . what else? We would sit for hours in the senseless game of singing out in chorus 'one, two, three'e, fower' imitating with the chant the noise of the wagons as they lazily crossed the rails of 'the way'. These wagons ran on a direct haulage system from Springwell and Kibblesworth colliery. It was officially called 'the Pontop and Jarrow Railway', I believe. The noise of the rope on the rollers whipped them into a loud hum which always fascinated us. One of our more stupid games was crawling under the wagons whilst they were stationary, and out the other side, knowing full well that at any minute they might re-start.

DINTING Taking up portions of the floor either to make better height and to make a solid base (setting props on a soft floor is dangerous since they are likely to slide or sink). I personally hate dinting and would rather shift twice as much going forward than half as much digging downwards. A Doncaster word, I think. We have no real equivalent for it in Durham except 'bottom caunch', possibly, or simply 'taking up bottom'.

DIRL The noise produced when something is hit, a resounding vibration like a bell, heard especially from the blacksmith's cabin and the fitters' workshop at the surface. Thomas Wilson in *The Pitman's Pay* uses it thus:

> Thy tongue runs like wor pulley wheel,
> And dirls me lug like wor smiths hammer!

DIRT Name given to the stone on the caunch or lip at the edge of the face. Also name given to a kind of shale found between the seams or in the seam. Dirt is also called 'muck' in Yorkshire. Greenwell gives it a different meaning – 'foulness or fire damp'.

DISTRICTS The name given by miners to the sub-divisions of the seam. A district might have two or three or more units. Cavilling teams were assigned to a particular unit within a district. In a district you will probably find the same height of coal for all the units, the same degree of warmth and the same type of equipment will be in use so the different cavil groups understand each other's problems even though they might not have the same degree of water or gas, or anything like that.

DOGGIES A kind of charge-hand employed to assist and direct the haulage lads. The doggie would also assist when tubs came off the road. He might walk round the whole area of the haulage at pit bottom and all the districts and if a tub was off the road the young worker would bang on the pipes with the locker which would echo through the pit and inform the doggie that there was a tub off the road. He would then follow the noise in the pipes up to where the tubs were off the road. With the innovation of skip winding in the last fifteen years, 'doggies' are disappearing and the number of haulage lads becoming fewer. In Durham the doggie was the wagonwayman who repaired and looked after the haulage districts and roads; he was not really an official but more like the chargehand in the factory system. His name, I believe, came from the dog nails which he used in repairing the roads.

DOG NAILS These are used by the road layer. He clears the road with a shovel and pick making it as level as he can and throwing the waste into a tub. He then takes a big, heavy diesel rail for the diesels to run on and lays them on the sleepers. These are hammered into place with dog nails which are nails with a thick head for securing the rail.

DOWNCAST The name given to the shaft which carries the air down to circulate around the mine. You can always identify the downcast shaft as it is the one exposed to the air, the upcast shaft having a concrete 'chimney' around it to conduct the gases and bad air away.

DRAG North-east Durham; on Wearside 'dreg', and in Yorkshire 'locker'. A metal or wooden rod which the haulage lads thrust into the spokes of a tub or mine car to slow it down or to stop it 'getting amain'. When the tubs are coming past, they can pick up quite incredible speeds, and it's quite an art slinging these one after another into the spokes as they go by. In Doncaster they are given the slang name of 'dicks'. The expression 'dick' was taken quite literally by a Hatfield doggie who used to take delight in whittling or cutting these lockers in the shape of a penis – very polished and very professional – and chasing young haulage lads around the shaft with them.

DRAWING OFF Term used in connection with withdrawing or 'drawing off' the props or supports. When the ripper is removing the props to advance them and allow the stone to fall behind him he shouts 'drawing off' to warn all around that a portion of the old roof is now coming down.

DRIFT This is a tunnel which has been run out through huge areas of stone to get to new coal or to link up one unit with another. A drift will be known by various names, say '7's drift' or '6's drift' or it might be named after the man who heads the drift out ('McGarvey's drift'). Often roadways are driven by contract workers taken for the duration of the job, the 'big hitters' as they call them in Doncaster. These men are usually very big men and toil under tremendous strain. Contract gangs are often made up of foreigners – Poles, Germans and Yugoslavs – or Scotsmen.

DUFF The small particles of coal which are left as a residue after washing, a sandlike sediment. One also finds duff on the coalshed floor. During the 1926 strike, when the miners' coal bunkers were empty, they used to gather up the duff from round the pit heap and make 'duff balls', a kind of coal 'snowball' made by taking a lump of wet clay and rolling it round and round. When it was put on the fire, the lumps of clay and coal used to shoot off in all directions, and the fire wasn't bright, but it was better than no fire at all.

ELEPHANT FEET Great iron base plates for attachment to hydraulic props with the intention of stopping them sinking into soft bottoms. In fact they usually do sink all the same and are more difficult to get out again, with elephant feet beneath them. Knowing this the men often don't use them at all, but bury them illegally somewhere out of the way. When I started as a ripper I was told, 'never shovel what you can lift, and never lift what you can bury.' A maxim never found in the Coal Board's manual.

ENDLESS A rope haulage system. It consists of a wire rope going in a huge circle, empty tubs can be clipped on one side, thus pulling them in, while full ones are clipped on the other side, thus pulling them out. The ropes can be stopped to knock the clips off and disengage the tubs, or more usually the haulage lad is skilled enough to disengage the tub without stopping the rope.

FACE The coal face – the very extremity in the mine, the coal seam. Faces are all different, none of them any good. Some seams are 18in and even less, others tens of feet high. Some are red hot, others freezing cold; some are flooded, some smelly; the coal face advances at different levels – some are on gradients of 1 in 5, left to right, others are like big dippers, going first up then down.

FALL An unexpected collapse of the tunnels or the face. A fall can be anything from three lumps to half of Durham caving in. It can come in the main haulage ways or right in the centre of the face. It then becomes essential to support the fall as quickly as possible to stop it spreading further. As an old miner said to me in Doncaster 'it don't matter what fancy supports they invent, when Yorkshire puts her foot down hard something's got to go.' In other words the whole weight of the earth when it starts to move above us will get its own way in the end no matter how we try to prepare for it.

FAST Term used in the Doncaster coalfield for 'stuck'. If something is said to be 'fast' it means the reverse in fact, i.e. that it cannot move at all.

FAULT Arises where stone intersects the coal face. It is in fact a geological fault; sometimes the earth has 'rolled' and the whole seam is twisted around; sometimes it drops miles below leaving nothing but a blank face of stone where you thought a rich vein of coal would be. Step faults raise the coal in progressions (or drop it) so the same seam can be seen at three different levels, only a few feet apart. Faults can sound the death knell for a colliery. At Hatfield colliery so many faults were struck in the Hazel and north-east seams that the men started to call it 'Hatfield quarry', and propounded a theory that the manager was selling stone fireplaces to his friends. To the miner, a fault makes working conditions very unpleasant, especially with machine cutting; when the machine attempts to cut through the stone it creates clouds of dust which make it almost impossible to breathe. On some districts you literally couldn't see your hand in front of your face for the dust. Faults are commonly known in Durham, and in Yorkshire, as white walls. There is a lesser kind of fault known as a 'roll', a stratum of stone which intersects the seam for a short period and then disappears.

FEEDER This is a constant stream of either water or gas. It comes in through the strata of the coal or rock, from the floor, the roof or the coalface itself. Some of these feeders (especially water) run on for year after year no matter how big the pumps fetched in to deal with them. Underground streams are often responsible. As to gas feeders many experiments (some successful on a limited scale) have been carried out to tap the natural methane gas in the mine and use it for lighting surface buildings or else in the miners' cottages.

FETTLED Taken care of, 'that's fettled that', also as a greeting 'How duw?' 'Fine fettle' ('How are you?' 'Feeling fine') or else 'What fettle?' 'Champion'. Also a term describing a physical or verbal beating 'He syoon fettled him' (he soon took care of him).

FILLERS (HEWERS) All filling was originally done by shovel, but since the advent of the disc cutters most of it is done mechanically on to a long, moving chain. Nowadays the filler or, as he is called in Yorkshire, the stableman, will only work in the 'neuks' or stables – i.e at the end of the face, where the coal has to be hewn by hand before the face is ready for the machine to start its cutting. In tail gate stables small machines are coming in now to replace the workers. In the main gates there will be four or five stable men at work. They work very hard – shovel very hard all day and they still have their timbers to move in order to complete the job. I think they are the last shovel workers, except perhaps the rippers, who are still employed using the shovel – 'the idiot stick' – 'the banjo'.

FILL OFF AND SOD OFF Knocking off work. Until the coming of the power loading agreement, the face worker was paid by the piece. As soon as the team had finished the task, their money was assured and they could 'fill off and sod off' – i.e. go home. Certain caunch men were known to work themselves crazy, and finish their day in five hours. As one worker finished he pitched in to assist his marras with maybe harder places.

FIRE DAMP Miner's name for methane gas. This is the explosive one, the real killer. In very small quantities it is highly explosive. At 2½ per cent in the general body of the air all electrical equipment has to be switched off and the men withdrawn. From 7 to 14 per cent in the general body it is liable to spontaneous combustion – i.e without any spark or encouragement.

FLANKER To work a flanker comes I think from racing parlance. In its original form it refers (I think) to a horse coming up on the rails unexpectedly. In the pits it means to do something cute to the advantage of the person who has worked it.

FLAT SHEETS In the old days these were square sheets of iron about 1in thick and nailed on to planks to assist in the moving of tubs and corfs. Nowadays flat sheets are used by rippers to help them shovelling. This is a good example of how miners keep the same word for new inventions when the old thing it has described has died out. Flat sheets today are oblong sheets of varying length but normally about 5 or 6ft by 2½ or 3ft; these are put on the floor of the seam especially where the floor or bottom is soft. The ripper shovelling into a 20 or 30 ton heap of stones then has the boon of a solid level floor: his shovel can slide straight into the heap, instead of being caught on the uneven surface.

FLOOR The floor of the seam. Also widely known as 'bottom'.

FLOOR LIFT When the weight of the earth comes on the gateway of the mine it does not always force the roof down. Often if the roof is hard and the floor soft, the result is that the floor boils up: a day or two later you may not be able to walk up the gate, the floor will be nearly touching the roof. This is a job where 'dinting' is required i.e. the digging up of the 'floor lift' and the levelling of it so that a regular height is kept. This might have to be done regularly in seams with soft floors.

Floor lift is also, and more commonly, the result of 'bottom pressure', simply, while the coal is in the earth there is nowhere for the pressure forcing up from below to express itself. After the coal is removed the earth can boil up in an attempt to fill the vacant space.

FORESHIFT The first shift. The day shift; any shift starting after midnight.

FULL'UN Name applied to a tub or mine-car loaded with coal, as opposed to the 'chummin', or empty.

GALLOWA Pitmen's name for a pony underground.

GANNIN BOARD This is the main haulage road from which all the places cut off.

GATES (tunnels) On the conventional system of longwall face work there are three gates, or tunnels, the mullergate (mothergate) or main gate, which is the biggest of the three and carries the big belt which takes the coal from the face outbye. The mullergate also has all the mechanical and electrical equipment for the operation of the face-machines, the belts, chains, etc. The tail gates are at each side of the face and are smaller than the mullergate; up these gates come the men and materials necessary for production. Since a face could be 240yds long and 18in high a man in case of an emergency would have to crawl or be dragged a long way, probably too long for safety, therefore on long faces there is sometimes a 'dummy gate' or 'escape gate' which is half-way between the main gate and the tail gate. 'Dummy gates' are disappearing with more modern methods of face work.

GAUDY DAY In the earlier days of mining, this was the name given to any day which the workers made into a holiday. It might be the day the first cuckoo was heard, or the turnips were ready in the fields. If the truth were known the custom is still observed although on a more individualist basis. Often workers will say 'bugger it, lets have one off for the queen'. This seems to me a modern equivalent of the 'gaudy day' of old.

GOAF Also called the 'gob'. This is the section of waste ground behind the face and the working tracks. When the coal has been drawn out, the supports are moved forward leaving the roof to fall in behind into the 'goaf'. It is a planned cave-in.

HANDFILL In modern parlance the word refers to hand coal filling on the so called 'conventional system' (long wall, jib cutter, bored and fired) of face operation. The workers kneel (or lie) in a long line along the coal face shovelling or 'filling' the coal which has been previously fired down into the belt rolling along behind them.

HEAP The name given to all the surface installations of the mine.
HELPER UP Young boy employed to help the barrow men, or in later years putters, out of 'swally' or bad places. The Elliot family of Birtley, tell of a little lad sent to 'help up' a big hand putter. The putter being surprised at the idea of such a little lad being able to add any weight to the tub said: 'Thou's got ti help me up? Whey all reet then, here's a bunch of tokens, lash my legs with it when I start to push.' (This would be the practice if a pony was pulling the tub.)
HEWER In Durham this was the worker (before the advent of machine cutting) who by means of a pick, usually a hand pick, felled the whole area of coal and filled it into tubs. It required a very great deal of skill and a tremendous amount of strength. The hewer would hew out a long slice under the coal and temporarily sprag or support it while he crawled under it to hew further in. Later he had the help of powder and wedges to loosen it; still later with machine cutting he became a filler-hewer i.e. shovelling the bulk of the fired down coal on to the belts but fetching down the coal not loosened by the pick. Today when the machine fills the coal as well as hewing it (except in the 'neuks' or stables) the cuttermen perform the hewer's work. The pick is still an invaluable tool to the men working in coal or on the caunches, but it's not in such continual use. In the days of hand hewing, the men had also to pay towards a tradesman working in the blacksmith's shop (known as 'the pick shop'). The men had to buy all their own pick blades. At the end of each district or flat, there was a long steel rod on which the men placed their picks. These were sent to bank every day to be sharpened, ready for the next shift. My father remembers as many as three pick blades being used in a single shift. Also the hewers had a lamp crook, for hanging their oil lights on; they stuck the crook into a prop whilst working.
HINGER (hanger) Today it can be found at the gate ends of the face. The hanger is connected to an iron girder, sticking out of it at right angles will be a small roller. When lowering a neighbouring girder, it will fall into this roller and one can simply push it forward to advance it rather than having to lift and push. 'Dinnat let yer hinger dingle' is a well known farewell, having both pit and sexual connotations. Certain 'hingers' were nicknamed 'Budgie' or 'Canary' because of the way they clung to the supports.
HOGGER Three meanings for this term are all in use today. (A) Stockings with the feet cut out (or sleeves from sweaters) which are worn over the tops of normal pit stockings. Their purpose is to prevent bits of small coal or stone falling between the boot and the stocking, thus

necessitating cessation of work (very costly on contract particularly if one has built up a steady pace of shovelling). (B) Most rubber and leather pipes, but usually the rubber ones and those used in pumping operations. Metal pipes wouldn't be known as hoggers. (C) The pit shorts which the miner wears underground, he will usually wear these 'hoggers', his boots, belt battery and knee pads and nothing else: sometimes only belt, boots and pads are worn.

HORNY TRAM A flat tub without sides having an upright on each corner rather like a four poster bed without a top. It is used in the conveyance of all kinds of timber and drums of oil to the coal face. In Durham I have seen the gate so low that the 'horns' have had to be sawn off to get the tram up to the face.

HOW! A form of greeting having many variants. 'How' is a shortened version of 'Howdo' (how are you?) to which you may reply 'alreet!' (all right) or 'ney se bad' (not so bad). The reply to 'how' is very generally 'champion'. Another version is 'hoi' a term meaning 'hallo' the return is always the same, 'hoi'. In Wearside pits they use 'cher' instead of 'marra' so a greeting from the Teesside and South Durham men is sometimes 'howcher' (how are you friend?), the same version of this in the north-east Durham and Northumberland collieries is 'ho ma' ('how are you marra/friend?'). Farewells are just as numerous and usually have a cheerful or sexual connotation which is linked to the mine. 'Look after yourself' in the Durham area becomes 'dinnat let thee dingle dangle' or 'keep thee prop up' or 'keep thee pooder dry' or, most common of all, 'keep had' (keep hold).

HUNKERS 'Squatting on the hunkers', squatting in a crouched position with the backside slightly off the balls of the feet while the weight of the body falls on to the toes. This is the relaxed posture which most miners down the pit or on the surface assume. There is of course nowhere to sit down on the face, one cannot sit on the flooded muddy ground so the next best thing, 'the hunkers', are relied upon.

INBYE In the direction of the face, away from the shaft.

INTAKE The name given to the shaft down which all the air travels; also called the 'downcast'; it is open at the surface to allow greater access to fresh air.

JOWL To sound the roof. Deputies today, and at one time face workers, carry a wooden rod, like a thin broomshank, and use this to 'jowl' or test the roof at the coalface, and from the sound (hollow or firm) to see whether it is safe. 'Jowling' is still carried on by all workers and many other miners further 'outbye'; today however the worker will pick the nearest long piece of wood or metal at hand to tap the roof, and in the process of tapping shakes down any little pieces of coal or rock which otherwise falling on to his back would give him a nasty shock, and (in anticipation of bigger lumps or a fall) might make him panic. The following song describes this very important operation.

> Jowl boys, jowl,
> Jowl boys and listen,
> There many a marra misten man,
> Because he couldn't listen.

JUD In the days of bord and pillar it was a measure of coal; it varied in tonnage and measure but was usually the amount which a strong man could bring down and fill in a shift.

> A got fifty oota the jud,
> Titty fallar, titty fallay,
> Ee by gob it was gud,
> Titty fallar, titty fallay,
> Ah, cum oot ti ger a shaft,
> The timmer give a crack,
> Yi bugga, a stone fell on me back!
> Titty fallar, titty fallay,
> Tral la lala la la,
> Ower the walls oot!

KEEKER This was the owners' 'weighman'. He looked over the tubs as they came to bank, checked and weighed them for the owners. He had also the task of laying them out, or confiscating the tubs which were not full to the brim or had stones in. The men would pay their own 'checkweighman': he was there to try and prevent the workers being excessively robbed; he was elected and paid by the men to look after their interests at bank. My grandfather became a keeker at Wardley, but in the first place he had not been a miner. He was a fish curer by trade, and was good at hawking fish around the doors, having his own little pony and flatcart. However he ended up at Follonsby. My father says:

> He worked on the stone heap as a teamer; this was very heavy work, consisting of tipping the tubs of stone over by hand, also shoving them along the gangway on to the stone heap; this was for

hours a day, all weathers, no sheltering, working all the time. He was also a very good time keeper, as well as a good workman, it was for these reasons that he was given the job as keeker, for which he received the huge sum of 1/- per day extra, also whenever the pit was idle through bad shifts (which was the thing in those days) he was paid. This was the only thing my father gained through his promotion. Of course there was also a head keeker over all the others.

KEPS Also known as 'catches'. Metal supports at the mouth of the shaft (two at each side), which come out to hold the cage once it is 'at bank' (on the surface). They are controlled with a handle operated by the banksman. With the advent of skip winding (i.e. feeding the coal directly into the cage without the use of tubs or mine-cars) the catches are no longer used. Previously men would refuse to 'ride' the cage if the catches were not operational.

KIRVING After the jib cutter has cut its slice into the face the small coals which it leaves in its wake are called kirvings. They lie piled right to the roof. Crawling along an 18in seam with tons of loose kirvings underneath you is painfully awkward work. One has to 'swim' or wriggle from side to side, as the kirving gives under you and you can't get any propulsion forward. Your back is flat against the top, your head lying on one side and the arms and legs get cut by a thousand little pieces of glass-like kirvings as you inch your way forward.

KIST (box) The kist is the place where the deputy sits and deploys the teams of men. It might be a cabin. At the start of the shift the deputy sits there, compares reports, and deploys the men. At the end of a shift men working overtime would go there to get what were called 'ready breks' which were notes given to men in Durham for some food and a drink of milk when they came out of the pit (pie notes, they call them in Yorkshire); here also the deputies give out wet notes allowing miners to ride out of the pit early if working conditions have been unusually bad. If there isn't a cabin, the kist can be just a corrugated sheet set up against the wind with a chock and a plank.

LAKING This is a Yorkshire expression for stopping off work voluntarily. There are many reasons why workers stop off work. In the case of older workers often they just can't do the work any more, the dust is getting worse all the time, and the noise of the equipment more unbearable. Four shifts of labour are usually enough for their needs and as much as they can physically take. The Friday gives them an opportunity to rest up for the week-end, or do things in the garden or about the house. For some Friday is the traditional night out, but on a four shift rota, the men on the last two Friday shifts lose their evening if they report for work, and those on the last shift even have their Saturdays ruined, which is unforgivable. If you have a regular face job and you know your own marras aren't coming in on a Friday, you might stop off in order not to have to work with a set of strangers who would come to fill the places of those absent.

LAMP CABIN Name given to the surface building in which the lamps are housed. It may be a separate building or part of a pit-head complex comprising baths, time office, token room and lamp cabin. After changing into your pit clothes you would progress to the time office board, and take down the check or token with your number on. Next you would go to the lamp room and take out the cap lamp and battery. One of your tokens would be hung on the lamp rack to show you had it out, the other would be kept until getting into the cage, when you give it to the onsetter who carries it back to the time office. The lamp is already lit when you take it from the cabin, but on a very low flame. This must be coaxed to life by gently blowing air into the top of the lamp and 'pricking' the wick a little higher.

LARE (learning) Self-gained knowledge.

LAY-OUT BOXES When a tub came to bank (surface) and was found to have a certain amount of foreign material in it (usually stone), it was 'laid out' or confiscated by the manager's weighman. The stone from the tub was placed into lay-out boxes of various capacities; when it was full or half full the miner might be made to pay a fine as well as having his tubs confiscated. At Felling colliery in the 1890s the lay-out box was 14 lb in capacity and a fine of 3*d* was inflicted when it was full.

LEADER (girder/timber) The girder or timber leader is the modern 'putter'. He no longer carries tubs to the hewers or fillers, instead he is in the haulage supply business and carries in materials. As in the case of our earlier 'putters' he works either with a pony or hand putting. He drives (if pony leading) a tub, a sledge, or a flat tram full of timber or straight iron girders. In certain pits there were gates and working places which had no rail for tubs to run on; here the pony would have to pull a sledge in the manner of Eskimos or Red Indians. Hand putters had the horrible job of pushing the tubs and trams by sheer brute strength in much the same style as those in nineteenth-century engravings.

LEAD 'UN A term used in putting. It relates to the tub which the putter leaves empty inbye the hewer's siding, while he flies out with the full 'un. The putter must never allow the hewer to 'wait on' or be short of tubs otherwise he would sharp find himself minus a job, and maybe a set of teeth as well. Just a few extra lines from my father about this: 'after the tub had been put into the siding the putter took his pony into the face and hung on to the full 'un, and came out behind it past the siding; he then stopped his pony and went back to the empty, hand putting it from the siding to the men at the face. He would push it until the front wheels dropped over the plate ends. Then he'd get back on the limbers with his full 'un and proceed outbye with it. When pushing the lead 'un we hung our oil lamp on a leather strap around our necks; this was dispensed with when we got electric lamps, not cap lamps, but big heavy lamps which weighed 7 lbs, however these were a great improvement to us putters.'

LIMMERS The shafts which attached the pony to the tub or tram. The driver sat on the limmers, sideways to the tub, with his shoulder hard against the tub and head crouched; if he lifted his head above the pony's height he was at a very deadly risk from the roof and its jutting obstructions. The limmers were connected to the tub by a cottle pin.

LINES-FACE In the days of wooden pit props and hand-set timber these lines were painted on the roof of the face to keep the setting straight.

LINES-GATE These lines were painted in the centre of the tunnel (what we call the 'gate') in order to keep the arches running in a straight line. Their position was established by a lines lad or a linesman, sitting with a plumbob (a weight on the end of a cord).

LINES-GRADE These lines are painted on the side of the rings in the 'tunnel'. They are about knee level and they run from the pit bottom right into the face. These are brought up to date every quarter by the surveyors. The object of these lines is to maintain a gradient – a consistent level – and to find what levels the gates are working. In Durham the 'lines lad' was the young lad who painted these lines on behalf of the surveyors. He would carry a little pot of whitewash to paint these lines on and you would usually find that whereas everybody else was black with coal coming out of the pit, he would be black with whitewash. The whole tunnel would have quite a lot of lines over it. I worked on this job when I first went down the pits in Durham and I progressed to putting on not only the grade lines in gates but also the face lines and it was a tremendously difficult thing to crawl along a very thin seam and try to keep the pot of whitewash up straight; you might find when you got to the end of the face that you had lost your brush or you had spilt all the water.

LOCO The locomotive down the pit. This term is applied to the battery powered 'locos' rather than the diesels. At Wardley colliery the mine-cars would be filled up at the loading points and the 'locos' would haul them to the shaft bottom, here the cars would be released and run down a self-acting gradient into the cage, which would be wound up to bank. My father in later life worked as a loco control officer in much the same capacity as a signal box man; he kept the locos co-ordinated from Wardley, Follonsby, and Usworth, and directed the mine-cars to each of the loading points.

LOWSE Is an expression for 'knock off' or finishing time. In the old days the shout would come down the shaft 'she's lowse', the onsetter would take up the shout in each of the galleries, 'lowse, she's lowse' and so it would go on, round the workings. It probably originated with the corfs being 'lowsed' or loosed off the rope, in order to allow the men to sit in the loop and be drawn up two at a time. The dialect expression used just before the shift ended was 'von ni lowse', or in Northumberland 'vorry near lowse' the 'r' being rolled in their peculiar way ('very nearly lowse' i.e. finishing time).

MAISTERS (masters) The old owners of the pits. Many of the workers who believed that nationalization would free them from the masters were quite surprised to find that these same masters, or their immediate servants, were given jobs on the Coal Board. Many miners believed that the owners were lucky to get away with their lives for the slaughter they had inflicted on the miners; instead they were paid £338 millions in compensation. Wardley pit was owned by the Bowes-Lyon family, who also owned Usworth pit and Felling pit as well as many others,

MANRIDING SET An underground passenger train. One version of it has a row of seats facing outwards, like a jaunting car, and another row facing in the opposite direction so that the men sit in two long lines back-to-back; this kind is generally worked by rope haulage. The other is like a big dipper train having room for perhaps four or six men, and pulled by a loco. In earlier times in Durham, before the use of special equipment, the men used to ride in by the tubs. Two men would sit in each one. This set would be pulled in by the big underground rope engines. At the back of the tub would ride a little laddie called a 'run rider'. He would ride on the chain end of the set beside the 'cow'. Part of his function was to jump off the back and signal to the winder inbye by means of the overhead signal wires.

MARK, BASE These were made at the dead centre of the tunnel or 'gate'. They were marked out by the surveyors when they came in with their dials and their theodolites. At this point a hole would be made in the roof with a chisel, and a wooden plug was banged in with a staple on it. From the staple a rope was hung which was used for sighting off the oil lamp. This would be a base mark and be marked with a triangle.

MARK, LINE PIN This would be set by the linesman, who had no instrument to measure with except his own eyesight and plumbobs. He would sight a centre from the base mark which had been established by the surveyors, and make another insertion into the roof. This would be his line pin. It was usually put into stone because stone very rarely moves in the roof, whereas the rings moved through pressure and were therefore unreliable. It was quite hard work standing in a lot of heat and with your arms fully outstretched trying to bang a hole into solid rock with a hammer and chisel. The part of the job that I quite enjoyed – kind of stupid pleasure – was in whittling lumps of wood into the plugs that were hammered into the top – and also in painting up the circle marked 'line pin'. Day after day I would think up new ways of painting the words 'line pin'. Often I would paint it as 'li-pi' or it would have a big L, or I would paint it in italics.

MARK, MEASURE A system of measuring the length of the tunnel from its start to the face. This measure mark was usually a vertical line painted on to the wall of the tunnel with the letters 'MM' (measure mark) beside it. I used to enjoy painting 'MM' – it made a change from 'line pins'.

MARKS, PAY These were generally made by the deputies, to measure the work of the caunchmen (who would be paid by the yard) – that is the stone men who worked at extending the tunnel. The caunchmen worked to yardage on contract, bargaining a price for each job. The deputy would come and measure what yardage they'd done and paint his payment on the side of the arch girder (probably it would be his initials) and a line as to how far they'd got. Sometimes he would paint his line on the 'dolly props' – the wooden props lagging the arches. If he was stupid enough to do this, the caunchmen could knock the dolly props backwards which would give them that much more yardage. It was through these systems of pay marks that the caunchmen's wages were worked out under the piece rate system. Most caunchmen worked piece rates – in fact a lot still do. However at the pit I went to in Yorkshire the contract system was abolished in favour of day rates.

MARRA This is a Geordie expression which means mate. A person's marra is the fellow whom he is cavilled with, whom he works with. In Yorkshire although marra has drifted down with the exiled Geordies, 'serry' will be the term used when speaking to one's mate. Under the butty system the chargeman had 'gobbers' working under him, the term 'gobber' also takes the place of marra. In some parts of Durham, particularly round Penshaw Monument way, the term 'cher' is used for 'marra': 'pass shul' cher' – pass the shovel, mate. Around Seaham 'sa' is used ('does tha see sa').

MASTER NOTE A form of payment. The team of colliers or rippers shared the task in a communal way; the tonnage they shifted was recorded on a big note called a 'master note': it billed all the work done by the team, the total money to be paid to them, the yardage advance, etc. At the end of the week (or fortnight, in the days of fortnightly pay) the cavil leader would take the master note into the pub and the men checked it and shared out the money, although this was normally paid a day later in order to have time to set to rights any wrong billing on the note. Actually the master note or 'big note' of my day was not the same as the earlier one. Then, the master note given to the team leader stated the total tonnage, total payments, extra payments etc.; and was shared out by the team leaders equally amongst the team. However in my day although the task was shared and the wages were a collective earning, the note had changed. On the 'big note' in my time the total tonnage would be billed as would the total wages, etc. But each man would also have an individual note, with his share already divided by the wages clerks. Each man would have his different tax deductions, rent payments, and off-takes, so that although all the wages would have started the same, they would all end up with a more or less individual payment, despite the fact that the task had been pooled. The team leader would still withdraw with his men into the canteen, to check the yardage paid on the big note and fight like hell if it was wrong, but it was not quite the same as it had been in the past.

MASTER SHIFTER The Master Shifter was the head man on the night shift and responsible for preparing the pit for the following day's coal work, etc. The head man on the *day* shift was the number one overman and the real power behind the throne.

MEETINGS The time when the cages meet each other in the shaft, one going up and the other coming down.

> When gannin up an doon the shaft
> The patent cage did threaten,
> For ti tak ivor old 'un's life if
> They chanced at meetings.

MELL (MALL) Name given to the sledge hammer used for knocking props out or smashing big stone/coal etc. Used for 'braying' timber in. It weighs between 3 lbs and 14 lbs.

MULLERGATE (the mothergate, main gate) The central gate (tunnel) out of which all coal comes, it has all the big belts and most of the electrical points. The tail gate did house certain of the equipment, the borer might have all his gear in that gate, maybe 20 yds of cable for his machine; today the tail gate might have the tank and pumps for the hydraulic chocks on the face.

NEUK (nook) The extreme ends of the coal face; in the first phase of mechanization it was the place where the cutter was turned around in order to jib in and set away cutting. With the invention of disc cutting the machine no longer has to be turned around, but it still has to have a start made for it, and the term 'neuk' is still applied to the areas of coal taken down and shovelled by hand at either side of the face. In Doncaster the neuk is called the 'stable'.

ODD 'UN A tub which the putter may have filled for himself before the hewers had got in to their place. More usually however it was an incentive to keep the hewer well supplied with tubs and not 'waiting on'. If the hewers got a good lad they might reward him or encourage him by filling an 'odd tub' for him, or else co-operate in filling the 'odd 'un' with the putter. The tub would then go out of the pit marked as having been filled by the putter. My father remembers that the only time they hewed odd tubs for themselves was after the last shift of men had gone outbye: 'we used to have about an hour or maybe three-quarters of an hour to make a little extra money. Of course this was after the 1926 strike when we had to work longer hours. We used to work 8 hours plus a winding time of 35 minutes.'

ONSETTER The onsetter has the same job at the bottom of the pit as the banksman on the top. He decides when you can ride out of the pit and when you cannot. He cannot really keep you in the pit after your time, but he can delay 'rapping' until the very last minute. Usually, if he is a decent fellow, he might let you out of the pit five or ten minutes early. But sometimes, if he is not and you've taken the mickey out of him, he can make you wait a long time before you get out of the pit, for he's in charge of the cage and lets you know it.

OUTBYE Places near the shaft bottom, and away from the face.

OUTCROP In some parts of the country coal crops out to the surface and leaves an exposed seam for some distance, on the sides of hills for example and near rivers. This is usually the start of 'drift mining' where the coal is worked from the side of the hill, and no shaft is dug out. Miners used to go outcropping to find themselves coal when they were out on strike.

OVERCAST Sometimes air courses and roadways will pass over the top of one another; where this is so it is called an overcast: that is a section of the tunnel where air is channelled over the top by means of another smaller 'tunnel'.

OVERMAN The overman is the official immediately above the deputy and immediately below the under managers. These are the people the workers have to deal with in their day-to-day struggles. These overmen can make life difficult or, if not pleasanter, easier for the workers, depending on how willing they are to make concessions. Wardley pitmen, for some obscure reason, always call them 'overmens', even when they are talking about one person. At one time the overman had a whole district of face units to cover walking all around and directing all the men. Now there is an overman for every unit.

PACK A stone wall built near the face to help take the weight of the earth. A pack is built like a dry stone wall, the same as one would see in the Northumbrian hills or Scotland. One stone is placed upon another to form a kind of stone 'corral', small stone is shovelled into this until a solid pillar up to the roof is constructed. Packs vary in length but are normally 2ft 9in to 3ft wide and 5yds long.

> So, we fill up the pack, with the stones thats in the gate,
> And wi'l get them neet and tidy, weel afore bait,
> For there's girders to hump and girders ti set,
> Side ti tak' off that ah' dare bet,
> Doon thu' Brockwell seam in the north of number five west.
> (*Stoneman's Lament*, Johnny Handle)

In the days of conventional long wall filling (by hand), three packs other than the gate packs would be constructed. These would be 3ft wide and 4yds wide. Nowadays the only packs are those put on by caunchmen at the gate ends.

PEGGY This is the term that Durham colliers applied to the handpick. When the face was hand hewed the hewers used to have to pick the coal down in preparation for filling. The handpick is still used in the 'neuks' or stables, at the end of some faces, but the pick is dying in its usefulness. The rippers use a pick more than anyone else on the face, but even with them it is used less than in the past.

PICK-WINDY In Doncaster called a 'jigger'. A small version of the pneumatic drill used in hewing down coals and stone and breaking up big coals and stone. It is held much like a tommy gun being of the same sort of length.
PIG TAIL Durham name for a haulage clip. It clips the tub chain to a haulage rope. There are many types of clips, varying in size according to particular haulages or roads. The hambone clip I used to hate; it was a large awkward object with a head like a hambone shank; if you happened to get a clout off one of these as it moved along the rope, it took some time to recover. Bottom rope clips are also difficult to deal with; you have to put one foot on the arm of the clip to tighten the jaws round the rope, similar to starting a motorbike. I remember once my foot lost its grip on the handle and it sprang back to hit me in the knee cap, it knocked me cold across the line, and had the clip been connected with sufficient strength to pull its load I probably would have been minus a foot or two.
PILLS This is the name given to the powder charges that are used in shot-firing operations – perhaps 6in long. Various forms of explosives would be used, dynagex being the general one. Stronger powder is used for stone than for coal, which is obviously softer. Although there is a stipulated amount meant to be put in each hole, if the shot-firer is a decent bloke he'll put a lot more than that in the hole in order to blow the coal into dust practically. Which isn't much value to the pit but makes the operation of shovelling the stone and the coal a lot easier. 'You can myek yer arms gaan like beeswings.' 'Pills' are stored in the powder house underground at some pits (although not in Durham) and they will be carried in bags by the workers working on the job – in little bags like those used for carrying bowls. They contain sufficient powder only for the day's operations. For carrying in the powder the workmen might be paid 1*s.* 8*d.* per bag extra. If other work had been done by workers for which there was no payment the deputy might book them 'powder money' even though they'd carried no powder at all. This is a job I would never do, whatever they paid me – it's too much 'clart-on' (trouble). 'Pooder' was never known as 'pills' in Durham.
PIT HEAP The slag heap – all the waste which is taken from the mine. This is also the name given to all the surface buildings, etc., near the shaft.
PIT SENSE A kind of instinct, and intuition coupled with actual experience of many underground happenings. An ability to take notice of minute warning signs, sounds and smells. A miner with pit sense will

immediately recognize something as different (like the absence of mice from a gate or working place) and know at once that something is the matter. I knew a miner down Wardley pit who could to all intents and purposes 'smell' fire damp although it is known to be odourless. Another miner could tell the strata and seams above him by the taste of chemicals in the water dripping through the roof.

PIT WHIP Pick shaft carried by the putter. It is about 6in long with a leather thong at the end. It was hung on the belt with the thong wrapped round the belt. When it was carried in the hand, the thong was wrapped round the thumb in a loop. The putter would carry his lamp and his whip in the same hand and drive with the other. The major purpose of the whip was not that of whipping the pony, although some did, but to act as a lifting aid when the tub got off the way.

PLACING The planning of the day's work. The overman arranges the work of the day and advises the miners as to the jobs outstanding and what is necessary to be done, e.g. he may say to the loco driver 'there are five tubs of waste materials inbye to come out and nothing can go in until this is done.'

POODER (powder) The gunpowder used in blasting operations.

POST A type of very hard stone which makes a good roof.

PRICKER This is a long thin iron rod with a flat hook on the end and is used by the shot-firer. After a hole is bored, he puts this into the hole and tests the hole inch by inch for cracks to see if any gas is leaking through into the hole which might cause an explosion when the shots are fired.

PROP BOBBY This is a term that is used in the Doncaster coalfield and it applies to a worker who goes around counting props (a dataller). This was done because if props were difficult to get out, or if they were dangerous, the workers just used to bury them or knock them flying back into the waste where they were buried. To offset this the management employs a datal worker who walks up and down the face counting how many props are in the section and how many are missing. He usually has a little counter which he depresses with his thumb. In order not to lose count he points to every prop as he clicks his counter. It looks as if he is blessing them. There isn't much antagonism to this particular worker, in fact you never know which one has reported you, because there might be one on every shift. If you can try and suggest that the prop was gone before you got there you may get away with it, but usually it results in a fine of about £5.

PULLERS This is a Geordie expression which is used to describe the workers who move the conveyors over after a cut of coal has been taken.

The pullers also withdraw props and set certain new ones but generally their work was to bump up the conveyor and the box ends on either side of the conveyor, setting whatever supports are necessary. The task of these men was designated 'pulling and drawing' so this group of workers became 'pullers and drawers'. Perhaps six or eight men would operate in a team. They 'pulled' in the belt in sections and drew off the wooden timber. This timber was chopped out since the weight upon it did not allow it to be knocked out. For this work the drawer's axe (pronounced 'ayex') would have to be razor sharp, and the worker might spend half an hour sharpening it every day. The timber was chopped into a point at the top then knocked. Drawers and pullers would withdraw the wooden chocks and reset them. The disengaging device called the 'trip' would be used in this operation. My father recalls the pan pullers at Wardley when he was a young miner: 'When I was 17 years old, I worked in a place called the east drift, we had three faces down there each 80yds long. We had five hand fillers on these faces filling on to "jigger pans" which had a face engine which caused the pans to "jig" the coal down into the tubs. The name pan pullers came from this conveyor, which was used before the advent of belts. The "pans" were hotched forward as the face progressed.'

PUTTERS A term used in Tyneside (they are called 'trammers' in Yorkshire). These workers are usually young lads who haven't gone onto the face though some are older. They bring the 'chummins' or empty tubs to the face, and carry the 'fullins' away, either by means of a pony or else by hand (which is incredibly hard work). A putter might have four or five places to supply and never be able to please everybody because all the workers want their tubs fetched back there first. My Dad was a putter for a lot of years underground. He tells me that it was very hard work, sitting cramped upon the limmers all day, wearing nothing but little hoggers. Your head would be bent down below the level of the tub, oil lamp on the thumb, while twisting the body all ways round turns. He writes to me as follows: 'Remember son, only four foot plates were used, also wherever the face turned off, a swape turn was put in, either left or right hand over. The putter's backside had to drop inside the limmers, and as the tub came round the turn, his backside between the limmers and the tub pushed it round the turn; then the handles used to stick into your back. By the end of the shift, maybe the putter may have 30 or 40 tubs put (it would have to be good putting to get 40 by the way).' Where the worker had to push the tub out by hand it was very hard work especially in the lower places – pushing the tubs out just with his own muscle power and still having to try to keep up with the workers who were shovelling in their places. They worked the same as in the pictures of nineteenth century mine drawings of boy or woman labourers pushing tubs. Even six years ago pushing trams of girders and timber along by hand was not uncommon.

Nuw, ah'm just a smallie laddie,
Hardly ould enough ti hew,
But a'v held me a'n at puttin
Wiv th' best a'h iver knew,
Giv us pleanty bait and bottle
Pleanty beef and baccy chows,
And a'hl bet me bunch 'o tokens that from
Gannin doon ti lowse.

A'hl be runnin for the lead 'in
Runnin for the lead 'in,
Runnin for the lead 'in
From gaanin doon ti lowse.

(*The Putter . . .*)

PUTTER-HEWERS These were the putters next in turn to take their place at hewing (putters graduated to hewing after preparing themselves for the work by building up their muscles putting). As a rule these weren't cavilled like the others, but acted as 'spare' men. If a putter was off one of them filled his place; the same was true if a hewer was off. If there were no vacant places, the overmen would try and find a place where they could hew for a full shift. My father tells me that the putter-hewers always did all right for themselves as regards money.

RAMBLE A very thin 'parting' of muck or shale, often black as coal, which lies between the top of the coal and the roof. Hewers were fined for having 'ramble' in their tubs; trying to avoid such a thing took the colliers time and (in the days of piece-rate) lost them earnings. Later on prices were negotiated to take account of 'ramble'. In the Follonsby colliery *Prices and Conditions* (19 February 1931, p. 3) we read the following entry for hewers: 'Ramble scale $\frac{1}{4}d$ per ton for each complete inch above two inches. Average of three measurements.'

READY BREK Ready brek is a note which is given by the deputy to a workman who does more than two hours' overtime; it entitles him to a free purchase of pies, a glass of milk or anything else to a certain value. In

Yorkshire it is called a 'pie note'. This has recently been restricted to specific things; previously the miner could have anything he wanted. He may come out of the pit with the ready brek and not wanting anything he might get some chocolate to take home for the children, for instance. The men objected very strongly in Yorkshire when they were told they could only use it to eat in the canteen – and then only for pies (I used to bring mine home for the dog). Geordies are very keen on having this half-a-crown's worth of food and it was said that when the Americans first landed on the moon the first thing they saw was a dozen Geordie pitmen running for a ready brek.

RIDE To travel in the cage up or down.

ROLL A roll is a small geological fault which comes into the face as a length of stone and might stay for a few weeks and then get shorter and just disappear. Wardley colliery was said to have been sunk into a nest of faults.

ROONDY COALS Shiny hard coal about the size of a rounded house brick. Roondy was considered the right size for domestic use. Often great lumps of coal would be in the miner's load of house coal. One of the jobs the boy scouts would get in Wardley was smashing up big coals into roondy, since the coal in Durham which the miners got was never screened; nowadays you are lucky to get an odd piece of big coal. Roondy coals were also known as 'Billy Fairplay'.

RUBBING THE SAW This was the putters' version of equalizing chances, similar to the cavilling system. My father, who spent more of his time underground as a putter than at anything else, explained the system thus:[3]

> You ask me about rubbing for places, well we putters were cabled to a flat; say the South Flat, Six Pillar or West Flat for example. These places were just like 'zones' as you knew them at Wardley, only they were known as 'flats'. Flats were split up into districts – 2nd South, or 1st North and so on. Now, there may have been 30 hewers in the flat and 5 putters. We putters were cabled to these flats, having drawn our cables at the special cabling meeting. First out of the draw no. 1, having first claim on that particular flat, then no. 2 and so on. Now the reason for the rubbing is very simple: 30 men, two to each board or wall, which meant 15 places for the putters to supply. Some of these places could be a long way away, as much as 14 pillar (a pillar, if I remember correctly, was 44 yards long) while others were close at hand. Some of them would be standing rough, maybe only tub height

and width and as such very bad for making good money. To equalise all the chances the putters would rub for places. This was by the deputy placing numbers on his saw or a piece of wood. They were written down in this fashion 2, 5, 1, 4, 3; of course the putters didn't know in which order the numbers had been placed. No. 1 would rub first, then the others follow. The next day it was the turn of no. 2 to rub first so that by the end of the week each putter had had a first rub . . . a very fair system don't you think?

The Putter has a verse about 'rubbing the saw':

> There's a half a dozen gaanins
> In the flat that a'm at nuw,
> An' if a'd me ain a choosen a'w
> Wad be in '1' or '2',
> But some huw or other, huw it
> Comes a'w divinaar,
> As sure as rub me cavil its
> The worst 'un on the saw.

RUNNER When a tub has got 'amain' or out of control and is free-wheeling down a gradient or along a flat it is said to be a 'runner'.
SCABBY-ROOF A bad roof, where the coal does not break easily from the top and leaves an uneven 'bitty' roof with lumps of coal clinging to it. This makes it difficult for timbering and painful for crawling, because of the sharp bits sticking down.
SCAFFOLD Numerous scaffold exist at the coal face. (A) the rippers' scaffold at the tail gate. This is a 'wall' of wooden planks erected facing the coal face 'lip'. It is erected to prevent the stone flying into the gate after it is fired. It also keeps the heaps of stones compact and facilitates shovelling. (B) Boring scaffolds. These are erected in the gates in order to stand on them to bore the lips, also to stand on when setting the 'arch girders' (tunnel supports) into the roof. (C) Main gate scaffold. A kind of chute to direct, control and dispose of the stone after it has been fired at the lip. When talking my glossary over with other miners we all found different examples of the use of scaffolds. For example my first instance was totally new to George Hodgson of Murton Colliery and later of the Staffordshire coalfield. Perhaps the reader should realize that many of our implements at the coalface are in fact innovations of 'pit sense' i.e. they are inventions of the workers, and not always standard implements

of every colliery or coalfield. We have often been surprised to see a worker from another colliery or district come up with a totally new construction of 'scaffold' to ease the work.

SCORE The number of tubs upon which the putters' wages were calculated. Strangely enough the score was not the same number of tubs at every colliery. Most of the Tyneside collieries, however, reckoned on 20 tubs per score (with one or two exceptions of 21).

SCRATTIN Casual shovelling. Hens can be described as 'scrattin about' when they are digging up soil with their feet. In the mine the term describes those carefree workers who take the odd shovelful and then a drink of water, another shovel and then stop to fix their belts or knee pads.

SCREENS The surface installation at which all of the coal is brought for washing, sorting and grading. It is here that all the stones and foreign material must be extracted from the coal. In the old days it was here where the tubs came after they had passed from the hewing places and out of the mine. The contents of these tubs was closely inspected by the keeker, to see that any foreign material in the tub was extracted and the hewer who had filled it, punished, usually by fine, confiscation of the tub or both. My grandfather worked as a keeker on the screens. His duties were to see the screens kept moving and that the coal was clean. He also had about thirty men working under him – but he had to do the same work as them. My father writes as follows:

> Now in those days, David, we didn't have washers for the coal: all the coal coming to the surface was hand hewed and was tipped on the screens by means of a 'tippler'. The pit heap at Follonsby had five tipplers. The lads used to push the tubs in, lift a lever and the coal would fall on to the screens. I can tell you that because my father was a keeker; he was not very well liked by the hewers. They thought (wrongly) that keekers were to blame when the tubs of coal were laid out. Well, son, I started work on the screens, and seen what went on. The procedure was as follows: an adult would be placed at the start of the screens, and as the coals came along the stones were taken out by hand. In the case of the large coal having stone on it the men utilized a tool called a 'snap' with which the stone could be sliced off. Now, the other stone was known as 'band'. If the tub of coal was very dirty (with band) then the man at the head of the screens 'rapped on them', then stopped the screens in order to 'lay the tub out'. As you state, the 'lay out boxes' held about 14 lbs; now if over three boxes were filled

out of one 10 cwt tub it was known for the hewer to lose that tub. Up to three boxes and *6d* was kept off his price, which he would receive for the coal. This *6d* was a lot of money in those days to be taken from a rate of maybe 2/- per ton which the hand hewers were being paid at that time. Now, the keekers (including my father) always worked from the tub at the other end of the screen in order to prevent bother, and as such had nothing to do with tubs that were laid out. Of course on occasions the coal would reach his position very dirty and then he would have to stop the screens himself to make sure the coal was laid out and all the stones taken from it. You see David, in those days, to keep a contract it was important to have clean coal, hence the layouts. Some men kept their coal very clean while others just shovelled everything in. If I can explain about the hewers; in the 'board and walls', the coal was about 2ft 3in thick this on the top of the 'band' which could be anything up to 2ft 6in thick, under that maybe 10in of bottom coal. You see, if you can follow me, the hewers had to hew the coal, then shift this band and throw it backbye; but some of the hewers wouldn't clean it up properly and just threw it into the tubs along with the coal. Mind you, some of the stone couldn't be avoided. Different today, eh? I have seen stones as long as 4ft and 2 or 3ft thick come out on top of mine cars, we haven't any lay outs today of course.

On our heap, when I was a boy, David, it was quite true the men had a checkweighman, but the coalowners had their man, in the weigh office, working shifts. These men were on the staff and their job was checking on each of the hewers' tokens, how many tons of coal each man had hewed, also keeping a strict count of putters' tokens, so as at the end of each shift they knew how many tubs of coal the men had hewed and how many tons. . . . We had a proper cabin in which all the hewers and putters tokens would be hanging, also a board on the wall stating the number of tubs hewed and the tons, as well as the putters' scores, which as you say was 20.

SCREENERS Name applied to all those workers who work on the screens. This was and often in older pits is back breaking work. Very young and very old and injured workmen work here with great hammers smashing and separating the coal and stone, the big and small.

SEAM Name given to the strip of coal. A seam can be an inch high or (so I've been told) 100ft in other lands. In Durham 18in was not an unusual height although 2ft would be about average: these were all thin seams which made it impossible for the miner to stand up.

SET A 'run' or train of tubs, full ones or empties.
SHAFT A vertical passage into the depth of the mine from the surface; nowadays there can be no less than two shafts, often there are more. In the old days one shaft was quite typical. Shafts can vary in depth from a few yards to over three-quarters of a mile. In certain other countries shafts are said to be far deeper than this. In many areas of Northumberland and Durham the coal cropped out to the surface, especially at river banks and valleys. In these cases the coal would be drifted in from the hillsides, until it got too distant whereupon a vertical shaft might be sunk from the surface, leaving the drift entrance as a day hole.
SHOT-FIRING The name given to the blasting of coal or stone. The man who fires the shots is known as the shot-firer or shot-lighter (in Yorkshire 'shottie'). Nowadays the shot-lighter is a 'grade 2' deputy.
SHOT STICK (stemming rod) The long wooden pole used by the shot-firer to ram his shot and 'stemming' into the hole before blasting.
SKELP To hit.
STAITH The constructions on river banks for tranferring coal into the waiting colliers (coal-carrying boats). In my time the Tyneside staiths were chutes which directed the coal from a height into the boats. In earlier years there were also 'drops' which swung the whole tub over and lowered it into the keels. Perhaps some fifteen years ago, our expeditions as lads from Wardley colliery would lead us down the 'dilly lines' which ran from Heworth colliery down a steep incline past the back of the new Ellen Wilkinson Estate, through Bill Quay and out at the staiths. We would follow the line till it broke out on the banks of the Tyne just opposite Lesley's naval yard; here the colliers would be filled up with what seemed like an endless stream of coal shooting right out of the staiths and falling some feet into the waiting boats below. Certain of the colliery staiths of my childhood were in fact very old sites long since established by the collieries, particularly those mentioned by John Gibson in his surveys in 1787, as Usworth, Felling (Brandling Main staith) and Washington.[4] In 1834 the staith on Felling shore, serving Messrs Grace and Co.'s colliery at Felling, was in full production. Heworth shore held the staith of John Hutchinson and Co. and served the distant Sheriff Hill colliery. Just below Bill Quay staith (named on very ancient documents as Bill Point) was Pelaw Main Staith, this belonging to Messrs Perkins and Co. the owners of Ouston colliery. Close to this was the Coronation staith, the property of the affluent Brandlings and Co. of Heworth colliery.[5]

STAPLE A small shaft which runs from one seam to another, it may or may not have a winding contraption; in most cases it has ladders and a chute for materials or coals.
STAPLE SHAFT The staple shaft is a shaft running from one seam of coal to another. This is to give easy access so that the main cage from the surface doesn't have to be used to take men from one seam to another. This staple shaft has on many occasions saved people's lives from explosions and cave-ins, giving them an escape route to a higher seam level.
STEEL MILL The miners' light before the coming of the Davy lamp. It consisted of a flint and a steel 'catch' which struck the flint when a wheel was turned. This produced a stream of low heat sparks. These mills were almost certainly less dangerous than the candle but still very dangerous. They caused many, many terrible explosions. The mill was normally hung around the neck, one person wearing it and turning the wheel, the other person working it. A young boy normally turned the wheel for the hewer.
STEM The operation of ramming the 'stemming' in tight after the shots.
STEMMING The material rammed behind a shot to contain its explosive force when a shot is fired; it would simply fly back out of the hole unless there was some obstruction pushed up behind it. 'Stemming' is rammed up hard behind the powder so that the full force of the blast is dispersed into the rock or coal. The best stemming is the clay which is left lying in the gates; it is soaked and then moulded into long 'pills'. Modern stemming can be either a jelly substance contained in a long sausage-like bag, or long plastic tubes, full of water.
STOPPING The fresh air which comes down a mine has to circulate in every unit and district. In order to make it cover everything, certain 'short cuts' it would otherwise use have to be blocked off. This is done by erecting 'stoppings' which are air-tight doors or barriers. There are several kinds of 'stoppings'. It could simply be a brattice cloth sealed down at the edges with 'hard stop' (a kind of quick-setting cement that goes rock hard) or sand. It might be a system of air-tight doors. Certain of these stoppings have also to be fire proof, and explosion proof; these of course are much more elaborate constructions, and are usually erected when sealing a face off that is worked out. When it is necessary to stop the supply of air to the old district, stoppings are erected in each gate; portions of the roof and sides are extracted until there is a solid rock or stone base to work from; sand is then packed tight, until there is a solid

wall perhaps 20yds long and 15 or 20ft wide, the whole lot is then sealed up with 'hard stop' and cement blocks. After an explosion, to stop the supply of air to any fires, or to render spontaneous combustion impossible, a stopping is put on and the affected districts sealed off. In the days of private enterprise, the coal owners' chief concern was with getting his mine operational again as quickly as possible. The stoppings therefore were put on practically immediately. On some occasions the villagers tried to stop such a thing happening, because of the possibility of miners trapped inbye, the stopping effectively cutting off their air and suffocating them.

STRETCH A stretch of coal is the Durham term for a stint. It's a man's allotted task on the face. Stints may be more or less difficult to fulfil but the various stretches on the face would be shared by the team: workers in the good places would pitch in to give a hand to those on difficult stretches when they had finished their own stints.

SUMP KEY This is a non-existent key which young lads when first going down the pit (in Doncaster) are told to get. The sump is under the shaft – under the resting place of the cage – and of course you cannot get it.

SWALLY In Yorkshire called a 'swilly'. This is a dip in the way caused by earth movements and often filled with water. Swallies are also caused by the undulations of the seams as they dip and rise. They nearly always buckle the rails. This is one of the difficulties faced by the haulage lads of today as by the putters of old; if they miss seeing the buckle, the tub comes off the way. The story was told to me of a pony who was being worked very hard, and used to run full pelt into a swally every time he felt tired in order to get the tub thrown off the road; the men waiting for the tubs used to come out and hit his legs but he used to take no notice of this. There is a Durham song about putters, and the difficulties made by swallies:

Nuw, there's a hitch'in and a swally,
Filled wi water like a ford,
And a lots a ways, a'l twisted and a clarty ganin bord,
There's raw planks, and raggy canches,
And yi sometimes get a smack,
Wey it meks yer twist your gizzard if yi'v chance ti
Catch yor back.

(*The Putter . . .*)

SWITCH Name given to the equipment in the main gate which runs the conveyor chain or belt for the coal and carries it away from the face outbye. It also contains the telecommunications between sections of the face, the caunch, the surface and various other centres of activity. I believe in the South Midlands it is the name given to the middle of the coal face.
TEAMING BYE This was the term used when all the trucks had been filled, there were no skips to load, so all of the coal drawn out of the pit would be shoved along the gangway, and then tipped over into a heap so as to keep the pit going as long as possible.
TOKENS Discs bearing the hewer's number; after he had filled the tub, he hung on his token to show the officials at bank it was his tub. Of course marras pooled their earnings and their tokens. The putter also had tokens, these were hung on every tub which he 'put' or drew outbye, this to identify whom to pay and how much, according to the number of tubs carrying his token. Putters in many cases would also 'hang up' or pool their tokens in order to equalize the work and prevent senseless competition. If the putter was keen to earn better money (and they all were), he would charge inbye, to try and get in before the hewers and fill a tub for himself. This done, he would not only hang his token on but also a 'butterfly' of twisted paper, to show that he had both filled and put the tub. My father remembers that the hewers' tokens were round in shape and bore an initial and number. The putters' were oblong in shape and were put on a piece of marlin about 12in long. The hewers put their token on a small hook in the bottom of the tub, to which the token of the putter had already been placed. Sometimes this was in the middle of the side of the tub. These were taken off as the tub was 'teamed' into the screens, in this way a check was kept of the number of tubs hewed and put. He also says that: 'When we putters hewed any coal during out shift we split the string on our token, and placed a roll of paper through it. This was known as a putter-hewer's tub and token.'
TRAM Originally the detachable chassis on which rode the corf or basket of coal. This would be detached at the shaft, slung on to the loops of the shaft-rope, and drawn up to the pit bank. These were replaced by the cage which carried the tubs to surface, and this in turn is being replaced by skip-winding. Tram was also the term exclusively given to an underground vehicle ('jotty') which was approximately two tub lengths long, and had a side which was held up by a detachable cottle pin. It carried the long timber and when the pins were removed the side could drop down to allow their discharge.

TRAPPER A young boy who used to be employed at the air doors, opening and shutting them for the putters coming in and out with their tubs. In Jarrow colliery the owners introduced self-operating doors which worked on a hinge; when hit by the tubs they would open by themselves. The miners complained that the tubs knocked the doors out of line and thus affected the current of the air; the owners were accused of putting the men's lives in jeopardy, simply to save the pittance paid to the trappers as wages.

> . . . the trapper,
> That's the name they give the door boy,
> And it's a queer job,
> A'l on me tod . . .
> (BBC Radio Ballad *The Big Hewer*, Ewan MacColl)

TUBBING An old term for the 'walling' which was put round the shaft to keep back water; in earlier times it was planks but eventually metal was used, then concrete.
TWYNE Name given to the operation of the putter in switching the weight of his body to different positions to assist the tub round the turns. The backside, the elbow and the knee were all skilfully (and painfully) deployed in easing the tubs round the blunt corners of the 'way'. A better description of 'twyning' would in all probability require a physical demonstration since the contortions which the putter forced his body into to keep his tub on the way just could not be described by words alone.
UNIT A team after cavilling would be sent on a unit. The unit under conventional working, would consist of two sections and three gates, the main gate running up the middle of the area, and tail gates at each side. All of the backbye workers, the timber lads, putters, etc. would be assigned to one unit, as well as the hewers.
VIEWER Name given to the manager in the old days. Usually the viewer was the manager of both surface and underground work.
WAILERS In the early days of Durham mining boys employed to pick out slate pyrites and other foul admixtures from the coal.
WASTE All the old workings and gates, usually abandoned or little used. These were generally the return airways where little air passes and few men ever have need to pass through. Wastemen are workers who travel the pit making sure these old places aren't too bad, doing minor

repairs to the supports and keeping enough air circulating to stop the accumulation of gas.

WATER NOTE (or wet note) A piece of paper written out by the deputy which allows the miner to ride out of the pit early, sometimes by as much as an hour. A man may have been working in water a foot deep, or with water running in on the roof, really foul filthy work and the tools and coal wet through, possibly lying flat in 6in of water in an 18in seam, but even in a 3 or 4ft seam kneeling in a foot of water with it streaming in on your head. Miners try to get away from it early. The note originated in the days before pit head baths (Wardley's opened in February 1958) when the man had to walk home in his pit clothes, soaked right through; when the baths opened he was still allowed out early so that he could wash in comfort without having to jostle with the crowds of men who have all ridden out together.

WINNING A winning might be a new drift to make a gate, or it might be drifting out a new face – developing a new face. This work can sometimes be harder than working on the face although you might only have a small number of men employed in it, the hardship coming from the fact that the workers are using only temporary supports. Even today they have to use temporary supports as they cannot get the proper operating equipment in (also there is no proper current of air). You might get an air blower which directs the air in a bit further, but it might be very hot on a new winning, although equally it could be freezing cold. It is hideous work.

WOODY COAL Coal which will not cleet or separate.

WORKIE TICKET This term is applied to the person who pushes his luck with other workers too far, who talks too much, or takes the mickey too often. It's said he is 'working his ticket'. I've been termed a workie ticket myself at times through talking too much about people's subservience to the management and getting to the stage where people have used violence because I've got on their nerves. People don't usually remain workie tickets very long; they soon get cured of their ailment.

WORKING THE ROOF 'The roof is working!' is a danger signal. It signals the fact that the weight is starting to come on the face. It starts to groan, cracks (breakers) start to appear, little lumps, medium lumps and then big lumps start to fall. Weight bumps start to sound like thunder. When little lumps start to trickle from the roof it is said to be 'tinkering on'. The name given to each report of the roof is 'bonk' (Greenwell calls it 'bowk' – perhaps it was in his day). In most cases of course it is the 'goaf' or waste areas which collapse. Unfortunately parts of the face 'slip' and give way at times, even in our 'modern' mines today.

YARD STICK All officials from the deputy up to the gaffer carry one of these heavy walking sticks. Today used partially as a measuring instrument, in part as a hoist to put an oil lamp into high areas of the roof, but first and foremost as Lord Mayor's chain: a symbol of authority. It is carried almost entirely by officials, although there is no earthly reason why an ordinary miner should not carry one. Certain workers carry them to assist in travelling long distances while walking; in Durham, when I was a line boy, I used to tie my whitewash brush to one while painting lines on the high roofs. In most of the collieries today a long tube with a bulb at the end is put up into the roof. The bulb is squeezed and the air in the roof enters into the bulb. This 'test' is then redirected into the oil lamp where the portion of fire damp can be calculated. This process saves the tired arms of the official hoisting his lamp on a stick into the roof.

YOKEN The name of an extra tub which the putter who tried to go it alone – rather than working for shares – would fill. If he got out quick with his 'fullin' he could fill an extra tub before the other putters got out. If he succeeded he was said to have a 'gaynor' on the other putters; if he failed his tub would block the way of the other putters and he was then said to have a 'yoken'.

Notes

1 A reprint of the third edition (1888) has been recently published by Frank Graham.

2 John Trotter Brockett, *A Glossary of North Country Words*, Newcastle, 1846.

3 It will be seen that my father refers to 'cavils' as 'cables', this is simply because the word cavil is pronounced by most Durham pitmen 'cable'.

4 John Gibson, *Plan of the Collieries on the River Tyne and Wear*, 1787, Institute of Mining Engineers, Newcastle.

5 William Fordyce, *The County Palatine of Durham*, vol. I, p. 114.

Subject index

Index of places